AF412828

Soils, Ecosystem Processes, and Agricultural Development

Shinya Funakawa

Editor

Soils, Ecosystem Processes, and Agricultural Development

Tropical Asia and Sub-Saharan Africa

 Springer

Editor
Shinya Funakawa
Graduate School of Global Environmental Studies
Kyoto University
Kyoto, Japan

ISBN 978-4-431-56482-9   ISBN 978-4-431-56484-3   (eBook)
DOI 10.1007/978-4-431-56484-3

Library of Congress Control Number: 2017930415

Printed on acid-free paper

This Springer imprint is published by Springer Nature
The registered company is Springer Japan KK
The registered company address is: Chiyoda First Bldg. East, 3-8-1 Nishi-Kanda, Chiyoda-ku, Tokyo 101-0065, Japan

Preface

The negative impacts of agricultural development on our environment are rapidly growing, yet we are increasingly dependent on the agricultural sector for food and energy. The situation is similar in the tropics, where subsistence agriculture with low–input management has long comprised most agricultural systems. The main objective of this book, therefore, is to integrate environmental knowledge observed in local agriculture, based on the understanding of soil science and ecology, and to propose possible technical solutions and a more integrated approach to tropical agriculture.

We first start with a rather classical pedological issue, i.e., an investigation of weathering processes of soil minerals under different geological and bioclimatic conditions, which could be regarded as the cumulative result of biological activities and, in turn, may control natural biological processes. Traditional agricultural practices with different resource management strategies are then analyzed in terms of their modification of natural biological processes. The goal of this section is to compile and describe local knowledge accumulated in traditional agriculture under different natural constraints for resource utilization. Finally, we focus on the present situation of tropical agriculture, that is, resource utilization in modern agriculture after application of technical innovation (increased application of chemical fertilizers as well as agricultural chemicals). Here, possible technical approaches to resource management that reasonably support agricultural production while mitigating environmental degradation are discussed.

In summary, we will describe and analyze the ecological and technical countermeasures available for mitigating environmental degradation due to the increasing agricultural activities by humans, based on our scientific understanding of traditional agriculture in the tropics. This is an effective approach, as such ecological and technical tools previously involved in traditional activities are expected to be easily incorporated into present agricultural systems.

Kyoto, Japan   Shinya Funakawa

Contents

Chapter 1
Introduction

Shinya Funakawa

Abstract This volume describes soil and ecosystem processes and agronomic practices that are well-suited for sustainability in tropical regions. In Part I, pedogenic processes on different continents, i.e., in humid Asia and equatorial Africa, are discussed in detail, with special reference to our recent findings regarding clay mineralogy and soil fertility in these regions. Part II provides a comparative analysis of pedogenic acidification processes under different geological and climatic conditions in humid Asia where Ultisols predominate and follow with a comparative analysis of Oxisols in Cameroon. This analysis is presented within the context of an "ecosystem strategy for resource (nutrients) acquisition," as a driving force of mineral weathering and soil development. Parts III and IV provide an analysis of "low-input" agricultural methods and a description of trials for establishing "minimum-loss" agriculture, respectively, in tropical Asia and Africa. Firstly, shifting cultivation and/or slash-and-burn agriculture in different regions, i.e., in the forest–savanna boundary of Cameroon, monsoonal forests in Thailand, and the miombo woodland of Zambia, are analyzed. These agricultural systems are then comparatively analyzed in the context of on-site re-accumulation and/or temporal redistribution of nutrients. In addition, the interaction between agricultural and pastoral activities in the Sahel is analyzed from the viewpoint of the spatial redistribution of resources, i.e., transport of resources to the farm from outside. Agricultural practices consistent with "low-input" and "minimum-loss" management have been characterized, and they provide a knowledge base and technical framework for scientifically based sustainable soil management. We then describe several trials undertaken for the control of soil erosion by either wind or water in different regions, characterized by different tropical climates and soils, e.g., semiarid savannas in Niger, Ultisols in monsoonal forests in Thailand, and on Ultisols and Oxisols in savannas and forests of Tanzania and Cameroon. A trial conducted on a cropland in Tanzania—for the purpose of minimizing losses of nitrogen due to leaching—that utilizes soil microbial biomass as a temporary pool is also described.

S. Funakawa (✉)
Graduate School of Global Environmental Studies, Kyoto University, Yoshida-Honmachi, Sakyo-ku, Kyoto 606-8501, Japan
e-mail: funakawa@kais.kyoto-u.ac.jp

© Springer Japan KK 2017
S. Funakawa (ed.), *Soils, Ecosystem Processes, and Agricultural Development*,
DOI 10.1007/978-4-431-56484-3_1

Keywords Clay mineralogy • Ecosystem strategy for resource (nutrients) acquisition • Low-input agriculture • Minimum-loss agriculture • Tropics

1.1   What Is Soil?

Five essential soil-forming factors, i.e., parent material, climate, biological activity, topography, and time, were initially specified by Dokuchaev more than 100 years ago. This idea is a classical in soil science, and many researchers have explained the present soil distribution pattern based on these soil-forming factors. In my opinion, during the formation of a particular soil, the contribution of biological activity is crucial, since the parent material of soil could develop to the "actual" soil only after undergoing interactive reactions with biosphere. More specifically, the role of biological activity and time in actual soil formation differs from that of the other three soil-forming factors, as shown in Fig. 1.1.

Parent material, topography, and climate are considered to be inherent conditions that are essentially a result of the earth's activity. The parent material of a specific soil is determined either directly by the movement of earth's crust, which distributes different kind of rocks with variable chemical composition to different parts of the land surface, or indirectly through the mode of deposition, forming different types of soils such as residual, colluvial, alluvial, or aeolian. This factor could govern soil formation through mineral weathering and supply of mineral nutrients to the biosphere. Topography is primarily the result of orogenesis, which controls the direction of water movement and the hydraulic environment of soil. Climate is primarily controlled by the distribution pattern of land and ocean on the earth's surface and the rotation and revolution of the earth, which are mostly determined by the earth's (and solar) activities.

In contrast, biological activity is not the initial condition. Rather it is the driving force of soil development and is also regulated by the inherent conditions of earth's surface similarly as soil. Time may relate to the cumulative biological activity and is not independent from climate and/or topographic factors. For example, in the soil under rainforest climate, soil and biological reactions proceed more rapidly; thereby soil development would be faster than that under dry climatic conditions. In the soil on a stable flat terrain, the rate of soil erosion is low and the soil is aged and strongly weathered; therefore, it has low residual mineral content.

In Part I of this volume, the soil-forming processes in different continents, i.e., in humid Asia and equatorial Africa, have been discussed in detail, with special reference to our recent findings regarding the clay mineralogy and soil fertility in these regions. First, based on the soils in humid Asia, the importance of parent material and climate factors for the development of upland soils have been discussed, with special emphasis on the process of vermiculitization of illitic minerals from sedimentary rocks and/or felsic plutonic rocks under a udic soil moisture regime as a determining factor for certain chemical properties of soils, such as cation exchange capacity and level of exchangeable aluminum, and also for

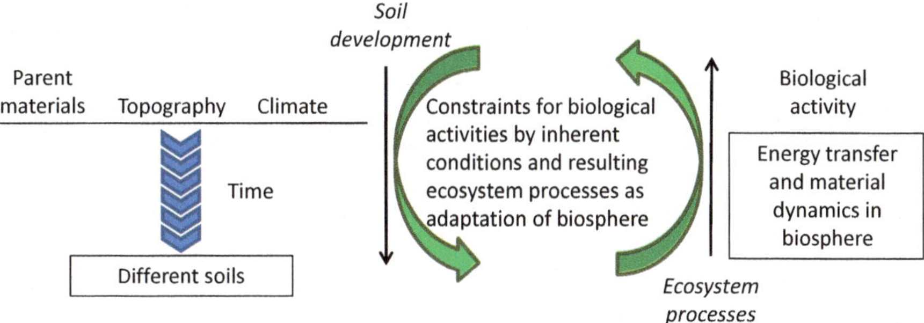

Fig. 1.1 Relationship between biological activity and soil development

the formation of strongly acidic Alisols under udic soil moisture regime (Chaps. 2 and 3). This is followed by a comparative analysis of soil properties of equatorial Africa, taking into account both chemical and mineralogical characteristics (Chaps. 4 and 5). In these discussions, clay mineralogical characteristics, especially of crystalline layer silicates, are highlighted as a key element that largely influences the physicochemical properties of the soil. In addition, the role of amorphous oxides and/or hydroxides of aluminium and iron is emphasized as a controlling factor for the retention of soil organic matter and phosphate in tropical soils (Chap. 6).

1.2 How Do Soil Development and Natural Ecosystem Processes Correlate?

The interaction between terrestrial ecosystem processes and soil development is an essential process. Biological activity is a driving force of soil development, while soil conditions largely derived from parent material and climate can either restrict or enhance biological activities and ecosystem processes. Several processes are involved in the interactions between soil and ecosystem processes: for example, nutrient uptake by plants and soil acidification/mineral weathering, microbial respiration, and redox reactions in soil. I believe that the former combination above is one of the most crucial processes under humid climatic conditions, where precipitation exceeds evapotranspiration, and, consequently, mineral weathering proceeds only in one direction due to the consecutive removal of by products from the soil system under leaching conditions.

The essential interplay between soils and biological activities observed in the nutrient uptake by plants and soil acidification/mineral weathering is represented below:

1. Different kinds of acids are generated during the process of energy transformation in the ecosystem processes, which are then added to the soil. For example,
   excess uptake of cations over anions into forest biomass, dissolving CO_2 derived

from respiration of soil microbes and plant roots, transformation of nitrogen and its leaching loss, and release of organic acids from plant roots and soil microbial activities could all be listed as ecosystem internal acid load (van Breemen et al. 1984).

2. It is known that mineral weathering is accelerated under the presence of acids. For example, weathering reaction of K-feldspar to form kaolinite is represented below:

$$2KAlSi_3O_8 + 2H_2CO_3 + 9H_2O \rightarrow Al_2Si_2O_5(OH)_4 + 4H_4SiO_4 + 2K^+ + 2HCO_3^- \tag{1.1}$$

As mentioned in the examples, mineral weathering reaction is represented as follows:

$$[\text{Thermodynamically less stable mineral A}] + nH^+ + [\text{Anion}^{n-}] \rightarrow [\text{Thermodynamically more stable mineral B}] + H_4SiO_4 + [(\text{soluble}) \, \text{Cation}^{n+}] + [\text{Anion}^{n-}] \tag{1.2}$$

3. Many cations, such as K^+, Mg^{2+}, and Ca^{2+}, released during mineral weathering, would be utilized by plants and other organisms as essential nutrients.

Based on the above facts, the relationship between biological activities and mineral weathering is possible taking into consideration the following: (1) biological activities accelerate mineral weathering through ecosystem internal acid loads, and (2) living organisms utilize the essential nutrients, which are released through mineral weathering reactions. Thus, information on the interaction between biological activities and soil processes under different bioenvironmental conditions (differing in climate, parent material, and topography/water conditions) is important to understand the processes of ecosystems or soils.

To analyze the relationship between soil formation and ecosystem processes in detail, we applied the proton budget method, which was originally used in studies of acid deposition and/or soil acidification (van Breemen et al. 1983), to a soil horizon levels. In Part II of this volume, pedogenetic acidification processes under different geological and climatic conditions in humid Asia, where Ultisols (Acrisols and Alisols in World Reference Base for Soil Resources [IUSS Working Group WRB 2014]) predominate, have been comparatively analyzed (Chap. 7). This analysis is presented within the context of an "ecosystem strategy for resource (nutrients) acquisition" and as a driving force of mineral weathering and soil development (described in Chap. 3). Similar processes in Oxisols/Ferralsols in equatorial Africa have been analyzed following the same methodology used to compare Acrisols and Alisols in Asia (Chap. 9). Ecosystem processes involved in the differentiation of forest and savanna vegetation in central Africa are discussed in Chap. 8, with special reference to soil processes, which might contribute to clarify the interaction between ecosystem and soil processes.

1.3 How Do Human Agricultural Activities Adapt and/or Modify Natural Ecosystem Processes? – Future Perspective of Development on Tropical Agriculture

Agricultural production inevitably removes ecosystem resources such as mineral nutrients from the ecosystems as crop yield. It indicates that the replenishment of nutrients once removed is essential for sustaining agricultural productivity for a long time period. Before the establishment of modernized agriculture, input of external resources from outside was limited. Farmers had to depend on natural processes such as mineral weathering and litter input for replenishing mineral nutrients for next cropping. There are essentially two strategies for nutrient replenishment. One is the on-site re-accumulation of mineral nutrients or temporal redistribution of nutrients. This approach is typically observed for shifting cultivation, in which both cropping and fallow phases are incorporated in land rotation. It is believed that, in traditional shifting cultivation, sustainable crop production requires added nutrients via slash and burn after certain periods of fallow. The other possibility is the transport of resources from outside the farm, i.e., spatial redistribution of resources. Input of manure produced outside is such an example. When the impact of agricultural activities on the natural soil and ecosystem processes were limited, severe environmental constraints have driven adaptation of human beings to the respective natural environments (Fig. 1.2).

Such relationship between soil and ecosystem processes and human agricultural practices, characterized by heavy constraints of natural conditions and limited impact of human agricultural activities, was subjected to change considerably after introduction of modern technologies into agricultural practices since the early twentieth century, typically after chemical fertilizers had been available through development of Haber–Bosch process. The information and techniques for resource management and sustainable land use, which have been involved in classical, low-input agriculture, are presently not necessarily adopted; instead, modern agriculture, characterized by the use of heavy doses of agricultural chemicals and/or intensive irrigation, is the major approach for agricultural production.

It is a fact that extensive application of irrigation practices and/or nitrogen fertilizer have led to increased agricultural production. However, this has also caused negative impacts as well, for example, irrigation practice has caused serious expansion of secondary soil salinization (Middleton and Thomas 1992) and nitrogen contamination, especially into the water systems such as lakes and/or closed seas in downstream, is becoming a serious threat to the aquatic ecosystems (Rockström et al. 2009). On the other hand, low-input agriculture is considered to be still important in relation to the economic poverty in many countries and also to unstable and unpredictable conditions of agroecological infrastructures in near future such as climate, soil degradation, worldwide depletion of water, and nutrient resources (such as phosphorous). Notably, there are many examples of

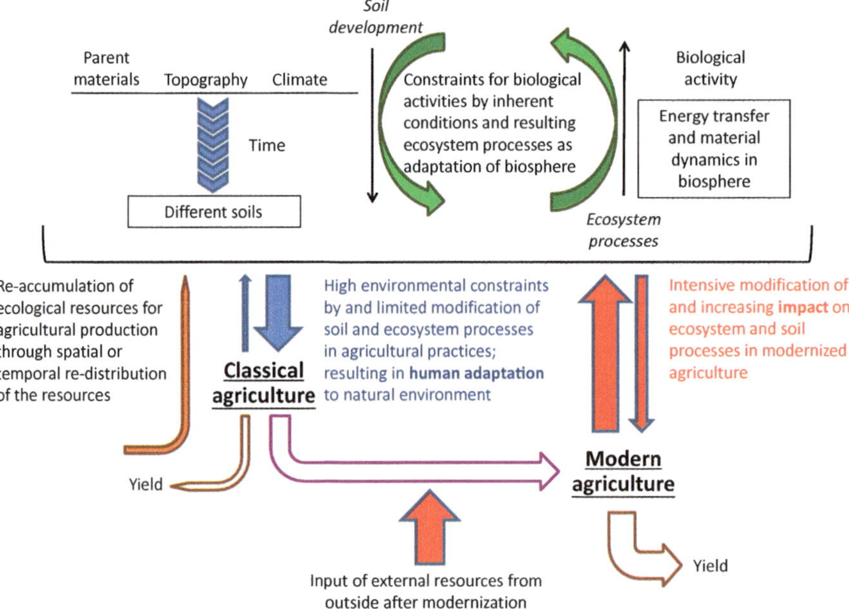

Fig. 1.2 Alteration of ecosystem and soil processes after establishment of modern agriculture

desertification, which has been largely enhanced by the extensive application of modern agricultural practices (Middleton and Thomas 1992).

In this context, I emphasize that we should maintain the knowledge of low-input agriculture with scientific basis, which has proved to be sustainable at least for a certain period of time and could be utilized for developing alternative agriculture that does not heavily depend on external input of resources. It is especially applicable in the cases of tropical agriculture, where the experiment of modern agriculture has only short history and economic and ecological uncertainty in near future has more reality than in the temperate zone. I consider that the main direction of the development of tropical agriculture in future could be set to as below:

1. It should decrease the loss of resources rather than increase their input.
2. It is important to control the fluxes of carbon, nitrogen, and mineral resources rather than to accumulate them in soils, since the proportion of flux relative to the pool size is usually much larger in tropical ecosystems than in temperate ecosystems.

Parts III and IV are composed of the analysis of "low-input" agriculture and trials for establishing "minimum-loss" agriculture in tropical Asia and Africa, respectively. In Chaps. 10, 11, and 12, shifting cultivation in different regions, i.e., in forest–savanna boundary of Cameroon, monsoonal forest in Thailand, and miombo woodland of Zambia, is analyzed, respectively. Then these agricultural systems are comparatively analyzed in the context of on-site re-accumulation of mineral nutrients and/or temporal redistribution of nutrients in Chap. 13. On the other hand, in Chap. 14, the interaction between agricultural and pastoral activities in the Sahel has been analyzed from the viewpoint of the spatial redistribution of resources, i.e., transport of resources from the outside to the farm. These agricultural practices are all equipped with technical background and knowledge enabling "low-input" and "minimum-loss" management.

Subsequently, in Chaps. 15, 16, and 17, several trials for controlling soil erosion either of wind or water in tropical agriculture in different regions under different climate and soils, i.e., in semiarid savannas in Niger, in monsoonal forest on Ultisols in Thailand, and in savannas and forest on Alfisols, Ultisols, and Oxisols in Tanzania and Cameroon, have been introduced. A trial for decreasing leaching losses of nitrogen, by utilizing soil microbial biomass as a temporary pool, in Tanzanian cropland is included in Chap. 18.

It is again emphasized that the main intention of all these analyses and trials, introduced in Parts III and IV, is derived from our primary concern that we should maintain the knowledge of low-input agriculture with scientific basis in order to develop alternative tropical agriculture, in which the loss of resources should be minimized by controlling the fluxes of carbon, nitrogen, and mineral resources. These directions would be justified both from the economic perspective related to the poverty of many farmers in the tropics and instability in the production circumstances specific to tropical countries in the near future, such as climate change, soil degradation, and worldwide depletion of water and nutrient resources.

References

IUSS Working Group WRB (2014) World reference base for soil resources 2014. International soil classification system for naming soils and creating legends for soil maps. World Soil Resources Reports No. 106. FAO, Rome

Middleton N, Thomas D (eds) (1992) World atlas of desertification, Second edn. Hodder Arnold Publication, London/Baltimore

Rockström J, Steffen W, Noone K, Persson Å, Chapin FS III, Lambin EF, Lenton TM, Scheffer M, Folke C, Schellnhuber HJ, Nykvist B, de Wit CA, Hughes T, van der Leeuw S, Rodhe H, Sörlin S, Snyder PK, Costanza R, Svedin U, Falkenmark M, Karlberg L, Corell RW, Fabry VJ, Hansen J, Walker B, Liverman D, Richardson K, Crutzen P, Foley JA (2009) A safe operating space for humanity. Nature 461:472–475

Van Breemen N, Mulder J, Driscol CT (1983) Acidification and alkalization of soils. Plant Soil 75:283–308

Van Breemen N, Driscoll CT, Mulder J (1984) Acidic deposition and internal proton in acidification of soils and waters. Nature 307:599–604

Part I
Distribution of Clay Minerals in Tropical Asia and Africa with Special Reference to Parent Materials (Geology) and Climatic Conditions

Chapter 2
Parent Materials and Climate Control Secondary Mineral Distributions in Soils of Kalimantan, Indonesia

Tetsuhiro Watanabe and Supiandi Sabiham

Abstract Secondary mineral distributions in soils from Kalimantan, Indonesia, were investigated to examine the effects of the parent materials and climate at different elevations on the distributions. B horizon soils were sampled at 60 sites on gentle slopes at different elevations (20–1700 m altitude). Each major parent material (sedimentary, felsic, and intermediate to mafic) was represented at different elevations. The soil samples were classified from their total elemental compositions using cluster analysis. Secondary minerals were measured by X-ray diffraction and selective extractions. The samples were divided into ferric (high Fe contents), K&Mg (high K, Mg, and Si), and silicic (high Si) groups. The ferric soils were derived from mafic parent materials, whereas the others were derived from felsic or sedimentary parent materials. The K&Mg soils had higher total base contents (suggesting primary minerals) and were less weathered than the silicic soils. Secondary minerals in the ferric soils were characterized by high contents of Fe oxides and gibbsite. The K&Mg and silicic soils had similar secondary mineral (kaolinite and vermiculite) contents, but more mica was found in the former. Only the silicic group soils had secondary mineral contents that changed as the elevation changed (the kaolinite content increased and the vermiculite and poorly crystalline Al and Fe contents decreased as the elevation decreased). Higher temperatures at lower elevations may cause minerals to be altered more. Secondary mineral distributions were primarily controlled by the parent material (mafic or felsic/sedimentary) and, secondarily by the climate, which varied with elevation.

Keywords Borneo • Weathering • Soil mineralogy • Clay minerals

T. Watanabe (✉)
Graduate School of Agriculture, Kyoto University, Kitashirakawa-Oiwakecho, Sakyo-ku, Kyoto 606-8502, Japan
e-mail: nabe14@kais.kyoto-u.ac.jp

S. Sabiham
Faculty of Agriculture, Bogor Agricultural University, Jl. Meranti, Kampus IPB Darmaga, Bogor 16680, Indonesia

© Springer Japan KK 2017
S. Funakawa (ed.), *Soils, Ecosystem Processes, and Agricultural Development*,
DOI 10.1007/978-4-431-56484-3_2

2.1 Overview

The distribution of secondary minerals is fundamental to understanding the elemental cycles in natural and agricultural ecosystems because secondary minerals control the chemical and physical properties of soil (Schulze 2002). Vermiculite and smectite contribute to the cation exchange capacity (CEC), and interlayer Al hydroxides decrease the CECs of vermiculite and smectite (Barnhisel and Bertsch 1989; Funakawa et al. 2008). Al and Fe oxides/hydroxides contribute to the adsorption of anions such as phosphate (Parfitt 1978), and this is particularly the case for poorly crystalline Al and Fe, which are important even though they are present at low contents (Huang et al. 2002; Watanabe et al. 2015). Organic matter dynamics are also affected by secondary minerals (Bruun et al. 2010; Percival et al. 2000). Information on the distributions of secondary minerals is therefore essential to understanding nutrient dynamics in ecosystems and to managing agricultural and forestry land appropriately.

The parent material influences the quantities and types of secondary minerals that are present in a soil (Brady and Weil 2002). Minerals in the parent material can retain their frameworks when they are transformed into secondary minerals. This process, called transformation (Nahon 1991), is important in the weathering of 2:1-type minerals, especially the weathering of mica to vermiculite and then smectite (Churchman and Lowe 2012; Watanabe et al. 2006). Minerals in the parent material, e.g., feldspars, micas, and amphiboles, dissolve in the soil solution, from which secondary minerals can precipitate through the neoformation process (Nahon 1991). Gibbsite, kaolinite, and smectite are formed via neoformation (Huang et al. 2002; Churchman and Lowe 2012), although kaolinite and smectite also form as authigenic sedimentary minerals (Reid-Soukup and Ulery 2002; Murray and Keller 1993) and may be contained in sedimentary rocks.

The minerals in the parent material can be characterized by their weathering resistances. Felsic rocks (e.g., granite) and sedimentary rocks (e.g., mudstones and sandstones) are rich in quartz, alkali feldspars, and micas, which are more resistant to weathering than amphiboles, pyroxenes, and olivines, and have relatively low dissolution rates (Nagy 1995; Brantley and Chen 1995; Lasaga 1998; Rai and Kittrick 1989). More mica, which is the starting material for the transformation of 2:1-type clay minerals, is found in felsic and sedimentary rocks than in mafic rocks. In contrast, intermediate to basic rocks (e.g., andesite and basalt), which are found in some parts of Kalimantan, are rich in amphiboles, pyroxenes, olivines, and Ca feldspar, which weather rapidly (Nahon 1991; Lasaga 1998). More weathered soil is therefore expected to be present over mafic parent materials. The studies so far have been focused only on soils from limited areas of Kalimantan, mostly in East Kalimantan. This area contains large areas in which sedimentary rocks (e.g., mudstones and sandstones) are the parent materials.

The climate varies with elevation and affects the distributions of secondary minerals. Lower temperatures at higher elevations cause lower evapotranspiration rates, leading to higher leaching rates in the soil (Brady and Weil 2002). A high leaching rate affects the composition of the soil solution, such as giving low pH and $H_4SiO_4^0$ activity (Norfleet et al. 1993; Huang et al. 2002). Lower temperatures at higher elevations decrease the rate at which organic matter decomposes, therefore, inhibiting the crystallization of Al and Fe oxides and hydroxides by organic matter (Huang et al. 2002; Schwertmann 1985).

Thermodynamic analyses of the equilibria between soil solutions and minerals can allow a theoretical understanding to be gained of the distributions of secondary minerals and the conditions under which the secondary minerals formed (van Breemen and Brinkman 1976; Karathanasis et al. 1983; Rai and Kittrick 1989; Norfleet et al. 1993; Karathanasis 2002; Watanabe et al. 2006). The neoformation and transformation of secondary minerals are both affected by the composition of the soil solution (Karathanasis 2002; Watanabe et al. 2006). The $H_4SiO_4^0$ activity in a soil solution mainly controls the relative stabilities of gibbsite, kaolinite, and smectite (Rai and Kittrick 1989). Gibbsite is most stable under strong leaching conditions and low $H_4SiO_4^0$ activities, whereas kaolinite is stable at moderate $H_4SiO_4^0$ activities, and smectite is stable at high $H_4SiO_4^0$ activities. The pH is decisively important to transformation processes. Mica weathers to vermiculite at low pH values (Karathanasis 2002; Watanabe et al. 2006). The parent material and climate affect the composition of the soil solution. Si released from a parent material is a source of $H_4SiO_4^0$, and strong leaching conditions caused by high precipitation and low evapotranspiration cause the $H_4SiO_4^0$ activity to be low (Huang et al. 2002). The compositions of soil solutions in soils formed from a range of parent materials and under different climatic conditions need to be analyzed so that the conditions under which the secondary minerals formed can be theoretically supported.

In this chapter, the effects of the parent materials and variations in the climate with elevation on the distributions of secondary minerals in Kalimantan are discussed where hilly landscape without active volcanoes is characteristics and different from other tropical regions on shields.

2.2 Soil Samples

A total of 60 samples of B horizon soils with a range of parent materials and formed at a range of elevations were collected from across Kalimantan (Fig. 2.1). The sampling sites were at elevations of between 20 and 1700 m a.s.l. The mean annual temperatures at the sites were 18–27 °C, and the mean annual amounts of precipitation at the sites were 1970–4040 mm (www.worldclim.org). Sedimentary rocks (e.g., mudstone, sandstone, shale, and slate) are the most widely distributed rocks in Kalimantan, and 40 out of the 60 sites were on these rocks. The other sites were on felsic to mafic igneous rocks. Five sites were on felsic or felsic to intermediate rocks, six were on intermediate (andesitic) rocks, and nine sites were on mafic or mafic to intermediate (basaltic to andesitic) rocks. Sites at different elevations with each of the major parent materials found in Kalimantan (i.e., sandstone, mudstone,

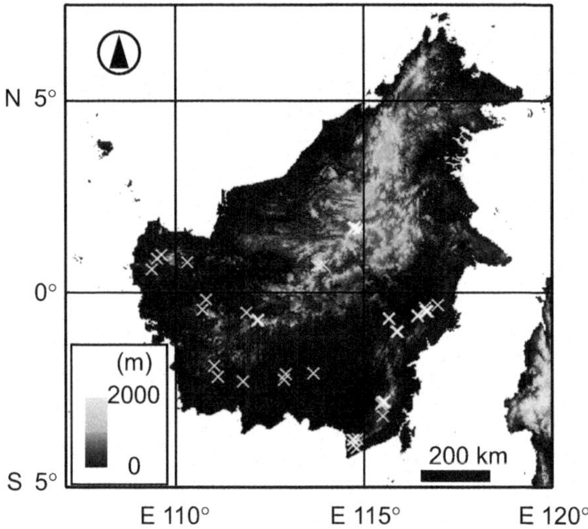

Fig. 2.1 Sampling sites (*white crosses*) on an elevation map of Kalimantan, Indonesian Borneo. The map data were from the WorldClim database (http://www.worldclim.org)

and slate; basalt and andesite; and granite) were included. Most of the sites were on gentle upper slopes that had been little disturbed (e.g., in forests or rubber plantations). The soils were found to be Alisols (25 sites), Acrisols (5 sites), Cambisols (26 sites), Ferralsols (3sites), and Luvisols (1 sites) (IUSS Working Group WRB 2015).

Climatic data (the mean annual temperature, mean annual precipitation, and mean monthly temperature at each sampling site) were obtained from the WorldClim database (www.worldclim.org). Evapotranspiration at each site was estimated from the monthly temperatures at the site using a previously developed method (Thornthwaite 1948).

2.3 Analytical Procedure

The chemical and mineralogical properties of the samples were analyzed: $pH(H_2O)$; $pH(KCl)$; CEC; exchangeable Ca, Mg, K, and Na; total C, Si, Fe (Fe_t), Al, Ca, Mg, K and Na contents; clay, silt, and sand contents; sodium dithionite and citrate extractable Fe (Fe_d) (Blakemore et al. 1981); and acid ammonium oxalate extractable Al and Fe (Al_o and Fe_o) (Blakemore et al. 1981; McKeague and Day 1966). Al_o and Fe_o include poorly crystalline aluminosilicates and Al and Fe oxides/hydroxides (Shang and Zelazny 2008). The Fe_d include crystalline iron oxides and the fractions extracted by the acidic ammonium oxalate (Fe_o) (Blakemore et al. 1981). The total reserve in bases (TRB) was defined as the sum of the total Ca, Mg, K, and Na contents.

The mineral composition of the clay fraction of each sample was determined using an X-ray diffractometer. The clay minerals were semiquantified from the relative areas of the X-ray diffractogram peaks for layer silicates (kaolinite (0.7 nm), mica (1.0 nm), and minerals with 1.4 nm layers). The amounts of gibbsite and kaolinite in the clay fraction of a sample were determined by performing differential thermal analysis (DTA).

Thermodynamic analyses of the stabilities of the secondary minerals were performed on 43 out of the 60 samples. The samples that were analyzed were selected because of the mineral compositions of their clay fractions, their parent materials, and their elevations. A soil water extract of each sample was collected by standing the soil for 1 week in water at a soil/water ratio of 1:2 at 25 °C and 1 atm. The solution composition (pH and inorganic monomeric Al (Driscoll 1984), Si, Fe, Ca^{2+}, Mg^{2+}, K^+, Na^+, NH_4^+, F^-, Cl^-, NO_3^-, SO_4^{2-}, and inorganic carbon concentrations) were determined. Then the ion activity in each sample was calculated. The stabilities of the minerals in the samples were evaluated using stability diagrams, as has been described elsewhere (van Breemen and Brinkman 1976). The thermodynamic data used in the diagrams were taken from a publication by Karathanasis (2002) for gibbsite and kaolinite and from a publication by Lindsay (1979) for muscovite, quartz, and amorphous SiO_2.

The soil samples were classified by their parent materials and the degree of weathering that had occurred by performing cluster analyses on the total Al, Fe, Si, Ca, Mg, K, and Na contents of the samples. The total Al, Fe, and Si are not very mobile in soil, so are considered to reflect the nature of the parent material, whereas the total Ca, Mg, K, and Na are very mobile (Hudson 1995), so are considered to reflect both the degree of weathering (Hudson 1995; Price and Velbel 2003) and the nature of the parent material. The total element contents of the samples were standardized and subjected to principal component analysis to give factor scores for each sample. Cluster analysis was then performed on the factor scores, using Euclidean distances and the maximum (complete linkage) method.

The Shapiro–Wilk test was used to determine if a group of data followed a normal distribution. The measured values for the soil groups determined by cluster analysis were subjected to analyses of variance and then multiple comparisons analyses. Each multiple comparisons analysis was performed using the Tukey test if the data followed a normal distribution and using Dunn's method if the data did not follow a normal distribution. Spearman rank correlation tests were used to identify relationships between the total element contents and between the elevations and mineralogical and physicochemical properties because most of the data were not normally distributed.

2.4 Identifying Clusters in the Soil Samples from the Total Elemental Compositions

The results of the principal component analysis on the total elemental compositions are shown in Table 2.1 and Fig. 2.2. The factor loadings for the first three principal components, which accounted for 81 % of the total variances, for Si, Fe, Al, Ca, Mg, K, and Na are shown in Table 2.1. The first component coefficients were high for Al, Fe, and Si, suggesting that this component could be used to determine whether a soil was derived from an iron-rich or silicon-rich parent material. This factor is later referred to as the "Fe vs Si" factor. The second component coefficients were high for K and Mg, and this component is referred to as the "K&Mg" factor. The third factor coefficients were high for Na and moderate for Ca.

Performing cluster analysis on the factor scores produced by the principal component analysis allowed the 60 samples to be divided into three groups. The three groups of samples were characterized by the principal components described above, and the distributions of the three groups of samples on a plot of the Fe vs Si and K&Mg component scores are shown in Fig. 2.2. A photograph of a representative soil profile for each group is shown in Fig. 2.3. One group consisted of ten soil samples with high positive Fe vs Si scores, and these samples had high total Fe contents (Table 2.2). This group is later referred to as the "ferric group." Another group consisted of 13 samples with high K&Mg scores. These samples were silicic in nature and had relatively high K and Mg contents, so this group is later referred to as the "K&Mg group." The other group consisted of 37 samples that were silicic in nature and had low K&Mg scores, and this group is later referred to as the "silicic group." In the cluster analysis results, the distance was larger between the ferric group samples and the other two groups than between the K&Mg and the silicic group samples.

The total Si contents were higher in the samples in the K&Mg and silicic groups than in the samples in the ferric group, whereas the total Fe and Al contents were higher in the samples in the ferric group (total Fe contents >210 cmol kg^{-1}) than in the K&Mg and silicic groups (Table 2.2). The average total Al, Fe, and Si contents in the K&Mg group were similar to the contents in the silicic group. The total K and Mg contents were highest in the samples in the K&Mg group.

2.5 Physicochemical Properties of the Soils

The physicochemical properties of the samples in each group are shown in Table 2.3. The clay contents decreased in the order ferric group (74 ± 13 %), K&Mg group (53 ± 12 %), and silicic group (36 ± 14 %), whereas the sand contents were higher in the silicic group (38 ± 18 %) than in the other two groups. The pH(H$_2$O) values were low in all of the groups, and the means were between pH 4.6 (for the silicic group) and pH 4.9. The ΔpH value was greater for the K&Mg

Table 2.1 Results of the principal component analysis of the total elemental compositions of the samples

	PC1	PC2	PC3
Factor loading			
Si	−0.95	−0.04	0.01
Fe	0.94	−0.05	0.17
Al	0.90	0.18	0.07
Ca	0.08	0.49	−0.59
Mg	0.08	0.90	0.18
K	−0.28	0.82	−0.01
Na	0.22	−0.15	−0.85
Eigenvalue	2.7	1.8	1.1
Total variance explained (%)	39	26	16

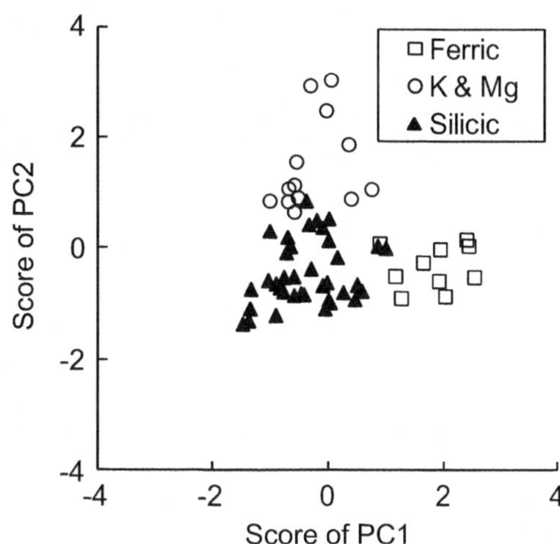

Fig. 2.2 Principal component scores (PC1 is the Fe vs Si factor and PC2 is the K&Mg factor) for the three soil groups (the ferric, K&Mg, and silicic groups)

group (−1.1) and the silicic group (−0.7) than for the ferric group (−0.3). The TRB was higher for the K&Mg group than for the other groups, and no statistically significant difference was found between the ferric and silicic group TRB values. The CEC was higher for the K&Mg group than for the silicic group. The CEC/clay was higher for the silicic group than for the ferric group. The CEC/clay value for the K&Mg group was between and not significantly different from the CEC/clay values for the silicic and ferric groups. These properties are referred in the following sections.

Fig. 2.3 Profiles of the soils in the ferric, K&Mg, and silicic groups. The elevations are shown in *parentheses*

2.6 Effects of the Parent Materials and Degree of Weathering on the Distributions of Secondary Minerals in Kalimantan

2.6.1 Ferric Group

The ferric group samples were concluded to be derived from mafic parent materials. Mafic rocks contain large amounts of colored rock-forming minerals that are rich in Fe and Mg and are poorly resistant to weathering (Lasaga 1998; Nahon 1991). Soils formed from mafic rocks will therefore have high free iron oxide and clay contents (Schaetzl and Anderson 2005a; Gray et al. 2009). The total Fe contents of the ferric group soils were >210 cmol kg^{-1}, higher than the average Fe contents of basalt and andesite (149 and 98 cmol kg^{-1}, respectively) (Best 2003). The Mg contents were lower, at less than 13 cmol kg^{-1}, in the ferric group samples than in the K&Mg group samples, probably because intense weathering of the ferric group soils caused much of the Mg originally present to be leached out. The ferric group samples came from areas marked on the geological map as having intermediate to mafic geology (Supriatna et al. 1993, 1994), supporting the conclusion that the ferric soils were formed from mafic parent materials. The ferric group soils had high secondary iron oxide (Fe_d) and clay fraction contents (Table 2.3). The ferric group soils had high

Table 2.2 Total elemental composition (means and standard deviations) for each of the soil group

Group	Number of samples	Si cmol kg^{-1}			Al cmol kg^{-1}			Fe cmol kg^{-1}			Ca cmol kg^{-1}			Mg cmol kg^{-1}			K cmol kg^{-1}			Na cmol kg^{-1}		
Ferric	10	580	±240	b	520	±120	a	260	±30	a	0.9	±1.3	6.3	±3.9	b	2.3	±2.2	c	3.9	±6.1	a
K&Mg	13	1200	±200	a	335	±81	b	80	±30	b	1.9	±3.2	15	±4	a	28	±8	a	1.7	±1.0	a
Silicic	37	1200	±200	a	281	±104	b	70	±40	b	0.9	±1.3	3.5	±3.0	b	11	±9	b	6.5	±8.7	a

Different letters indicate the data were statistically different at $P < 0.05$

Table 2.3 Mean values and standard deviation of physicochemical and mineralogical properties for each soil group

Group (number of samples)		Ferric (10)		K&Mg (13)		Silicic (37)	
Total C	g kg^{-1}	11 ± 5	a	7 ± 3	ab	6 ± 3	b
pH(H$_2$O)		4.9 ± 0.3	a	4.9 ± 0.5	a	4.6 ± 0.3	b
pH(KCl)		4.6 ± 0.5	a	3.8 ± 0.3	b	3.9 ± 0.3	b
ΔpH		−0.3 ± 0.5	a	−1.1 ± 0.5	c	−0.7 ± 0.3	b
Clay	%	74 ± 13	a	53 ± 12	b	36 ± 14	c
Silt	%	15 ± 6	b	26 ± 6	a	22 ± 11	ab
Sand	%	11 ± 9	b	20 ± 11	b	41 ± 18	a
CEC	cmol$_c$ kg^{-1}	16 ± 10	ab	19 ± 8	a	13 ± 6	b
CEC/clay	cmol$_c$ kg^{-1} clay	24 ± 17	b	36 ± 13	ab	37 ± 16	a
Ex. bases	cmol$_c$ kg^{-1}	1.2 ± 1.5	ab	2.9 ± 2.8	a	0.6 ± 0.4	b
TRB	cmol$_c$ kg^{-1}	21 ± 8	b	64 ± 16	a	26 ± 17	b
Fe$_d$	g kg^{-1}	120 ± 20	a	37 ± 20	b	35 ± 20	b
Al$_o$	g kg^{-1}	3.0 ± 1.0	a	2.7 ± 1.5	a	1.6 ± 0.6	b
Fe$_o$	g kg^{-1}	3.1 ± 1.3	ab	3.6 ± 2.2	a	2.3 ± 1.8	b
1.4 nm[a]	%	4 ± 7	b	42 ± 24	a	24 ± 17	a
Kaolinite[a]	%	94 ± 9	a	47 ± 27	b	71 ± 22	b
Kaolinite[b]	g kg^{-1} clay	670 ± 120	a	530 ± 110	b	580 ± 140	ab
Gibbsite[b]	g kg^{-1} clay	88 ± 66	a	4 ± 11	b	30 ± 42	b

ΔpH pH(KCl) minus pH(H$_2$O), *CEC* cation exchange capacity, *CEC/clay* cation exchange capacity of the clay fraction, *Ex. bases* exchangeable bases, *TRB* total reserve in bases, *Fe$_d$* dithionite-citrate extractable Fe, *Al$_o$*, *Fe$_o$*, oxalate extractable Al, Fe
[a]Relative peak area in the X-ray diffractogram of the clay fraction
[b]Measured in the clay fraction from which iron oxide had been removed
Different letters indicate the data were statistically different at $P < 0.05$

ΔpH values (−0.3 ± 0.5), indicating that the samples were dominated by variable charges or had high oxide contents (Uehara and Gillman 1981) (Table 2.3). These results were consistent with soils from mafic rocks having high free iron oxide and clay contents (Schaetzl and Anderson 2005a; Gray et al. 2009). The ferric group samples contained little or no quartz or mica (Table 2.4), which are found in felsic rocks, mudstones, and sandstones, and little or no vermiculite (Table 2.4), which is formed through the weathering of mica.

The ferric group soils were strongly weathered and had low TRB values (21 ± 8 cmol$_c$ kg^{-1}; Table 2.3) and low iron oxide activities (Fe$_o$/Fe$_d$) (0.03 ± 0.02 g g^{-1}), indicating that the soils had low primary mineral contents (Delvaux et al. 1989; IUSS Working Group WRB 2015) and high crystalline iron oxide ratios (Shang and Zelazny 2008; Torrent et al. 1980), respectively. The secondary minerals contained in the soils were crystallized iron oxides, kaolinite, and gibbsite. The samples with high gibbsite contents (88 ± 66 g kg$_{clay}^{-1}$; Table 2.3) had low H$_4$SiO$_4^0$ activities, within the gibbsite stability field (Fig. 2.4). Poor resistance to the weathering of minerals in mafic parent materials has been found to result in low primary mineral

Table 2.4 Mineralogical properties of the soil samples

Site	Elevation	TRB	Selective dissolution					Peak detection by XRD[a]					Identified 1.4 nm minerals[b]	Kaolin	Gibbsite[c]
			Al_d	Fe_d	Al_o	Fe_o	Fe_o/Fe_d	1.4-nm	Mica	Kaolin	Gibbsite	Quartz			
	m	$cmol_c$ kg^{-1}	$g\ kg^{-1}$											$g\ kg_{clay}^{-1}$	
Ferric group – high elevation															
50 SK	1730	25	18	110	2.5	4.4	0.04	–	+	+	++	±	–	460	170
4 EK	700	24	14	74	4.5	4.4	0.06	–	–	++	+	–	–	750	71
6 EK	670	11	29	140	3.6	2.9	0.02	–	–	++	++	–	–	630	170
11 EK	650	16	20	120	2.6	2.5	0.02	+	–	++	+	–	V-HIV	660	72
55 SK	590	27	13	110	2.7	2.9	0.03	+	–	++	+	–	HIV	530	10
Ferric group – low elevation															
46 WK	70	34	25	120	5.0	4.9	0.04	–	–	+	++	–	–	830	48
31 CK	60	24	24	120	1.9	0.7	0.01	–	–	++	++	–	–	710	170
57 SK	60	19	16	120	2.5	2.3	0.02	+	–	++	±	–	HIV	640	31
58 SK	50	7	14	110	2.5	3.7	0.03	–	–	++	±	±	–	810	11
35 CK	20	20	26	150	2.7	2.1	0.01	–	–	++	++	–	–	690	120
K&Mg group – high elevation															
2 EK	720	54	11	50	7.1	9.7	0.20	+	±	++	–	±	Sm	660	n.d.
9 EK	670	55	2.4	18	2.4	5.2	0.28	++	++	++	±	++	HIV-V	500	1.7
54 SK	610	51	6.4	46	2.6	5.7	0.12	–	–	++	±	++	–	570	n.d.
K&Mg group – low elevation															
12 EK	460	57	3.1	17	3.0	3.3	0.20	++	±	+	±	+	HIV	670	12
56 SK	310	69	13	79	2.0	2.6	0.03	+	–	++	+	–	HIV	710	39
18 EK	70	53	3.9	30	1.2	2.7	0.09	++	+	+	–	++	HIV-V	510	n.d.
28 EK	30	87	4.8	39	3.0	2.6	0.07	+	+	++	±	++	Sm	500	1.3
17 EK	70	52	4.5	41	1.9	2.2	0.05	++	+	++	–	+	V-HIV	420	n.d.
21 EK	60	84	4.3	34	1.4	2.7	0.08	++	+	++	–	+	V	350	n.d.

(continued)

Table 2.4 (continued)

Site	Elevation	TRB	Selective dissolution					Peak detection by XRD[a]					Identified 1.4 nm minerals[b]	Kaolin	Gibbsite[c]
			Al_d	Fe_d	Al_o	Fe_o	Fe_o/Fe_d	1.4-nm	Mica	Kaolin	Gibbsite	Quartz			
	m	$cmol_c\ kg^{-1}$	$g\ kg^{-1}$											$g\ kg_{clay}^{-1}$	
47 WK	60	46	6.5	36	2.6	2.2	0.06	++	+	++	±	+	HIV	460	0.5
49 WK	40	57	4.3	26	3.1	2.9	0.11	++	+	++	±	+	HIV	410	n.d.
16 EK	70	69	3.9	28	1.6	1.0	0.03	+	+	++	−	+	V	590	2.5
26 EK	30	94	4.8	38	3.5	3.8	0.10	++	±	++	±	+	Sm	540	n.d.
Silicic group – high elevation															
51 SK	1450	39	6.3	69	2.0	6.4	0.09	++	++	++	+	++	HIV	280	25
52 SK	1170	20	4.1	38	2.2	4.0	0.11	+	−	++	±	±	HIV	570	n.d.
36 WK	1030	49	6.1	60	1.4	3.1	0.05	+	+	++	++	+	HIV	330	130
37 WK	910	33	4.1	41	1.8	3.3	0.08	+	−	++	+	±	HIV	660	45
1 EK	900	19	3.6	20	2.6	7.1	0.36	++	−	++	+	−	HIV	680	40
53 SK	810	37	8.6	73	2.9	1.8	0.02	+	−	++	−	+	HIV	700	0.7
38 WK	760	2	7.1	40	1.5	1.8	0.04	+	−	++	+	+	HIV	580	91
3 EK	710	37	2.5	13	2.4	5.6	0.42	++	+	++	+	++	V-HIV	490	12
5 EK	680	20	3.7	19	1.7	2.6	0.13	++	−	+	+	+	HIV	400	12
8 EK	670	31	4.0	18	2.1	3.0	0.17	++	−	++	±	+	HIV	260	14
7 EK	670	75	3.7	18	2.4	6.3	0.35	+	+	+	±	++	HIV-V	400	2.0
10 EK	660	38	2.4	20	0.9	2.6	0.13	++	+	++	±	++	HIV-V	420	11
39 WK	600	29	7.1	35	1.6	1.1	0.03	+	−	++	++	±	HIV	690	120
Silicic group – low elevation															
40 WK	490	12	8.7	71	2.0	3.5	0.05	+	−	++	±	±	V	690	2.2
41 WK	430	16	18	87	1.9	2.2	0.03	+	−	++	+	+	HIV-V	480	33
13 EK	350	12	4.1	31	1.5	4.2	0.13	++	±	++	±	±	HIV	580	1.5
42 WK	340	49	5.3	59	0.7	0.6	0.01	+	+	++	+	+	HIV-V	470	74

43 WK	150	6	8.5	51	1.8	1.4	0.03	+	−	++	++	−	HIV	650	63
15 EK	100	45	5.8	34	1.7	1.8	0.05	+	+	++	−	+	HIV-V	510	3.6
44 WK	120	31	5.5	38	2.0	1.5	0.04	++	−	++	±	±	HIV	640	7.2
14 EK	100	34	3.7	24	1.8	2.6	0.11	++	+	++	−	+	V-HIV	480	n.d.
45 WK	90	27	6.9	43	1.5	1.6	0.04	+	−	++	+	−	HIV	490	32
30 CK	80	38	5.0	30	0.8	0.2	0.01	−	−	++	+	−	−	710	110
48 WK	60	12	5.1	40	1.9	1.3	0.03	++	−	++	±	+	HIV	640	n.d.
20 EK	60	14	1.8	10	1.2	2.1	0.21	+	+	++	±	+	V-HIV	530	n.d.
19 EK	60	42	5.7	37	2.1	1.3	0.03	+	+	++	−	+	V	510	n.d.
59 SK	50	62	10	60	1.8	1.1	0.02	+	−	++	±	−	HIV	700	7.9
22 EK	50	14	4.0	28	1.5	0.8	0.03	+	−	++	±	+	HIV-V	560	0.8
32 CK	50	25	3.1	55	1.3	0.6	0.01	±	−	++	++	−	−	710	80
27 EK	30	15	4.0	27	1.4	0.9	0.03	+	−	++	±	+	HIV-V	550	1.6
25 EK	40	1	1.5	2.9	0.2	0.1	0.05	−	−	++	+	+	−	650	53
60 SK	40	38	4.1	42	1.0	2.5	0.06	+	−	++	−	−	HIV	730	n.d.
24 EK	40	6	1.6	11	0.8	1.1	0.10	+	−	++	±	+	HIV-V	700	n.d.
23 EK	40	18	1.6	14	1.0	1.2	0.09	+	−	++	−	++	HIV-V	540	n.d.
33 CK	40	1	0.7	2.4	1.3	0.3	0.12	±	−	++	++	−	−	730	140
34 CK	20	2	1.3	16	0.2	1.0	0.06	−	−	++	−	−	−	730	n.d.
29 EK	20	16	4.6	23	1.1	0.9	0.04	±	−	++	−	−	−	850	n.d.

TRB total reserve in bases, Al_d, Fe_d dithionite-citrate extractable Al, Fe, Al_o, Fe_o oxalate extractable Al, Fe

[a]XRD, X-ray diffraction; ++, very clear; +, clear; ±, unclear; −, not found

[b]V, vermiculite; HIV, hydroxy-Al interlayered vermiculite; Sm, smectite; V-HIV and HIV-V denote a small and large part of the 1.4 nm peak remained after K saturation, respectively

[c]n.d., not detected

contents (TRB values) or in desilication, and advanced weathering has been found to result in low $H_4SiO_4^0$ activities, favoring the formation of gibbsite (Karathanasis 2002). Eight out of the ten ferric group samples had lower TRB values than the indicator of weatherable minerals in the ferric horizon (25 $cmol_c$ kg^{-1} (IUSS Working Group WRB 2015)), but three of the ten samples were classified as Ferralsols because they had high CEC/clay values. The ferric group soils were formed on mafic rocks rich in easily weathered minerals, and they were actually more weathered, with low active Fe ratios (0.03 ± 0.02 g g^{-1}) and TRB values (21 ± 8 $cmol_c$ kg^{-1}), than were the soils in the other groups. However, the ferric group soils had high CEC/clay values or contained high activity clays, and more weathering could occur. Some of the soils would not have matured enough to be classified as Ferralsols because of the hilly and undulating topography of Kalimantan (van Bemmelen 1970).

The distributions of secondary minerals in the soils derived from mafic rocks in Kalimantan and the formation of these secondary minerals have been described in few publications. 2:1-type minerals were found to be absent or were only present in small amounts, and gibbsite is present in the soils derived from mafic rocks in Kalimantan. This contrasts with the mineralogies of soils formed on mudstones or sandstones, which contain 2:1-type minerals and lack gibbsite (Koch et al. 1992; Watanabe et al. 2006). These soils lack 2:1-type minerals because of the absence of mica in the mafic parent materials and because of intense leaching preventing the formation of smectite, which has been found to occur in drier soils derived from mafic parent materials (Vingiani et al. 2004; Righi et al. 1999; Rasmussen et al. 2010).

2.6.2 K&Mg Group

The K&Mg group soils were concluded to be derived from felsic rocks or mudstones and sandstones and that they had been weathered relatively little. The mean Fe content of the K&Mg group soils was 84 ± 35 cmol kg^{-1}, which is lower than the typical Fe contents of basalt and andesite (149 and 98 cmol kg^{-1}, respectively) (Best 2003). Less-mobile Fe in the parent material will become more concentrated during the soil formation process, so the soils were concluded to be derived from rocks with even lower Fe contents, i.e., felsic or sedimentary rocks. These soils were collected from areas that are shown on the geological map to have felsic or sedimentary rocks (Supriatna et al. 1993, 1994). The low Fe_d and clay contents of the K&Mg group soils and the presence of mica, vermiculite, and quartz indicated that the soils were derived from felsic or sedimentary rocks (e.g., mudstones, sandstones, and slate) that contained mica. The high total K and Mg contents (Table 2.2) of the soils may have been derived from 2:1-type minerals (mica and smectite). The K&Mg group soils had high mica and vermiculite contents (Tables 2.3 and 2.4), and the mica was not completely weathered into vermiculite, unlike in the silicic group soils. The soils were from sites that were close to rivers

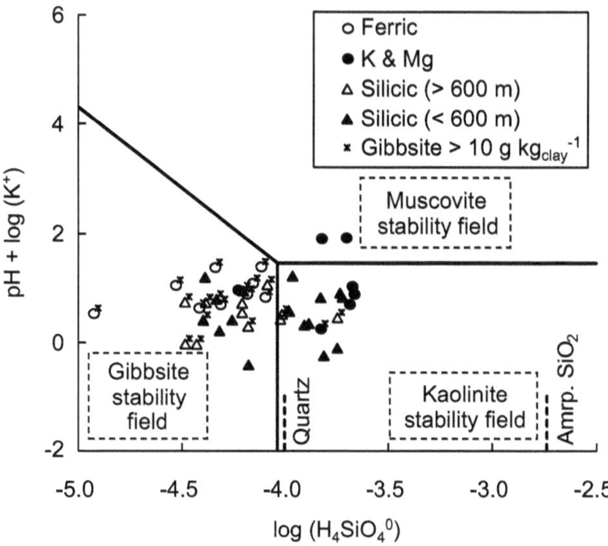

Fig. 2.4 Compositions of the soil solution extracts plotted on a diagram representing the relative stabilities of gibbsite, kaolinite, and muscovite. Samples with higher gibbsite contents (>10 g kg_{clay}^{-1}) are marked with *asterisks*. The short *dashed vertical lines* indicate the solubilities of quartz and amorphous SiO_2

(where deposition of river sediment can be assumed to occur) or from rather undulating landscapes (where relatively high levels of erosion may have occurred recently), so the soils may have been formed over a short period of time. The K&Mg group soils were less mature than were the soils in the other groups, and half of the K&Mg group soils were classified as Cambisols.

The K&Mg group soils were characterized by the presence of 2:1-type minerals (mica and vermiculite) and low gibbsite contents and occurrence rates (Tables 2.3 and 2.4). Most of the K&Mg group soil solution compositions were in the kaolinite stability field (Fig. 2.4), and some were in the muscovite stability field. The solution compositions corresponded with the mineralogies of the clay fractions, that is, the presence of mica, and the absence or presence of only small amounts of gibbsite. The samples had high TRB values, indicating that primary minerals remained in the soils. The high primary mineral contents would cause the $H_4SiO_4^{0}$ activities to be high, meaning that the formation of kaolinite would be favored over the formation of gibbsite. Mica was found to be less stable than kaolinite in the samples (which had low *y*-axis (i.e., pH + log (K^+)) values in Fig. 2.4), and the mica was weathered to form vermiculite. The low pH values and K^+ activities found in aqueous extracts of the soils would have been caused by the high levels of precipitation and strong leaching effects that occur in Kalimantan.

2.6.3 Silicic Group

The silicic group soils were concluded to be derived from felsic or sedimentary rocks but were more weathered than the K&Mg group soils. Felsic or sedimentary rocks were concluded to be the parent materials for the same reasons as the conclusions for the K&Mg group soils were drawn, i.e., because the total Fe contents were low and because of the information provided by geological maps (Supriatna et al. 1993, 1994). The silicic group soils had low Fe_d and clay contents and contained vermiculite and quartz (Tables 2.3 and 2.4), supporting the conclusions about the parent materials. Some of the silicic group soils contained no or only a small amount of quartz, but the soils did contain 2:1-type minerals or had low Fe_d contents. The soils had low TRB values, and the mica peak in the X-ray diffractogram of each silicic group soil was absent or small (Table 2.4), indicating that the soils were more weathered than were the K&Mg group soils. The silicic group soils were mainly classified as Alisols, which are the most common upland soils in Kalimantan (FAO et al. 1998, 2003; Funakawa et al. 2008).

The clay fractions in the silicic group soils were characterized by abundant kaolinite and the presence of gibbsite and vermiculite with or without interlayer Al hydroxides (Tables 2.3 and 2.4). The soil solution extracts were in the gibbsite or kaolinite stability fields (Fig. 2.4), which agreed with the conclusion that the mica had been weathered to form vermiculite. It has previously been verified that vermiculite can form from mica (Malla 2002; Churchman and Lowe 2012), and vermiculite has been found to form from mica under udic soil moisture regimes (Watanabe et al. 2006; Nakao et al. 2009). This contrasts with the limited degree to which vermiculite is transformed in regions under ustic moisture regimes (e.g., Thailand and Tanzania) (Watanabe et al. 2006; Nakao et al. 2009; Araki et al. 1998). The high pH values in such areas will mean that mica is stable (Watanabe et al. 2006). The Al_o contents were lower in the silicic group soils than in the other soils (Table 2.3). The Al_o contents of the silicic group soils may have been low because of low clay contents. Clay contents have been found to strongly correlate with Al_o (Ohta and Effendi 1993). The clay fraction may have been translocated to lower horizons (Ohta and Effendi 1993). Al_o in soil may also dissolve when the soil pH is low (Funakawa et al. 1993; Watanabe et al. 2008; Fujii et al. 2009). The Fe_o contents were lower in the silicic group soils than in the K&Mg group soils. Fe_o has been found to be more associated with the total C content than with the clay content (Ohta and Effendi 1993). However, the differences between the total C contents of the different groups were not significant in this study. Fe_o can also be translocated with dissolved organic carbon (Fujii et al. 2009).

2.7 Distributions of Secondary Minerals at Different Elevations

The secondary mineral distributions were found to be related to the elevation for the silicic group soils but not for the ferric and K&Mg group soils. For the silicic group, the kaolinite, 1.4 nm minerals, interlayer Al hydroxides, gibbsite, Al_o, and Fe_o distributions were all related to the elevation.

Most of the ferric group soils did not contain 2:1-type minerals, so no trend in layer silicates (kaolinite, 1.4 nm minerals, and interlayer Al hydroxides) could be found. It is not clear what other factors could explain there being no trend in the secondary mineral distributions at different elevations for the ferric and K&Mg group soils. The ferric group soils were presumably highly weathered, and all of these soils had started to lose certain mineralogical variations, i.e., all of the soils contained gibbsite and had lost most of the poorly crystalline Al and Fe (Al_o and Fe_o). The K&Mg soils were less mature than the soils in the other groups, and the K&Mg soils had higher primary mineral contents (including of mica), as suggested by the TRB values (Table 2.3), than did the soils in the other groups. These high primary mineral contents would have disturbed the possible mineral distributions at different elevations because the primary minerals will have provided 2:1-type layer silicates and Al and Fe (which could have been the source of poorly crystalline Al and Fe).

A trend was found in the secondary mineral distributions in the silicic group soils with elevation. The soils from high elevations contained more gibbsite and more 1.4 nm minerals (vermiculite), had higher Al_o and Fe_o contents, and contained less kaolinite than did the soils from lower elevations (Table 2.5). More interlayer Al hydroxides were found in the soils from higher elevations than in the soils from lower elevations (Table 2.4). The soils from high elevations also had high TRB values. The soils from low elevations were more weathered than the soils from high elevations in terms of the secondary minerals (the soils from low elevations had high kaolinite contents and low 2:1-type mineral, Al_o, and Fe_o contents) and TRB values. The higher kaolinite contents and lower 2:1-type mineral contents and TRB values at lower elevations may indicate that kaolinite formed at the expense of the 2:1-type minerals and other primary minerals because of the higher temperatures (Schaetzl and Anderson 2005b) at lower elevations than at higher elevations. Less annual precipitation and excess precipitation (i.e., precipitation minus potential evapotranspiration) occurs at the low elevation sites than at the high elevation sites. The moisture contents at both elevations would be sufficient for mineral weathering to occur, and it was not limited by the amount of water available (Rasmussen et al. 2007). The Al_o and Fe_o contents may have been higher at high elevations than at low elevations because of crystallization being retarded by organic matter (Schwertmann 1985; Huang et al. 2002) at the relatively low temperatures the soils at higher elevations would have been exposed to. The gibbsite contents and the amounts of interlayer Al hydroxides between the 2:1 layers were also higher at higher elevations than at lower elevations. The abundance

Table 2.5 Spearman's rank correlation coefficients for the relationships between elevation and the mineralogical and physicochemical properties

Group	Number of samples	Relative peak area in XRD analysis			Gibbsite	Kaolinite	Selective dissolution			TC	Clay	CEC	TRB
		1.4-nm	mica	kaolin			Fe_d	Al_o	Fe_o				
All	60				0.33*	−0.30*		0.31*	0.59***			0.34**	
Ferric	10											0.76***	
K & Mg	13												
Silicic	37	0.51**		−0.54***	0.37*	−0.49**	0.36*	0 62***	0 75***			0.40*	0.37*

Fe_d dithionite-citrate extractable Fe, Al_o, Fe_o oxalate extractable Al, Fe, *CEC* cation exchange capacity, *TRB* total reserve in bases. Only statistically significant values ($P < 0.05$) are shown
*, $P < 0.05$; **, $P < 0.01$; ***, $P < 0.001$

of Al hydroxides (i.e., gibbsite and interlayer Al hydroxides) in the soils from high elevations would have been caused by the low $H_4SiO_4^0$ activities of the soils at high elevations (Fig. 2.4) retarding the loss of Al through the formation of kaolinite. The low $H_4SiO_4^0$ activities might have been caused by the leaching (Huang et al. 2002) that would be caused by the high levels of excess precipitation found at high elevations. The stability diagram shown in Fig. 2.4 indicated that the soils from low elevations may have been in transitional states between kaolinite-dominated soils with high $H_4SiO_4^0$ activities and gibbsite-dominated soils with low $H_4SiO_4^0$ activities.

Other possible reasons for there being trends in the secondary mineral distributions in the silicic group soils as the elevation changed are given next. Higher contents of 2:1-type minerals in the parent materials at high elevations would affect the relative amounts of kaolinite and 2:1-type minerals in the soil. Higher erosion intensities at high elevations would continually provide new material for soils to form from and increase the 2:1-type mineral and poorly crystalline Al and Fe (Al_o and Fe_o) contents. The 2:1-type mineral contents of the parent materials were not examined in this study, and differences in erosion intensities were not apparent, at least through field and profile observations. These possibilities cannot, however, be ruled out.

Higher occurrence rates of gibbsite and interlayer Al hydroxides between 2:1 layers at high elevations in Kalimantan have not been described before, although the possibility that this is the case has previously been suggested from data for soil from one site (Watanabe et al. 2006). Gibbsite has been found at high elevations in other tropical regions (Betard 2012; Herrmann et al. 2007). In Thailand, gibbsite is found at high elevations (>1000 m), where more leaching is expected than at low elevations (Herrmann et al. 2007). The gibbsite contents of soil and the amounts of interlayer Al hydroxides present increase as elevation increases in the Sierra Nevada in central California (Dahlgren et al. 1997). Weathering and intense leaching at high elevations may cause gibbsite and hydroxyl Al interlayered vermiculite to dominate in that area (Dahlgren et al. 1997). The weathering of 2:1-type minerals is assumed to generally follow the pathway mica $\rightarrow$ vermiculite $\rightarrow$ hydroxyl Al interlayered vermiculite (Barnhisel and Bertsch 1989; Churchman and Lowe 2012). In this study, however, the more weathered soils found at low elevations (suggested by the soils having low TRB values, low 2:1-type mineral contents, low Al_o and Fe_o contents, and high kaolinite contents) contained less interlayer Al hydroxides than did the less weathered soils at high elevations (Table 2.4). Higher $H_4SiO_4^0$ activity in the soils at low elevations may promote kaolinite formation and retard the formation of Al hydroxides (Huang et al. 2002). Furthermore, interlayer Al hydroxides dissolve at low pH values (Watanabe et al. 2006), so less interlayer Al hydroxides will be present when the $pH(H_2O)$ is lower than 5.0 (Ohta and Effendi 1993). However, interlayer Al hydroxides are presumably replenished at high elevations by primary minerals dissolving, as has been found in less weathered Cambisols in Japan (Watanabe et al. 2006).

2.8   Neoformation of Secondary Minerals Containing Al

The $H_4SiO_4^0$ activity was concluded to control the neoformation of secondary minerals containing Al (i.e., kaolinite, gibbsite, and interlayer Al hydroxides). Gibbsite was formed at low $H_4SiO_4^0$ activities in highly weathered ferric group soils and less weathered silicic group soils containing 2:1-type minerals from high elevations. Gibbsite is generally assumed to be the ultimate product of weathering, but it was still found in the soils containing 2:1-type minerals that were subjected to strong leaching conditions (Dahlgren et al. 1997, Herrmann et al. 2007). Gibbsite forms in the initial stages of weathering under strong leaching conditions with low $H_4SiO_4^0$ activities (Vazquez 1981; Herrmann et al. 2007), even from highly weatherable tephra (Churchman and Lowe 2012; Ndayiragije and Delvaux 2003). In the study area, gibbsite formed in the ferric group soils because of strong weathering and in the silicic group soils at high elevations because of strong leaching conditions. The $H_4SiO_4^0$ activity was low in both cases. On the other hand, for the K&Mg group soils and silicic group soils at low elevations, insufficient weathering or insufficient leaching to lower the $H_4SiO_4^0$ activity occurs, and kaolinite rather than gibbsite was dominant. The anti-gibbsite effect (Jackson 1963; Ndayiragije and Delvaux 2003), the inhibition of the formation of gibbsite by the precipitation of Al between 2:1 layers, was not apparent in the silicic group soils, but gibbsite and interlayer Al hydroxides seemed to coincide, as has been found previously (Watanabe et al. 2006). The gibbsite and interlayer Al hydroxide trends could have been similar because kaolinite formed at high $H_4SiO_4^0$ activities inhibited the formation of polynuclear Al (Huang et al. 2002).

2.9   Conclusion

In this study, the effects of the parent materials and the elevation on the distributions of secondary minerals in upland soils in Kalimantan were examined. The primary factor controlling the distributions of secondary minerals was the parent material (mafic or felsic/sedimentary), and the secondary factor was the climate at different elevations (Fig. 2.5).

The parent material affected the occurrence of 2:1-type minerals (i.e., mica and vermiculite) and the neoformation of gibbsite and kaolinite. The parent materials of the ferric group soils were mafic rocks, which lack 2:1-type minerals and are poorly resistant to weathering. The soils were therefore strongly weathered, and the $H_4SiO_4^0$ activities in the soils were low. The low $H_4SiO_4^0$ activities favored the formation of gibbsite. The parent materials of the K&Mg and silicic group soils were felsic rocks (e.g., granite) or sedimentary rocks (e.g., mudstone, sandstones, and slate). Mica derived from these parent materials weathered to form vermiculite at the low pH values of the soils. The relatively high primary mineral contents of the

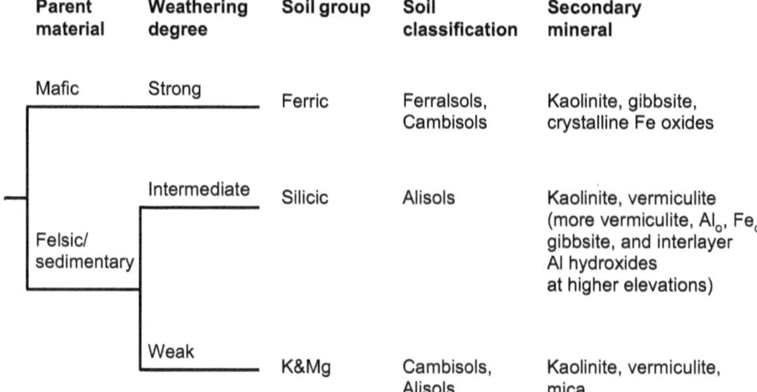

Fig. 2.5 Schematic diagram of the conditions leading to the secondary mineral distributions found in upland soils in Kalimantan, Indonesia. The soil classifications and secondary minerals shown are not exhaustive. Al_o and Fe_o are the acid ammonium oxalate extractable Al and Fe, respectively

K&Mg group soils (indicated by high TRB values) may have caused the soils to have high $H_4SiO_4^0$ activities, favoring the formation of kaolinite.

Effects of the climate on the secondary mineral distribution were only apparent in the silicic group soils (Fig. 2.5), most of which are the most common soils (Alisols) in Kalimantan. Low temperatures at high elevations would have retarded the crystallization of oxides, resulting in the Al_o and Fe_o contents being relatively high. The decrease in vermiculite and increase in kaolinite with decreasing elevation would have been caused by the weathering or dissolution of 2:1-type minerals at the high temperatures found at low elevations. High gibbsite contents were found at high elevations, and this was caused by the $H_4SiO_4^0$ activities being low at high elevations because of the high leaching rates caused by low evapotranspiration rates.

The observed trends in the distributions of secondary minerals have implications for differences in soil properties (e.g., CEC, P sorption capacity, and accumulation of organic matter) and therefore nutrient dynamics in soils in Kalimantan.

Acknowledgement This chapter is derived from Watanabe et al. (2017) published in Soil Science Society of America Journal, Soil Science Society of America, doi: 10.2136/sssaj2016.08.0263.

References

Araki S, Masanya BM, Magoggo JP, Kimaro DS, Kitagawa Y (1998) Characterization of soils on various planation surfaces in Tanzania. In: Proceedings of the World congress of soil science, Montpellier, France

Barnhisel IB, Bertsch PM (1989) Chlorites and hydroxy-interlayered vermiculite and smectite. In: Dixon JB, Weed SB (eds) Minerals in soil environments. vol SSSA Book Series 1. 2nd edn SSSA, Madison, pp 729–788

Best MG (2003) Composition and classification of magmatic rocks. In: Igneous and metamorphic petrology. 2nd edn. Blackwell, Boston, pp 16–50

Betard F (2012) Spatial variations of soil weathering processes in a tropical mountain environment: The Baturite massif and its piedmont (Ceara, NE Brazil). Catena 93:18–28. doi:10.1016/j.catena.2012.01.013

Blakemore LC, Searle PL, Daly BK (1981) Methods for chemical analysis of soils. New Zealand Soil Bureau Scientific Report 10A. DSIR, New Zealand

Brady NC, Weil RR (2002) Formation of soils from parent materials. In: The nature and properties of soils, 13th edn. Prentice Hall, Upper Saddle River

Brantley SL, Chen Y (1995) Chemical weathering rates of pyroxenes and amphiboles. In: White FA, Brantley SL (eds) Chemical weathering rates of silicate minerals, vol 31. Reviews in Mineralogy. Mineralogical Society of America, Washington, DC, pp 119–172

Bruun TB, Elberling B, Christensen BT (2010) Lability of soil organic carbon in tropical soils with different clay minerals. Soil Biol Biochem 42(6):888–895. doi:10.1016/j.soilbio.2010.01.009

Churchman GJ, Lowe DJ (2012) Alteration, formation, and occurrence of minerals in soils. In: Huang PM, Li Y, Sumner ME (eds) Handbook of soil science properties and processes, 2 edn. CRC Press, Boca Raton

Dahlgren RA, Boettinger JL, Huntington GL, Amundson RG (1997) Soil development along an elevational transect in the western Sierra Nevada, California. Geoderma 78(3–4):207–236. doi:10.1016/s0016-7061(97)00034-7

Delvaux B, Herbillon AJ, Vielvoye L (1989) Characterization of a weathering sequence of soils derived from volcanic ash in Cameroon – taxonomic, mineralogical and agronomic implications. Geoderma 45(3–4):375–388

Driscoll CT (1984) A procedure for the fractionation of aqueous aluminum in dilute acidic waters. Int J Environ Anal Chem 16(4):267–283

FAO, EC, ISRIC (2003) World soil resources. FAO, Rome

FAO, ISRIC, ISSS (1998) World reference base for soil resources. 84 World soil resources reports. FAO, Rome

Fujii K, Uemura M, Hayakawa C, Funakawa S, Sukartiningsih KT, Ohta S (2009) Fluxes of dissolved organic carbon in two tropical forest ecosystems of East Kalimantan, Indonesia. Geoderma 152(1–2):127–136. doi:10.1016/j.geoderma.2009.05.028

Funakawa S, Hirai H, Kyuma K (1993) Speciation of Al in soil solution from forest soils in northern Kyoto with special reference to their pedogenetic process. Soil Sci Plant Nutr 39:281–290

Funakawa S, Watanabe T, Kosaki T (2008) Regional trends in the chemical and mineralogical properties of upland soils in humid Asia: with special reference to the WRB classification scheme. Soil Sci Plant Nutr 54(5):751–760. doi:10.1111/j.1747-0765.2008.00294.x

Gray JM, Humphreys GS, Deckers JA (2009) Relationships in soil distribution as revealed by a global soil database. Geoderma 150(3–4):309–323. doi:10.1016/j.geoderma.2009.02.012

Herrmann L, Anongrak N, Zarei M, Schuler U, Spohrer K (2007) Factors and processes of gibbsite formation in Northern Thailand. Catena 71(2):279–291

Huang PM, Wang MK, Kampf N, Schulze DG (2002) Aluminum hydroxides. In: Dixon JB, Schulze DG (eds) Soil mineralogy with environmental applications. SSSA Book Series 7. SSSA, Madison, pp 261–289

Hudson BD (1995) Reassessment of Polynov ion mobility series. Soil Sci Soc Am J 59 (4):1101–1103

IUSS Working Group WRB (2015) World reference base for soil resources 2014, Update 2015. World Soil Resources Reports No. 106. FAO, Rome

Jackson ML (1963) Aluminum bonding in soils: a unifying principle in soil science. Soil Sci Soc Am Proc 27(1):1–10

Karathanasis AD (2002) Mineral equilibria in environmental soil systems. In: Dixon JB, Schulze DG (eds) Soil mineralogy with environmental applications. vol SSSA Book Series 7. SSSA, Madison, pp 109–151

Karathanasis AD, Adams F, Hajek BF (1983) Stability relationships in kaolinite, gibbsite, and Al-hydroxyinterlayered vermiculite soil systems. Soil Sci Soc Am J 47(6):1247–1251

Koch CB, Bentzon MD, Larsen EW, Borggaard OK (1992) Clay mineralogy of two Ultisols from Central Kalimantan. Soil Sci 154:158–167

Lasaga AC (1998) Rate laws of chemical reactions. In: Kinetic theory in the Earth sciences. Princeton Series in Geochemistry. Princeton University Press, Princeton, pp 3–151

Lindsay WL (1979) Aluminosilicate minerals. In: Chemical equilibria in soils. Wiley, New York, pp 56–77

Malla PB (2002) Vermiculites. In: Dixon JB, Schulze DG (eds) Soil mineralogy with environmental applications. vol SSSA Book Series 7. SSSA, Madison, pp 501–529

McKeague JA, Day JH (1966) Dithionite- and oxalate-extractable Fe and Al as aids in differentiating various classes of soils. Can J Soil Sci 46:13–22

Murray HH, Keller WD (1993) Kaolins, kaolins, and kaolins. In: Murray HH, Bundy WM, Harvey CC (eds) Kaolin genesis and utilization. The Clay Mineral Society, Boulder, pp 1–24

Nagy KL (1995) Dissolution and precipitation kinetics of sheet silicates. In: White FA, Brantley SL (eds) Chemical weathering rates of silicate minerals, vol 31. Reviews in Mineralogy. Mineralogical Society of America, Washington, DC, pp 173–233

Nahon DB (1991) Introduction to major geochemical processes of weathering. In: Introduction to the petrology of soils and chemical weathering. Wiley, New York, pp 1–49

Nakao A, Funakawa S, Watanabe T, Kosaki T (2009) Pedogenic alterations of illitic minerals represented by Radiocaesium interception potential in soils with different soil moisture regimes in humid Asia. Eur J Soil Sci 60 (1):139–152. doi:10.1111/j.1365-2389.2008.01098.x

Ndayiragije S, Delvaux B (2003) Coexistence of allophane, gibbsite, kaolinite and hydroxy-Al-interlayered 2: 1 clay minerals in a perudic Andosol. Geoderma 117:203–214. doi:10.1016/s0016-7061(03)00123-x

Norfleet ML, Karathanasis AD, Smith BR (1993) Soil solution composition relative to mineral distribution in Blue Ridge Mountain soils. Soil Sci Soc Am J 57(5):1375–1380

Ohta S, Effendi S (1993) Ultisols of lowland dipterocarp forest in east Kalimantan, Indonesia, III. Clay minerals, free oxides, and exchangeable cations. Soil Sci Plant Nutr 39:1–12

Parfitt RL (1978) Anion adsorption by soils and soil materials. In: Brady NC (ed) Advances in agronomy, vol 30. Academic Press Inc., New York, pp 1–50

Percival HJ, Parfitt RL, Scott NA (2000) Factors controlling soil carbon levels in New Zealand grasslands: is clay content important? Soil Sci Soc Am J 64(5):1623–1630

Price JR, Velbel MA (2003) Chemical weathering indices applied to weathering profiles developed on heterogeneous felsic metamorphic parent rocks. Chem Geol 202:397–416

Rai D, Kittrick JA (1989) Mineral equilibria and the soil system. In: Dixon JB, Weed SB (eds) Minerals in soil environments. vol SSSA Book Series 1, 2nd edn. SSSA, Madison, pp 161–198

Rasmussen C, Dahlgren RA, Southard RJ (2010) Basalt weathering and pedogenesis across an environmental gradient in the southern Cascade Range, California, USA. Geoderma 154 (3–4):473–485. doi:10.1016/j.geoderma.2009.05.019

Rasmussen C, Matsuyama N, Dahlgren RA, Southard RJ, Brauer N (2007) Soil genesis and mineral transformation across an environmental gradient on andesitic lahar. Soil Sci Soc Am J 71(1):225–237. doi:10.2136/sssaj2006.0100

Reid-Soukup DA, Ulery AL (2002) Smectites. In: Dixon JB, Schulze DG (eds) Soil mineralogy with environmental applications. vol SSSA Book Series 7. SSSA, Madison, pp 467–499

Righi D, Terribile F, Petit S (1999) Pedogenic formation of kaolinite-smectite mixed layers in a soil toposequence developed from basaltic parent material in Sardinia (Italy). Clay Clay Miner 47:505–514

Schaetzl R, Anderson S (2005a) Soil parent materials. In: Soils: genesis and geomorphology. Cambridge University Press, Cambridge, pp 167–225

Schaetzl R, Anderson S (2005b) Weathering. In: Soils: genesis and geomorphology. Cambridge University Press, Cambridge, pp 226–238

Schulze DG (2002) An introduction to soil mineralogy. In: Dixon JB, Schulze DG (eds) Soil mineralogy with environmental applications. vol SSSA Book Series 7. SSSA, Madison, pp 1–35

Schwertmann U (1985) The effect of pedogenic environments on iron oxide minerals. In: Stewart BA (ed) Advances in soil science, vol 1. Springer, New York, pp 171–200

Shang C, Zelazny L (2008) Selective dissolution techniques for mineral analysis of soils and sediments. In: Ulery AL, Drees LR (eds) Methods of soil analysis Part 5 mineralogical methods. Soil Science Society of America, Inc., Madison, pp 33–80

Supriatna S, Djamal B, Heryanto R, Sanyoto P (1994) Geological map of Indonesia, Banjarmasin sheet. Geological research and development center, Bandung

Supriatna S, Heryanto R, Sanyoto P (1993) Geological map of Indonesia, Singkawang sheet. Geological research and development center, Bandung

Thornthwaite CW (1948) An approach toward a rational classification o f climate. Geogr Rev 38:55–94

Torrent J, Schwertmann U, Schulze DG (1980) Iron-oxide mineralogy of some soils of 2 river terrace sequences in Spain. Geoderma 23(3):191–208

Uehara G, Gillman G (1981) Chemistry. In: The mineralogy, chemistry, and physics of tropical soils with variable charge clays. Westview tropical agriculture series, vol 4. Westviwe Press, Inc., Colorado, pp 31–95

van Bemmelen RW (1970) The geology of Indonesia, vol vol. IA, 2nd edn. the Government Printing Office of the Netherlands, The Hague

van Breemen N, Brinkman R (1976) Chemical equilibria and soil formation. In: Bolt GH, Wruggenwert MGM (eds) Soil chemistry, A. Basic elements. Elsevier Scientific Publishing Company, Amsterdam, pp 141–170

Vazquez FM (1981) Formation of gibbsite in soils and saprolites of temperature-humid zones. Clay Miner 16(1):43–52

Vingiani S, Righi D, Petit S, Terribile F (2004) Mixed-layer kaolinite-smectite minerals in a red-black soil sequence from basalt in Sardinia (Italy). Clay Clay Miner 52(4):473–483. doi:10. 1346/ccmn.2004.0520408

Watanabe T, Funakawa S, Kosaki T (2006) Clay mineralogy and its relationship to soil solution composition in soils from different weathering environments of humid Asia: Japan, Thailand and Indonesia. Geoderma 136(1–2):51–63. doi:10.1016/j.geoderma.2006.02.001

Watanabe T, Hase E, Funakawa S, Kosaki T (2015) Inhibitory effect of soil micropores and small mesopores on phosphate extraction from soils. Soil Sci 180(3):97–106. doi:10.1097/ss. 0000000000000116

Watanabe T, Hasenaka Y, Hartono A, Sabiham S, Nakao A, Funakawa S (2017) Parent Materials and Climate Control Secondary Mineral Distributions in Soils of Kalimantan, Indonesia. Soil Sci Soc Am J (In press). doi:10.2136/sssaj2016.08.0263

Watanabe T, Ogawa N, Funakawa S, Kosaki T (2008) Relationship between chemical and mineralogical properties and the rapid response to acid load of soils in humid Asia: Japan, Thailand and Indonesia. Soil Sci Plant Nutr 54(6):856–869. doi:10.1111/j.1747-0765.2008. 00316.x

Chapter 3
Influence of Climatic Factor on Clay Mineralogy in Humid Asia: Significance of Vermiculitization of Mica Minerals Under a Udic Soil Moisture Regime

Shinya Funakawa and Tetsuhiro Watanabe

Abstract Geology and climate had a clear impact on the clay mineral composition for the soils in humid Asia. In soils derived from mica-free parent materials, such as mafic volcanic rocks, the clay mineral composition changes to the kaolin apex along the mineral axis of the 1.4-nm minerals. For other soils derived from the sedimentary rocks or felsic parent materials, mica and kaolin minerals dominated with lesser amounts of the 1.4-nm minerals in northern Thailand, while significant amounts of 1.4-nm minerals were formed in Indonesia and Japan. Based on these findings and a thermodynamic analysis using soil water extracts, the mineral weathering sequences in the soils from felsic and sedimentary rocks were postulated for each of the regions. In Thailand, under higher pH conditions associated with the ustic moisture regime, mica is relatively stable, while other primary minerals, such as feldspars, are unstable and dissolve to form kaolinite and gibbsite. Under the lower pH conditions of the udic moisture regime in Japan and Indonesia, mica weathers to form 1.4-nm minerals. Differentiated in this way, the soil mineralogical properties are considered to affect the chemical properties of soils, such as the cation exchange capacity (CEC)/clay and pH, and also the taxonomic classification of the soils. The CEC/clay of the soils derived from the sedimentary rocks or felsic parent materials showed a clear regional trend; that is, it was usually higher than 24 $cmol_c\ kg^{-1}$ (corresponding to Alisols if the argic horizon is recognized) under the udic and perudic soil moisture regimes in Indonesia and Japan. It was predominantly lower than 24 $cmol_c\ kg^{-1}$ (corresponding to Acrisols) under the ustic soil moisture regime in Thailand. In contrast, soils derived from the mafic volcanic rocks or limestones were more variable in clay mineral composition, the CEC/clay, or pH and were often high in Fe_d. The World Reference Base for Soil Resources (WRB)

S. Funakawa (✉)
Graduate School of Global Environmental Studies, Kyoto University, Yoshida-Honmachi, Sakyo-ku, Kyoto 606-8501, Japan
e-mail: funakawa@kais.kyoto-u.ac.jp

T. Watanabe
Graduate School of Agriculture, Kyoto University, Kitashirakawa-Oiwakecho, Sakyo-ku, Kyoto 606-8502, Japan
e-mail: nabe14@kais.kyoto-u.ac.jp

© Springer Japan KK 2017
S. Funakawa (ed.), *Soils, Ecosystem Processes, and Agricultural Development*,
DOI 10.1007/978-4-431-56484-3_3

classification is generally consistent with the regional trends of the chemical and the mineralogical properties of soils and describes successfully the distribution patterns of acid soils in humid Asia using the criterion of CEC/clay $= 24$ cmol$_c$ kg^{-1}.

Keywords Cation exchange capacity (CEC) • Humid Asia • Mineral weathering • Soil moisture regime • Vermiculitization

3.1　Introduction

In the previous chapter, the influence of both the geological and the climatic factors on clay mineralogy was analyzed using the soils from Indonesia and Cameroon. It was found that the major weathering pathways of the secondary minerals were different depending on the mineralogical composition of the parent materials, i.e., the transformation of dioctahedral mica regulated the clay mineralogy of the soils developed from the sedimentary rocks or the felsic igneous rocks, while the primary minerals rapidly weathered to form kaolinite and sesquioxides of Fe/Al among the soils developed from the mafic volcanic rocks. In this chapter, we analyzed the clay mineralogy of soils in humid Asia, i.e., Indonesia, Thailand, and Japan (for comparison), which were mostly derived from the sedimentary rocks or the felsic igneous rocks, with special reference to the influence of the soil moisture regime. Based on the analysis in Chap. 2, the expandable 2:1 minerals are expected to be one of the major components that would affect the soil chemical and physical characteristics through the appearance of high permanent negative charges derived from the isomorphic substitution in either the tetrahedral or the octahedral sheets. The soils in humid Asia exhibit relatively incipient mineralogical characteristics due to dominant steep slopes, crust movement, and volcanic activity on "young alpine fold belts compared with soils developed on stable plains associated with the 'Precambrian shield' in eastern South America or equatorial Africa (Driessen et al. 2001)." Therefore, it is expected that soils found under different development and weathering stages might be involved in this region, which would be favorable for assessing the weathering pathways of clay minerals.

On the other hand, a new classification scheme of the world's soils was recently proposed – the World Reference Base for Soil Resources (WRB) (ISSS-ISRIC-FAO 1998a; IUSS Working Group WRB 2014). One of the new concepts of the WRB was the introduction of clay activity (or the cation exchange capacity (CEC)/ clay) to classify soils in the highest category level (i.e., the reference soil groups [RSG]), which are used to distinguish soils that have an argic horizon into Lixisols and Luvisols (high base saturation [> 50 %] soils) or Acrisols and Alisols (low base saturation [< 50 %] soils). Upland soils in humid Asia are predominantly acidic due to excessive precipitation; and, therefore, Acrisols and Alisols are the component in the major RSGs if argic horizons are recognized (ISSS-ISRIC-FAO 1998b). The regional distribution patterns of the two soils are, however, still a controversial issue in this region, and there is scarce data upon which to base discussion.

Although both Acrisols and Alisols are acidic and agricultural production is strongly restricted with these soils (Driessen et al. 2001), the higher CEC/clay values in the Alisols indicate that the levels of exchangeable and soluble Al reach toxic levels and cause more serious constraints for agricultural use than in Acrisols. It is necessary to understand the regional distribution patterns of the two soils with reference to pedogenetic factors, such as geology and climate.

In the present study, the relationship between the mineralogical and the chemical properties of the upland soils in humid Asia was investigated further. In addition, the WRB classification scheme for classifying the soils in humid Asia in terms of regional distribution patterns of physicochemical and mineralogical properties of the soils was discussed.

3.2 Clay Mineralogical Characteristics of the Soils Studied

3.2.1 Soils Studied and Analytical Methods

The regions considered in this study are Japan, Thailand, and the Java, Sumatra, and East Kalimantan regions of Indonesia (Fig. 3.1). Japan is part of a volcanic belt and contains widely distributed felsic igneous and sedimentary rocks and tephra. In northern and northeast Thailand, the sedimentary and the felsic igneous rocks are the most common. Java and Sumatra are part of a volcanic belt and contain mostly tephra, andesite, and sedimentary rocks, with felsic rocks only found locally. Most of East Kalimantan is covered with sedimentary rocks where there are no active volcanoes, and soils are more weathered on stable rolling landscapes. All of these regions have a humid climate, although Thailand has a distinct dry season with lower annual precipitation. The temperatures in Indonesia and Thailand are higher than those in Japan, which strongly favors weathering. The soil temperature and the moisture regimes are, therefore, described as mesic-/thermic-udic in Japan, hyperthermic-udic in Indonesia, and hyperthermic-ustic in Thailand (Fig. 3.2). Among these, a moisture deficiency during the dry season is only expected in Thailand (Fig. 3.2). Considering the geology of each region, we collected upland soils from subsurface horizons at 204 sites (37 in Japan, 90 in northern Thailand, one in northeastern Thailand, 28 in Java, nine in Sumatra, and 39 in East Kalimantan) for the identification of the clay minerals. The soil samples were divided into four groups: Japan (JP), Thailand (TH), the sedimentary rock areas of Indonesia (ID-S), and the volcanic areas of Indonesia (ID-V). The Indonesian samples were subdivided on the basis of the parent material, as the volcanic rocks are usually andesitic or mafic and, therefore, quite different from the typically felsic sedimentary rocks. We use the following abbreviations for samples from the Indonesian sites: JV, Java; SM, Sumatra; and EK, East Kalimantan. For JP and TH, the parent rocks were mostly felsic igneous and sedimentary rocks.

All samples were well-drained residual soils. From these, we selected 53 soils from 38 of the 204 sites, which have the representative clay mineral composition

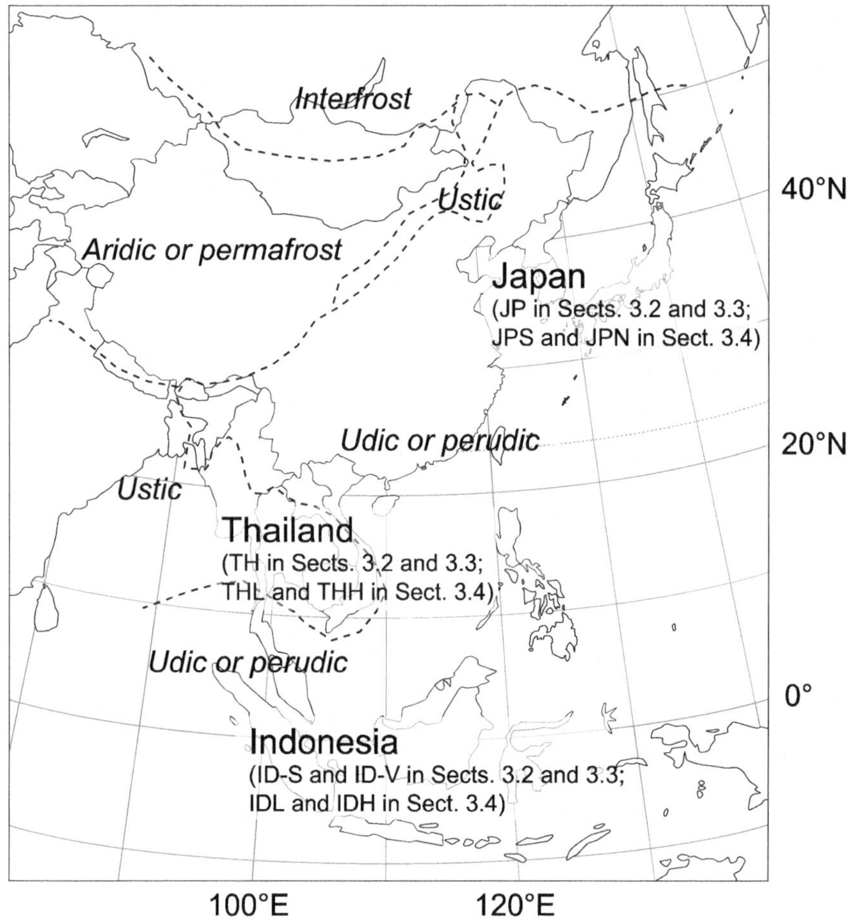

Fig. 3.1 Study area with soil moisture regimes

and the various parent materials for each region. The soil samples were then air-dried and sieved through a 2-mm mesh sieve, followed by general chemical analyses and a mineralogical analysis for the silt and the clay fractions using X-ray diffraction (XRD). The XRD analysis followed Mg saturation with glycerol solvation after the Mg saturation or K saturation with samples dried at 25, 350, or 550 °C. The clay minerals were semi-quantified by the relative peak areas of mica (1.0 nm), the kaolin minerals (1.0 and 0.7 nm), and the 1.4-nm minerals in the diffractograms.

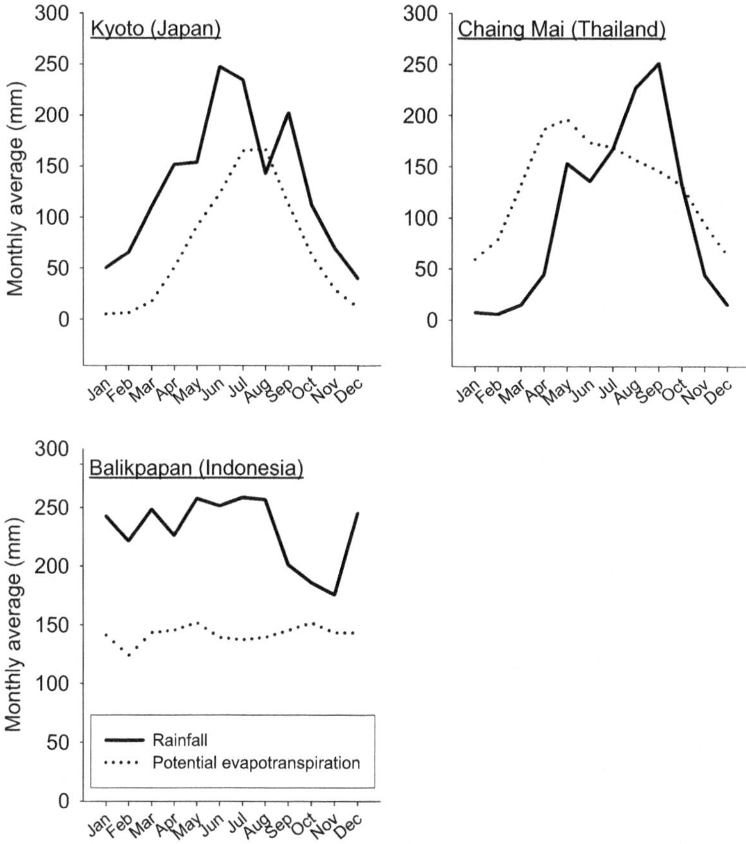

Fig. 3.2 Monthly distribution of rainfall and potential evapotranspiration for the three regions: southwestern Japan (Kyoto), northern Thailand (Chiang Mai), and East Kalimantan of Indonesia (Balikpapan) (The precipitation data were derived from the Global Historical Climatology Network, version 1 (GHCN 1), and the potential evapotranspiration was estimated with the Thornthwaite method (Thornthwaite and Mather 1957))

3.2.2 Mineralogy of the Silt and Clay Fractions

The X-ray diffractograms of the silt fraction indicated that mica was usually absent in soils with the parent materials of andesitic or mafic volcanic ejecta or the andesites that are characteristic of ID-V, such as gabbro or limestone. Mica was present in soils with parent materials of the sedimentary or felsic igneous rock (granite and rhyolite) and was characteristic of the JP, TH, and ID-S samples (Fig. 3.3). In ID-V4 (JV), ID-V5 (JV), and ID-S2 (JV), smectite was dominant in the silt fraction, which may be inherited from the parent rocks.

For the clay fraction from all regions, mica and vermiculite were absent when a mica peak at 1.0 nm was not detected in the silt fraction (Fig. 3.3). In contrast, when mica was present in the silt fraction, mica, vermiculite, and hydroxyl-interlayer

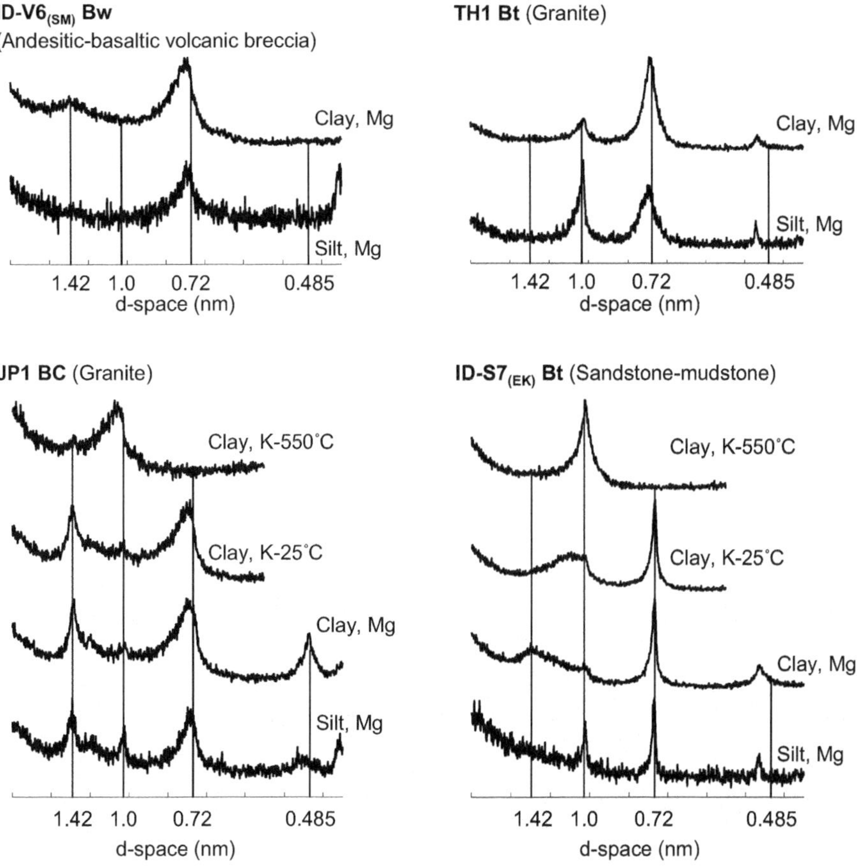

Fig. 3.3 X-ray (Cu–Kα) diffractograms from oriented specimens of silt and clay fractions (Parent materials are shown in parenthesis)

vermiculite (HIV) were common in the clay fraction. Mica was most abundant in 2:1 clay minerals in the TH samples, with vermiculite most common in the ID-S samples and HIV most common in the JP samples (Table 3.1, Fig. 3.3). Interlayered materials between the 2:1 layers were generally more common in the JP than in the ID-S samples, as the peak tended to remain with the heat treatment. In the JP and ID-S samples, the relative peak intensity of 1.4–1.0 nm was larger in the clay than in the silt fraction, suggesting that 1.4-nm clay minerals formed from mica (Table 3.1; Fig. 3.3). A broader peak at 0.7 nm among the JP and ID-V samples indicated the lower crystallinity of kaolin minerals compared to the ID-S and TH samples. In all JP samples, gibbsite was detected at 0.485 nm using XRD (Fig. 3.3). The exception was JP5, which consisted of short-range order minerals. With differential thermal analysis (DTA), gibbsite was usually detected in the JP and the TH samples, but not in the ID-V or the ID-S samples (Table 3.1). Soils from EK did not contain gibbsite, except for the ID-S10 (EK) sample. Levels of Al_o and Si_o indicated that most samples contain small amounts of amorphous minerals (Table 3.1).

Table 3.1 Mineralogical properties and selected cation activities of the soil water extract, expressed as negative logarithms

Sample group[a]	Sample name[a]	Horizon	Identified 1.4-nm minerals[b]	Gibbsite (g kg$_{clay}$$^{-1}$)	Kaolin minerals	Al$_o$ (g kg$_{soil}$$^{-1}$)	Si$_o$	Cation activities in soil water extract in negative logarithmic form				
								H$^+$	H$_4$SiO$_4$^{0}	Al^{3+}	Mg^{2+}	K$^+$
JP	JP1	Bw	HIV	44.7	691	2.8	0.4	4.5	3.3	4.8	4.5	3.8
		BC	HIV	74.2	693	3.7	0.7	4.6	3.7	5.1	4.8	4.1
	JP2	Cl	HIV	92.3	556	3.7	0.6	5.5	3.5	n.d	5.0	4.0
		C3	HIV	70.1	409	1.6	0.5	5.3	3.4	n.d	5.1	4.5
	JP3	AB	HIV-V	0.0	196	2.1	0.4	5.0	3.2	6.1	4.6	4.0
		BC	HIV-V	0.4	194	0.7	0.4	5.4	3.1	n.d	4.9	4.0
	JP4	AB	HIV-V	1.7	348	2.8	0.4	4.6	3.5	4.9	4.6	4.2
		BC	HIV-V	10.2	462	3.0	0.6	4.6	3.7	n.d	4.9	4.3
	JP5	AB	–	2.3	n.d.	46.6	13.1	5.4	3.6	7.9	4.8	4.2
	JP6	AB	Sm, HIV	3.3	721	5.4	0.5	4.4	3.5	4.4	4.1	4.1
		Bw	Sm, HIV	4.1	544	4.7	0.6	4.7	3.8	5.4	4.5	4.8
	JP7	E	V	3.8	41	2.3	0.3	3.7	3.3	4.1	4.4	4.0
		Bs	HIV-V	66.1	83	7.2	0.5	4.4	3.5	4.9	4.7	4.2
	JP13	Bwl	HIV	64.2	164	11.9	0.2	4.9	3.8	6.6	5.4	5.4
TH	TH1	AB	V	0.2	537	2.3	0.4	6.5	3.3	n.d	5.1	4.0
		Bt	–	0.1	609	2.0	0.5	6.2	3.6	n.d	4.9	4.2
	TH2	Bt	V	0.9	665	1.3	0.4	5.9	3.6	n.d	4.9	3.8
	TH3	Bt	Sm	3.7	531	0.9	0.4	6.4	3.6	10.6	4.8	4.0
	TH4	Bt	Sm, HIV	61.3	500	4.8	0.6	5.9	4.2	9.5	5.3	3.9
	TH5	Bt	–	0.6	425	1.8	0.4	5.8	4.0	n.d	6.3	4.1
	TH6	Bt	–	0.1	221	2.2	0.5	5.5	4.2	8.7	5.1	4.3
	TH7	AB	HIV	0.5	233	1.7	0.2	7.6	3.8	14.0	3.9	3.7
		Bt	HIV	1.0	287	1.7	0.2	6.5	4.1	11.0	4.6	4.8
	TH9	Bt	HIV-V	0.6	536	2.9	0.5	6.0	4.1	n.d	5.4	4.6
	TH10	Bw	Ch	0.3	326	1.3	0.3	5.4	3.8	9.1	5.2	4.2
	TH11	Bt	Ch	n.d.	439	1.6	0.3	6.0	4.3	0.0	5.8	4.9
	TH12	Bt	HIV-V	n.d.	561	2.1	0.4	6.0	4.2	n.d	5.1	4.5

(continued)

Table 3.1 (continued)

Sample group[a]	Sample name[a]	Horizon	Identified 1.4-nm minerals[b]	Gibbsite (g kg $_{clay}^{-1}$)	Kaolin minerals	Al_o (g kg $_{soil}^{-1}$)	Si_o	Cation activities in soil water extract in negative logarithmic form				
								H^+	$H_4SiO_4^0$	Al^{3+}	Mg^{2+}	K^+
	TH13	Bt	HIV-V	n.d.	363	2.5	0.4	5.7	3.9	8.7	6.0	4.8
	TH14	Bw	HIV-V	98.0	449	9.6	0.6	6.2	4.1	n.d	5.2	4.2
ID-S	ID-S1$_{(JV)}$	Bwl	–	5.7	792	5.1	1.4	7.0	3.5	12.8	4.3	5.3
		Bw3	–	4.3	714	4.6	1.4	7.0	3.2	n.d	4.7	5.6
	ID-S2$_{(JV)}$	A	Sm	n.d.	350	4.7	0.8	7.7	3.8	14.4	3.8	3.7
		R	Sm	n.d.	373	2.5	0.7	6.8	3.5	n.d	4.6	3.7
	ID-S3$_{(SM)}$	Bt	V-HIV	0.2	415	1.6	0.5	4.8	3.8	n.d	5.4	4.5
	ID-S4$_{(EK)}$	Bt	Sm, HIV	n.d.	674	2.1	0.4	4.6	3.9	6.4	5.6	4.6
	ID-S5$_{(EK)}$	Bt	V-HIV	n.d.	702	0.9	0.3	4.3	3.8	5.9	5.4	4.5
	ID-S6$_{(EK)}$	Bt	HIV-V	n.d.	578	2.1	0.3	4.6	3.9	6.2	5.9	4.4
	ID-S7$_{(EK)}$	BE	V	n.d.	419	1.6	0.1	4.9	3.5	n.d	4.4	3.8
		Bt	V-HIV, Sm	n.d.	512	2.1	0.1	4.4	3.5	6.4	5.0	4.2
	ID-S8$_{(EK)}$	Bt	V-HIV	n.d.	347	0.2	0.4	6.3	3.6	n.d	5.3	4.8
	ID-S9$_{(EK)}$	Bt	V-HIV	n.d.	380	1.8	0.2	4.5	4.0	6.0	5.4	4.2
	ID-S10$_{(EK)}$	Bt	HIV-V	10.4	375	1.5	0.3	4.6	4.0	6.4	5.6	4.7
ID-V	ID-V1$_{(SM)}$	BA	V-HIV	0.5	611	1.2	0.5	5.8	3.2	8.9	5.0	3.9
		Bw	HIV-V	n.d.	601	1.3	0.3	5.6	3.3	n.d	5.5	4.3
	ID-V2$_{(JV)}$	Bw	–	10.4	820	3.7	0.8	4.9	3.8	7.1	4.7	4.1
		Bt	–	11.1	952	3.7	1.0	4.7	3.9	n.d	5.2	4.9
	ID-V3$_{(JV)}$	Bw	–	n.d.	831	3.6	1.2	7.5	3.4	n.d	4.0	5.0
		Bt	–	n.d.	689	3.1	1.2	6.2	3.4	n.d	4.7	5.2
	ID-V4$_{(JV)}$	A	Sm	n.d.	273	0.9	0.6	8.2	3.2	n.d	3.4	4.4
		C	HIV/HIS, Sm	n.d.	354	1.2	0.6	7.1	3.3	n.d	5.1	5.5
	ID-V5$_{(JV)}$	Bw	Sm, HIV/HIS	n.d.	468	4.6	1.0	5.5	3.7	n.d	4.5	5.2
	ID-V6$_{(SM)}$	Bw	Sm	n.d.	721	2.9	0.3	5.7	3.5	n.d	4.9	4.8
	ID-V7$_{(SM)}$	Bt	Sm, HIV/HIS	n.d.	654	2.3	0.3	5.4	3.8	8.3	5.9	4.9

[a]JP: Japan, TH: Thailand, ID: Indonesia, S: sedimentary origin, V: volcanic origin, JV: Java, SM: Sumatra, EK: East Kalimantan, n.d. not detected
[b]HIV: hydroxy-Al interlayered vermiculite, V: vermiculite, Sm: smectite, Ch: chlorite, HIS: hydroxy-Al interlayered smectite

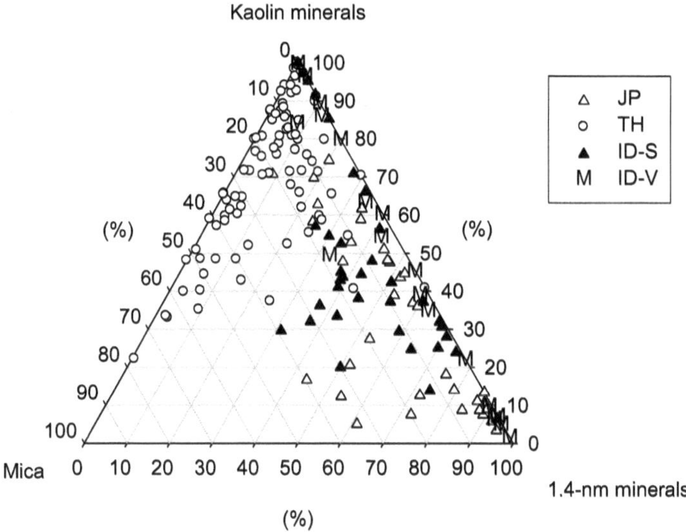

Fig. 3.4 Clay mineralogical composition of the soils determined based on relative peak areas of kaolin minerals, mica and 1.4-nm minerals in X-ray diffractograms

The regional trend of dominant clay minerals was summarized as follows (Fig. 3.4). The soils derived from mafic parent materials (mostly mica-free) had a clay mineralogy dominated by kaolin minerals with small amounts of mica. Under the higher pH conditions associated with the monsoon climate and the ustic soil moisture regime in Thailand, mica and kaolin minerals dominated with lesser amounts of the 1.4-nm minerals. In contrast, under the lower pH conditions of the udic and perudic soil moisture regimes in Japan and Indonesia, significant amounts of 1.4-nm minerals formed. Among them, most of the soils from Japan were dominated by 2:1–2:1:1 intergrade minerals that have appreciable amounts of interlayered materials, whereas those from Indonesia usually did not exhibit such clear interlayering (data not shown here). Our clay mineral composition results were generally consistent with previous reports for each region (Goenadi and Tan 1988; Matsue and Wada 1989; Araki et al. 1990; Koch et al. 1992; Supriyo et al. 1992; Yoshinaga et al. 1995; Yoothong et al. 1997; Prasetyo et al. 2001; Kanket et al. 2002).

3.3 Assessment of Mineral Weathering Conditions Based on Chemical Composition of the Soil Water Extracts

3.3.1 Background of the Thermodynamic Analysis

To assess inherent soil fertility for appropriate large-scale land management, it was important to understand how clay mineralogy relates to geological and weathering conditions. Important processes in the formation of clay minerals are the neoformation of gibbsite, kaolin minerals, and smectite and the transformation of mica. Neoformation is mainly controlled by $H_4SiO_4^0$ activity. Gibbsite forms when desilication is present and $H_4SiO_4^0$ activity is low (Huang et al. 2002). Kaolin minerals form under moderate $H_4SiO_4^0$ activity conditions, and smectite forms under high activity (Reid-Soukup and Ulery 2002). Mica, which is commonly present in the felsic and sedimentary rocks, weathers to vermiculite and smectite, with a decrease in the layer charge and the release of alkaline metals. The increased resistance to weathering of dioctahedral mica means that dioctahedral vermiculite is more common in soils than trioctahedral vermiculite (Malla 2002).

The distribution of clay minerals in the upland soils of humid Asia, namely, Japan, Thailand, and Indonesia, has been described previously. In general, hydroxy-Al interlayered vermiculite (HIV), mica, and kaolin minerals dominated in Japan (Matsue and Wada 1989; Araki et al. 1990), while kaolinite, mica, and smectite dominated in Thailand (Yoshinaga et al. 1995, 1997; Kanket et al. 2002). Kaolinite, vermiculite, and smectite were the primary clay minerals found in Indonesia (Goenadi and Tan 1988; Koch et al. 1992; Supriyo et al. 1992; Prasetyo et al. 2001). However, each of the above regional reports failed to compare the formation of local clay minerals with the formative processes in other regions. In addition, thermodynamic data on soil solution was lacking for humid Asia, and little emphasis had been placed on the thermodynamic analysis of the soil–water relationships. Soil solution data have been used successfully in the past to quantitatively predict and explain the distribution of clay minerals in soil and explain weathering trends (Kittrick 1973; van Breemen and Brinkman 1976; Karathanasis et al. 1983; Rai and Kittrick 1989; Norfleet et al. 1993; Karathanasis 2002). Such an approach may help to understand the distribution of different clay minerals throughout humid Asia. The objective of this section is to explain the distribution of clay minerals in humid Asia. To achieve this, we examine the composition of soil water extracts, as affected by the climate, the parent material of the soil, and the soil age.

3.3.2 Thermodynamic Analysis of the Soils Studied: Experiment

Soil water extracts were collected by continuously shaking the soils for 1 week at 25 °C and 1 atm with a soil (40 g) to water ratio of 1:2, followed by filtering through a 0.025-μm pore membrane filter (Millipore). For these samples, the pH and the concentration of Na^+, K^+, Cl^-, NO_3^-, SO_4^{2-} (via high-performance liquid chromatography), Si, Al, Ca, Mg, Fe, Mn (via ICP–AES), and F^- (via glass electrode) were determined. To eliminate the Al complexed with organic matter, the extracts were passed through a column filled with a partially neutralized (pH 4.2) cation exchange resin (Amberlite IR-120B(H); Hodges 1987). The amount of Al not adsorbed by the resin was determined by ICP–AES and assigned to Al complexed with organic matter, as opposed to the fraction retained that was assigned to the inorganic monomeric Al. The inorganic carbon concentration was measured with a total organic carbon analyzer (Shimadzu, TOC-V CSH). Ionic activities were calculated with the extended Debye–Hükel equation, using a successive approximation procedure (Adams 1971). Inorganic monomeric Al was distributed among Al^{3+}, $Al(OH)^{2+}$, $Al(OH)_2^{+}$, $Al(OH)_3^{0}$, $Al(OH)_4^{-}$, AlF^{2+}, and $Al(SO_4)^{+}$ according to the equilibrium constants of Lindsay (1979).

The clay minerals were assumed to be at or very near to the equilibrium with the soil water extracts. The stability of minerals was evaluated using stability diagrams and solubility diagrams. These diagrams represented the relative stability and solubility of minerals, as outlined by van Breemen and Brinkman (1976) and Karathanasis (2002), respectively. The thermodynamic mineral data used in these diagrams are from Karathanasis (2002) for gibbsite, kaolinite, quartz, and amorphous silica and from Lindsay (1979) for amorphous Al hydroxide, muscovite, and microcline.

The potential evapotranspiration for each region was estimated from the mean monthly temperatures recorded at meteorological stations near the sampling sites (Thornthwaite 1948).

3.3.3 *Thermodynamic Analysis of the Soils Studied: Results and Discussion*

The pH of the soil water extracts from the JP and the ID-S samples was low, mostly within the range 4.3–5.5; however, the pH is relatively high in the TH and the ID-V soils (pH 5.4–6.5) (Table 3.1, Fig. 3.5). The pH of the soil water extracts was generally higher than that determined for the soil suspensions, presumably due to the gradual dissolution of minerals that occurs during the 1-week preparation of the soil water extracts. The activities of the Al–OH species (Al^{3+}, $Al(OH)^{2+}$, $Al(OH)_2^{+}$, $Al(OH)_3^{0}$, and $Al(OH)_4^{-}$) are higher in the JP samples than in the ID-S samples, although the pH of the extracts from both regions was low (Fig. 3.5). The activities of the Al–OH species in the TH and the ID-V samples were low; and in some extracts, the Al concentration was below the detection limit due to higher pH. The $H_4SiO_4^{0}$ activity of the soil water extracts was highest in the JP (approx. $10^{-3.1}$–$10^{-3.8}$ mol L^{-1}) and the ID-V (approx. $10^{-3.2}$–$10^{-3.9}$ mol L^{-1}) samples, followed by the ID-S (approx. $10^{-3.5}$–$10^{-4.0}$ mol L^{-1}) sample. The lowest was in the TH (approx. $10^{-3.6}$–$10^{-4.3}$ mol L^{-1}) sample (Table 3.1).

In the stability diagram (Fig. 3.6), the compositions of the TH extracts are close to the line where both kaolinite and muscovite are stable, while those for the JP and the ID-S samples mostly lie within the field where kaolinite is stable and muscovite unstable. Exceptions to this trend are the ID-S1 (JV) sample, which is derived from limestone, and the ID-S2 (JV) sample, which contains significant montmorillonite. In the solubility diagram (Fig. 3.7), the compositions of the JP and the TH extracts mostly lie between the solubility lines of amorphous Al hydroxide and gibbsite,

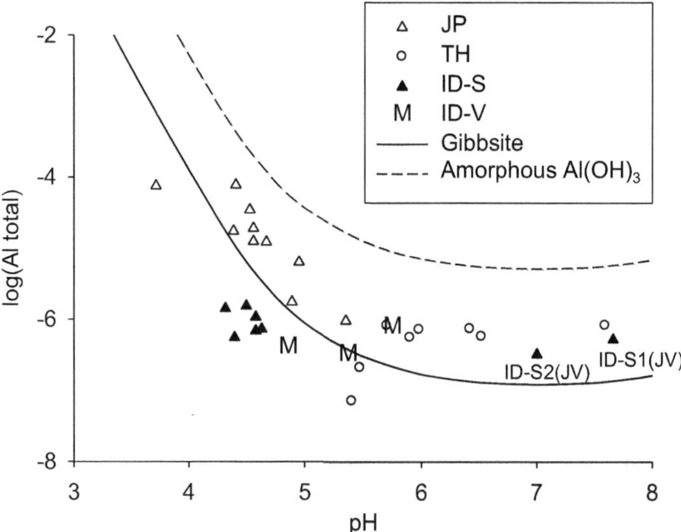

Fig. 3.5 The pH and sum of the activities of Al–OH species (Al^{3+}, $Al(OH)^{2+}$, $Al(OH)_2^+$, $Al(OH)_3^0$, $Al(OH)_4^-$) in the soil water extracts with gibbsite and amorphous Al hydroxide solubility lines

which indicates that gibbsite can precipitate. The ID-S samples and a few TH and ID-V samples lie under the kaolinite solubility line, indicating the dissolution of gibbsite and kaolinite.

3.3.4 Weathering Sequence in the Upland Soils of Humid Asia

The pH and Activities of Al–OH Species in the Soil Water Extracts The pH and activities of Al–OH species such as Al^{3+} were very important when considering mineral behavior, such as the stability of or the possibility of neoformation in soils. The pH of the extracts appeared to reflect the precipitation, the evapotranspiration, and the geology. The difference between the precipitation and the potential evapotranspiration in each region was 808 mm in Japan, −390 mm in Thailand, and 457–2709 mm in Indonesia. These trends may have explained the lower pH in the JP and the ID-S samples and the higher pH in the TH samples. The ID-V samples had a high pH despite intense leaching. This probably occurred because the ID-V soils contained easily weathered minerals, such as mafic minerals that could neutralize acidity when weathered rapidly. This was apparent at high $H_4SiO_4^0$ activity in the high pH extracts (Table 3.1).

The solubility diagrams (Figs. 3.5 and 3.7) for the JP and the TH samples indicated that the activity of the Al–OH species was controlled by the dissolution

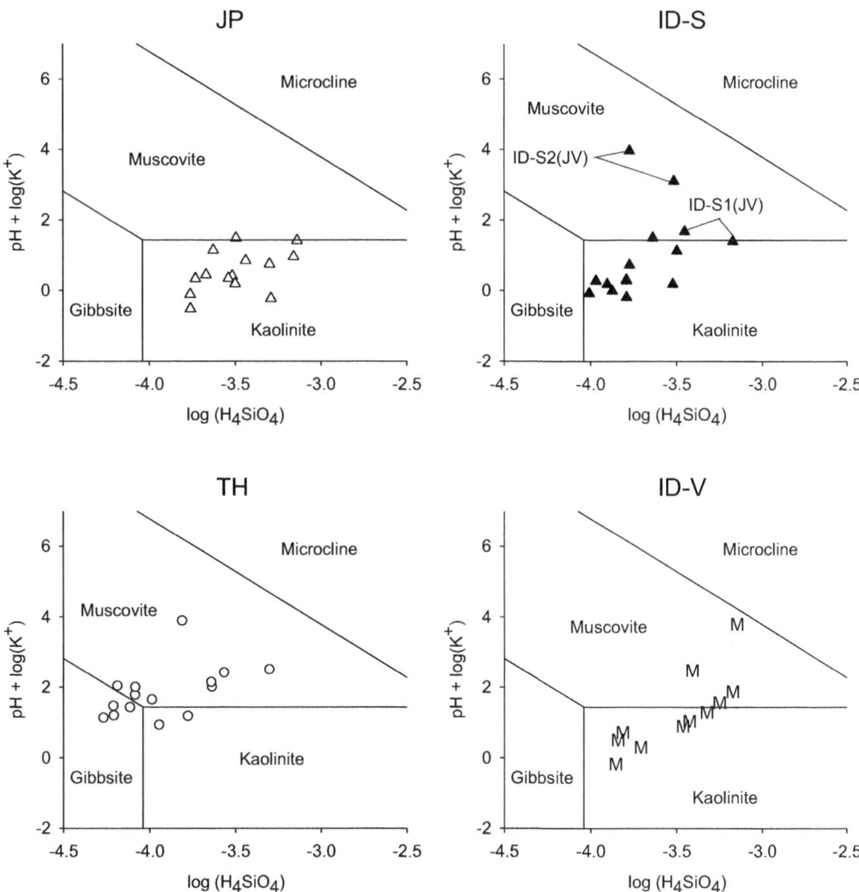

Fig. 3.6 Composition of the soil water extract from JP, TH, ID-S, and ID-V plotted on a stability diagram

and the precipitation of Al hydroxide, which had a crystallinity intermediate to gibbsite and amorphous Al hydroxide. In the ID-S samples, the dissolution of kaolinite controlled the lower activities. The Al–OH species recorded a higher activity in the JP samples than in the ID-S samples. As both regions are characterized by a low pH, the difference in activity may have reflected the freshness of the JP soils.

Neoformation and Dissolution of Gibbsite, Kaolin Minerals, and Smectite Gibbsite is usually present in soils from JP and TH (Table 3.1), consistent with the soil water composition (Fig. 3.5). The stability diagram of JP, however, indicated that kaolinite was more stable than gibbsite (Fig. 3.6). Huang et al. (2002) stated that the formation of gibbsite in saprolites and volcanic ash, where the $H_4SiO_4^0$ activity was high enough for the formation of aluminosilicates, was due to the rapid removal of the dissolved $H_4SiO_4^0$ from weathering zones or to

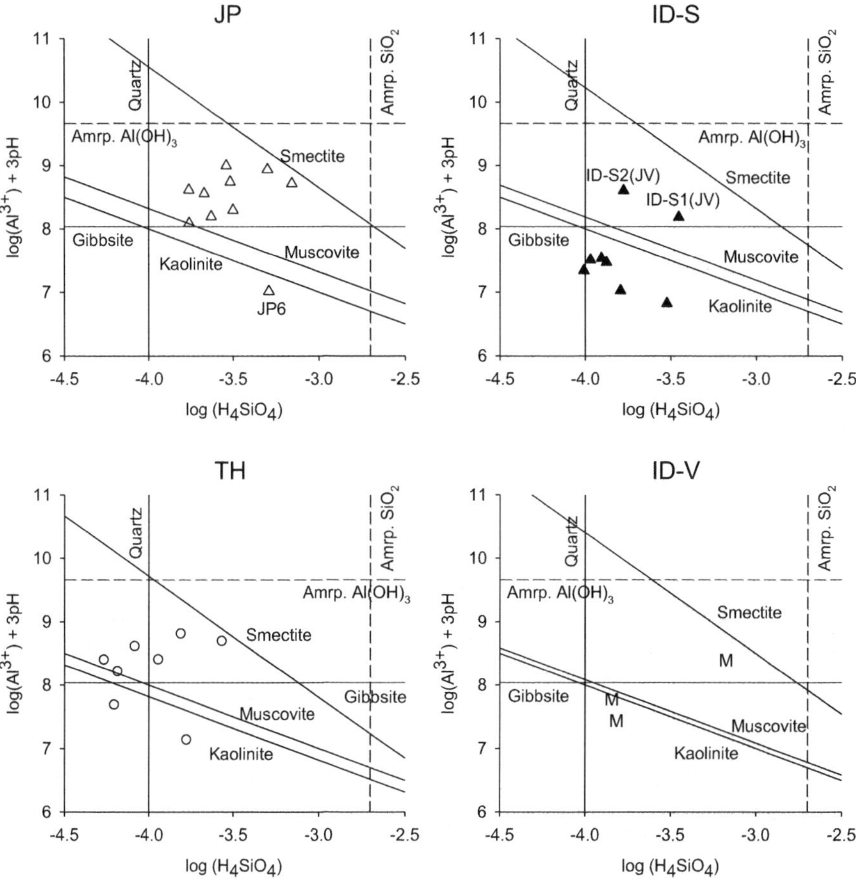

Fig. 3.7 Composition of the soil water extract from JP, TH, ID-S, and ID-V plotted on a solubility diagram

the local environment where the $H_4SiO_4^0$ activity was low. Gibbsite formation could also be kinetically favored over kaolinite in solutions where the $H_4SiO_4^0$ activity was high. Considering the gibbsite formation in this way and the freshness of the JP soils, gibbsite presumably formed together with the poorly crystalline Al hydroxide, the kaolin minerals, and the silica as unstable transitional minerals during an early stage of weathering. We considered that gibbsite formed in the same way in the TH samples, as more gibbsite and Al_o was found in the TH4 and the TH14 samples (Table 3.1), and there was no apparent relationship between the amount of gibbsite and the $H_4SiO_4^0$ activity. In the JP6, the TH3, and the TH4 samples, the coexistence of gibbsite and smectite, the latter of which is usually stable under high $H_4SiO_4^0$ activity, may have indicated the transitional presence of gibbsite. In the TH samples, the gibbsite formation and the stability may also have resulted from the very low $H_4SiO_4^0$ activity (Fig. 3.6).

In contrast, gibbsite was not present in the East Kalimantan samples except for in the ID-S10 (EK) sample (Table 3.1). This was consistent with the fact that the extract composition was undersaturated with gibbsite, indicating that gibbsite could not form or had been unstable and already dissolved. In the case of dissolution, the dissolution rate of gibbsite was much higher than that of kaolinite under low pH conditions (Nagy 1995; Lasaga 1998). We considered that sample ID-S10 (EK) was relatively unweathered and still contained gibbsite, as the sample recorded a larger peak area of 2:1-type clay minerals (Table 3.1) and had a broader peak at 0.7 nm in the X-ray diffractogram (data not shown) compared to other samples from East Kalimantan. In some acid soils, such as ID-S and JP7, kaolin minerals and gibbsite were present, but the trend was considered to be decreasing, as indicated by the fact that the solution composition was undersaturated with respect to the minerals in the solubility diagram (Fig. 3.7).

Easily weathered minerals in the ID-V samples, such as mafic minerals, must release Si very rapid upon weathering. The resulting high $H_4SiO_4^0$ activity may enhance the formation of kaolin minerals and smectite rather than Al hydroxide. It is widely accepted that smectite tends to form in solutions of high pH with high $H_4SiO_4^0$ and Mg^{2+} activities (Reid-Soukup and Ulery 2002). Such conditions are likely to exist in soils derived from the andesitic and the mafic parent materials.

Judging from the $H_4SiO_4^0$ activity of extracts, the most stable mineral, and, therefore, the most prevalent mineral given further weathering, was kaolinite in the JP and ID-V samples. The most stable and prevalent minerals were kaolinite and gibbsite in the TH samples (Fig. 3.6). In contrast, the most stable mineral (or kaolinite) was assumed to be decreasing in the ID-S samples (Fig. 3.7).

Transformation of 2:1-Type Clay Minerals In the case of the TH samples, both the mineralogical and the thermodynamic analyses indicated that mica was relatively stable and did not transform easily into 1.4-nm minerals (Figs. 3.4 and 3.6). Araki et al. (1998) investigated the weathering of Tanzanian soils, which experience a distinct dry season similar to that in Thailand, and reported the weathering of mica without the formation of 1.4-nm minerals. They assumed that it takes a long time, perhaps several million years, for mica to weather to kaolinite or gibbsite. According to the stability diagram, both pH and K^+ activity in solution were important for the weathering of mica. Mica was more stable than kaolinite with an increasing pH or K^+ activity (Fig. 3.6). For example, Rausell-Colom et al. (1965) demonstrated that the removal of K from micas was strongly dependent on the K concentration of the solution. In our experiment, however, K^+ activity in the TH extracts was almost the same as that in the JP and the ID-S extracts, ranging from $10^{-3.8}$ to $10^{-4.8}$ mol L^{-1} (Table 3.1). This indicated that pH plays an important role in the in situ weathering of mica. The apparent limits of $pH(H_2O)$ and pH of the soil water extract in terms of mica weathering to 1.4-nm minerals were 5.0–5.5 and 5.5–6.5, respectively. Mica was stable above these limits (Table 3.1).

In contrast, the mineralogical analysis of the JP and the ID-S samples showed the formation of the 1.4-nm clay minerals, such as HIV and vermiculite, from mica (Fig. 3.3). The stability and solubility diagrams indicated the dissolution of mica, which was interpreted in this dissolution process as 2:1 opened layers forming the 1.4-nm minerals, such as HIV and vermiculite, as transitional products (Figs. 3.6 and 3.7). In the JP samples, the low crystalline clay minerals and the easily weathered primary minerals dissolved and released Al to the soil solution. The released Al may have precipitated as Al hydroxide contaminants between the 2:1 layers as well as precipitating as gibbsite. The 2:1 layers of HIV were thought to be stable, probably because the buffering capacity of interlayered materials was such that it could react with acid more easily than the 2:1 layers. Such a preferential dissolution of the interlayered materials in 2:1 layers has been widely observed in the surface horizons of acidic forest soils (Ross and Mortland 1966; Bouma et al. 1969; Gjems 1970; Hirai et al. 1989; Funakawa et al. 1993). In the ID-S samples, small to negligible amounts of the easily weathered minerals were present in the soils from a tropical rainforest climate with relatively stable landscape conditions. This indicated limited potential for acid neutralization via mineral dissolution. Accordingly, the interlayered materials and the gibbsite may already have dissolved preferentially to 2:1 layers, with vermiculite forming. The dissolution of 2:1 layers or a reduction in layer charge may have resulted in the formation of smectite from vermiculite (Borchardt 1989).

The Weathering Sequence in Upland Soils of Humid Asia The mineral weathering sequence for each region was shown in Fig. 3.8. In Thailand, under higher pH conditions associated with the ustic moisture regime, mica was relatively stable; whereas, other primary minerals, such as feldspars, were unstable and dissolved to form kaolinite and gibbsite. Under the lower pH conditions of the udic moisture regime in Japan and Indonesia, the mica weathered to form the 1.4-nm minerals. The Japanese soils were young and had much Al–OH species in soil solution, resulting in Al hydroxide between the 2:1 layers and gibbsite, while the Al hydroxides and the gibbsite had not formed or were already removed from the more highly weathered Indonesian soils. In soils derived from mica-free parent material, which for clay mineral compositions are kaolin minerals and smectite, the clay mineral composition changed to the kaolin apex along the mineral axis of the 1.4-nm minerals (Fig. 3.8).

3.4 Regional Trends in the Chemical and Mineralogical Properties of the Soils: With Special Reference to the WRB Classification Scheme

3.4.1 Background

In Chap. 2 and previously in Chap. 3, a general distribution trend of the clay mineralogy in humid Asia was demonstrated with the possible weathering pathways of minerals under different geological and climatic conditions. Due to the definitive influence of expandable 2:1 minerals on the physical and the chemical

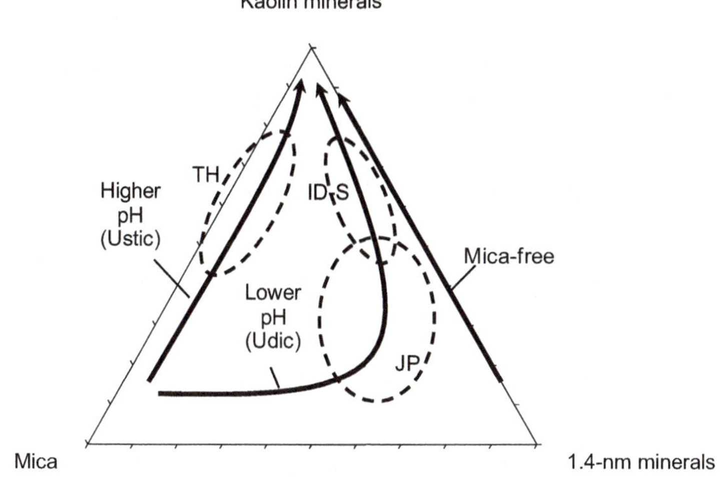

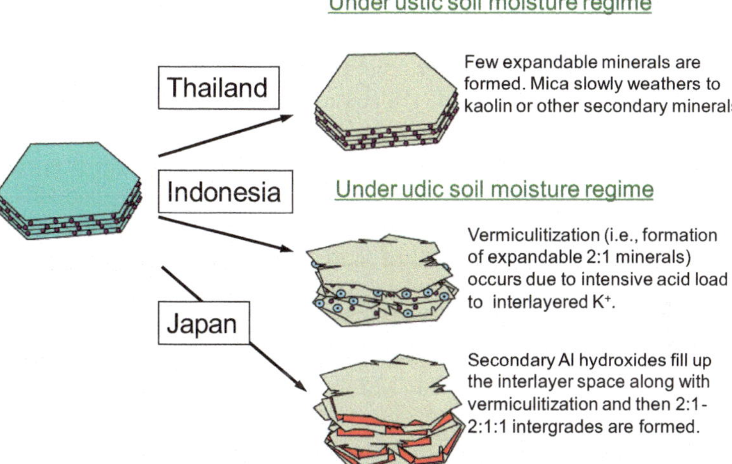

Fig. 3.8 Assumed mineral weathering sequence in the upland soils in humid Asia

properties of soils, such distribution patterns of clay minerals would control not only the physicochemical properties of soils but also the taxonomic distribution of the soils. As already introduced in the introduction to this chapter, one of the new features of the WRB is the introduction of clay activity (or CEC/clay) to classify soils in the highest category level (i.e., the reference soil groups [RSG]), which are used to discriminate soils that have an argic horizon into Lixisols and Luvisols (high base saturation [> 50 %] soils) or Acrisols and Alisols (low base saturation [< 50 %] soils). The upland soils in humid Asia are predominantly acidic due to excessive precipitation; and, therefore, Acrisols and Alisols are the components in the major RSGs if argic horizons are recognized (ISSS-ISRIC-FAO 1998b). The regional distribution patterns of the expandable 2:1 minerals, which were discussed in the previous sections, would also affect the regional distribution patterns of Alisols and Acrisols, which have not been clearly understood. In the present section, therefore, the relationship between the mineralogical and the chemical properties of upland soils in humid Asia is further investigated. In addition, we discuss the WRB classification scheme for classifying soils in humid Asia in terms of the regional distribution patterns of the physicochemical and the mineralogical properties of the soils.

3.4.2 Soils Studied and Analytical Items

A total of 186 B-horizon soils from the upland soil profiles with minimum disturbance (i.e., under forest or cropland with low-input management), a majority of which overlap with the samples used in the preceding sections, were used for the analysis in this section. These soils were grouped into seven categories based on pedogenetic conditions (Table 3.2). According to Fig. 3.1, a large part of East Asia is situated either under an ustic, an udic, or a perudic soil moisture regime; and, therefore, the present study was developed to consider the major soils in terms of the soil moisture conditions. As soils derived from the intermediate to the mafic volcanic rocks and limestone are considered to exhibit different properties and judging from the results presented in a previous section, they are separated from the other soils and grouped as "mafic." These soils were derived mainly from volcanic rocks in the Java and Sumatra Islands of Indonesia (ID-V soils in Sects. 3.2 and 3.3) and partly from the limestone in northern Thailand and the mafic intrusive rocks in Japan.

The sample groups of IDL and IDH were comprised of soils derived from the sedimentary rocks (mostly sandstone and mudstone) or the felsic materials in low ($<$ 600 m)- and high ($>$ 600 m)-elevation areas of Indonesia, respectively. Although both of the elevation ranges would be classified as having a hyperthermic soil temperature regime according to US Taxonomy (Soil Survey Staff 2014), they were divided into two groups in the present study since the percolation of organic matter into the B-horizon soils was apparently more extensive at elevations above 600 m, according to soil survey results. This was due presumably to extremely humid conditions (i.e., a perudic soil moisture regime) there.

The THL and the THH soils were collected from Thailand, mostly from the northern mountainous region with a monsoon climate. The border of the two groups was approximately 800 m a.s.l., above which there was predominately an evergreen forest, which succeeded because of decreasing water stress during the dry season. The soil temperature regimes of the THL and the THH sample areas were hyperthermic and thermic, respectively. Most of the parent materials of the soils were a wide variety of sedimentary rocks and granite.

The JPS and the JPN soils were collected from warm and cool temperate forests in Japan; and, therefore, soil temperature regimes of these soils were thermic and mesic. Japan has a humid climate, and the soils collected had a udic or a perudic soil moisture regime. The parent materials were predominantly sedimentary rocks (dominantly mudstone with minor occurrences of sandstone and shale) and partly felsic igneous rocks. Soils strongly affected by volcanic ejecta were not included in the samples. The analytical data on 18 soil profiles out of 50 were cited from Hirai (1995), in which almost the same analytical procedures were used as in the present study.

The soil samples collected from each pedogenetic horizon of the soil profiles were air-dried and passed through a 2-mm mesh sieve for the following chemical and mineralogical analyses: the soil pH, the cation exchange capacity (CEC), the contents of exchangeable bases (Na^+, K^+, Ca^{2+}, Mg^{2+} and NH_4^+), the exchangeable acidity (Al and H), the total C and N contents, the clay mineral composition (using XRD), and the contents of free oxides [DCB-extractable Fe (Fe_d) and Al (Al_d)]. The CEC solely derived from the mineral components (CEC_{min}) was calculated by eliminating the contribution of organic C using the following equation for each soil profile (4–7 soil samples):

$$CEC(cmol_c \ kg^{-1}) = a \times [total \ C(g \ kg^{-1})] + b \times [clay \ content \ (\%)] \qquad (3.1)$$

The coefficient, a, was determined by multiple regression, followed by the determination of CEC_{min} by subtracting the contribution of total C. The median values of a in different soil groups ranged from 0.16 to 0.22 $cmol_c \ g^{-1}$ C. In cases where the coefficients a and b could not be determined with reliability due to multicollinearity, the median value of a in each soil group was used instead. The CEC value per unit of clay content (CEC_{min}/clay, hereafter) was determined to be the CEC_{min} divided by the clay content.

For the analysis of the B-horizon soils, we used the data from one sample of each sub-horizon soil from the soil profiles – mostly involving profiles 30–50 cm in depth and corresponding to upper argic or a comparable horizon. The data obtained here were statistically analyzed using SYSTAT8.0 software (SPSS 1998).

3.4.3 Regional Trend in the General Physicochemical and the Mineralogical Properties of the Soils

The average values of selected chemical and mineralogical properties of the B-horizon soils from different regions, with analysis of variance (ANOVA) results, were summarized in Table 3.2. Although the clay mica was always the dominant component in the 1.0 nm minerals, judging from virtually no expansion of the 1.0-nm diffraction peak after glycerol solvation as well as a clear diffraction peak remaining after the 550 °C-heating in our samples, there was a possibility that halloysite could be considered as a minor component. As shown in Table 3.2 and

Table 3.2 Average values of selected chemical and mineralogical properties of the soils from different regions

Sample group	Number of samples	pH(H_2O) AVE	SE		Clay (%) AVE	SE		Total C[a] ($g\ kg^{-1}$) AVE	SE		CEC_{min}/clay ($cmol_c\ kg^{-1}$) AVE	SE		$Fe_d{}^a$ ($g\ kg^{-1}$) AVE	SE		Clay mineral composition 1.4 nm (%) AVE	SE		1.0 nm (%) AVE	SE		0.7 nm (%) AVE	SE	
Sorted by regions and parent materials																								
IDL	39	4.58	0.07	a	38.6	2.1	ab	5.4	0.5	a	32.4	1.5	bc	20.7	1.5	a	37	3	b	7	1	ab	56	3	cd
IDH	10	4.52	0.06	a	34.0	2.3	ab	6.5	1.0	ab	45.2	2.9	d	22.1	3.1	abc	37	5	b	18	4	abc	45	4	bc
THL	40	5.44	0.07	c	53.0	2.3	c	9.6	0.5	bc	22.2	1.3	a	37.3	2.6	cd	4	1	a	31	3	c	66	3	de
THH	24	5.50	0.09	c	47.2	3.2	bc	14.6	2.0	c	25.7	1.4	ab	40.8	5.0	bcd	12	2	a	17	4	b	71	4	ef
JPS	30	4.71	0.04	ab	35.1	2.4	a	12.9	2.6	bc	40.0	2.1	d	23.6	2.6	ab	58	4	c	11	2	ab	31	3	b
JPN	20	4.59	0.05	a	47.1	1.9	bc	33.4	3.0	d	39.9	2.2	cd	34.0	1.9	bcd	80	3	d	9	2	ab	11	2	a
mafic	23	5.02	0.16	b	66.5	2.3	d	11.2	0.9	bc	31.4	2.9	bc	65.4	6.7	d	14	5	a	4	3	a	81	5	f
Sorted by reference soil groups																								
Andosols	4	4.66	0.06	ab	50.9	1.7	ab	46.0	5.8	c	31.2	4.8	ab	30.4	2.2	a	83	3	c	9	3	ab	8	1	a
Podzols	6	4.38	0.09	a	43.7	2.1	ab	33.3	3.6	c	43.4	2.5	b	42.0	1.6	a	75	5	c	15	5	ab	10	1	a
Acrisols	32	5.28	0.09	b	56.4	2.6	b	9.4	0.7	ab	18.1	0.8	a	39.1	4.1	a	6	1	a	25	3	b	70	4	c
Alisols	64	4.76	0.06	a	45.7	1.8	a	6.8	0.5	a	33.2	1.1	b	31.6	2.5	a	33	3	b	9	1	a	57	3	bc
Luvisols and Lixisol	9	5.81	0.20	c	56.3	7.0	ab	9.9	1.3	ab	30.9	4.0	b	47.8	7.5	a	6	3	ab	15	7	ab	79	7	bc
Cambisols	71	4.95	0.07	ab	41.0	1.9	a	15.7	1.5	b	36.2	1.6	b	32.6	3.0	a	39	4	b	15	2	a	46	4	b
All	186	4.96	0.04		46.3	1.2		12.5	0.8		31.9	0.9		34.4	1.7		32	2		14	1		54	2	

AVE average, *SE* standard error

The values with the same letters are not significantly different by Tukey test ($p < 0.05$)

[a]ANOVA was applied after logarithmic transformation for normalizing dataset

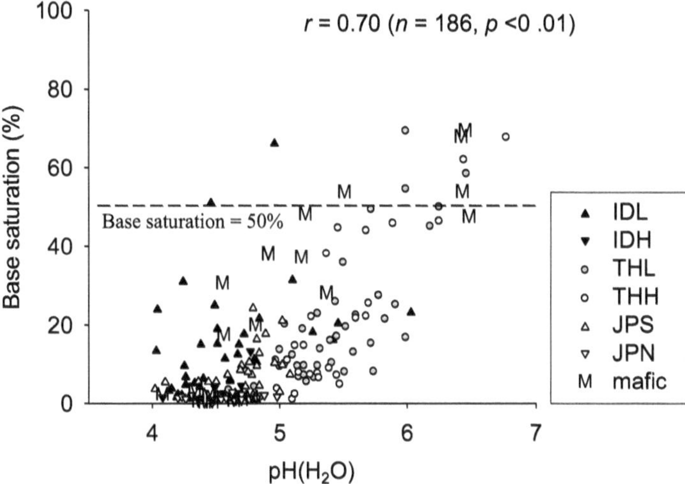

Fig. 3.9 Acidity and base saturation of the soils studied

Fig. 3.9, the majority of soils were acidic. The base saturation (by CEC at pH 7) was usually below 50 %. Among the B-horizon soils, the soils from the ustic soil moisture regime in Thailand showed significantly higher $pH(H_2O)$ values. Some soils from the mafic parent materials also exhibited higher base saturation above 50 % (Fig. 3.9). According to Table 3.2, the Alisols and the Podzols were strongly acidic, followed by the Andosols, the Cambisols, and the Acrisols. As for clay content, the soils from Indonesia contained less clay than those from Thailand or from the mafic parent materials. The total C content clearly increased as soil temperature decreased in the following order: IDL, IDH, THL (hyperthermic) < THH, JPS (thermic) < JPN (mesic). The regional trends of clay mineral composition discussed in Sect. 3.2 were also observed in the average values shown in Table 3.2, in which an increasing trend of 1.4-nm minerals in the order of THL, THH < IDL, IDH < JPS < JPN was clear.

The values of CEC_{min}/clay were plotted together with ECEC/clay (effective CEC [sum of exchangeable bases and Al] divided by clay content) or (exchangeable Al)/clay in Fig. 3.10a, b. The CEC_{min}/clay of the soils derived from the sedimentary rocks (excluding limestone) or the felsic materials showed a clear regional trend. It was usually higher than 24 $cmol_c$ kg^{-1} (corresponding to the Alisols if the argic horizon is recognized) under the udic or the perudic soil moisture regimes in Indonesia (IDL and IDH) and Japan (JPS and JPN); whereas, it was predominantly lower than 24 $cmol_c$ kg^{-1} (corresponding to the Acrisols) under the ustic soil moisture regime in Thailand (THL). The values of the THH soils were occasionally higher than 24 $cmol_c$ kg^{-1}, presumably because the percolating soil moisture conditions in the high mountains resulted in similar conditions as the udic soil moisture conditions. Most of the cation exchange sites represented by ECEC seemed to be occupied by exchangeable Al (Fig. 3.10b); and, therefore, the level

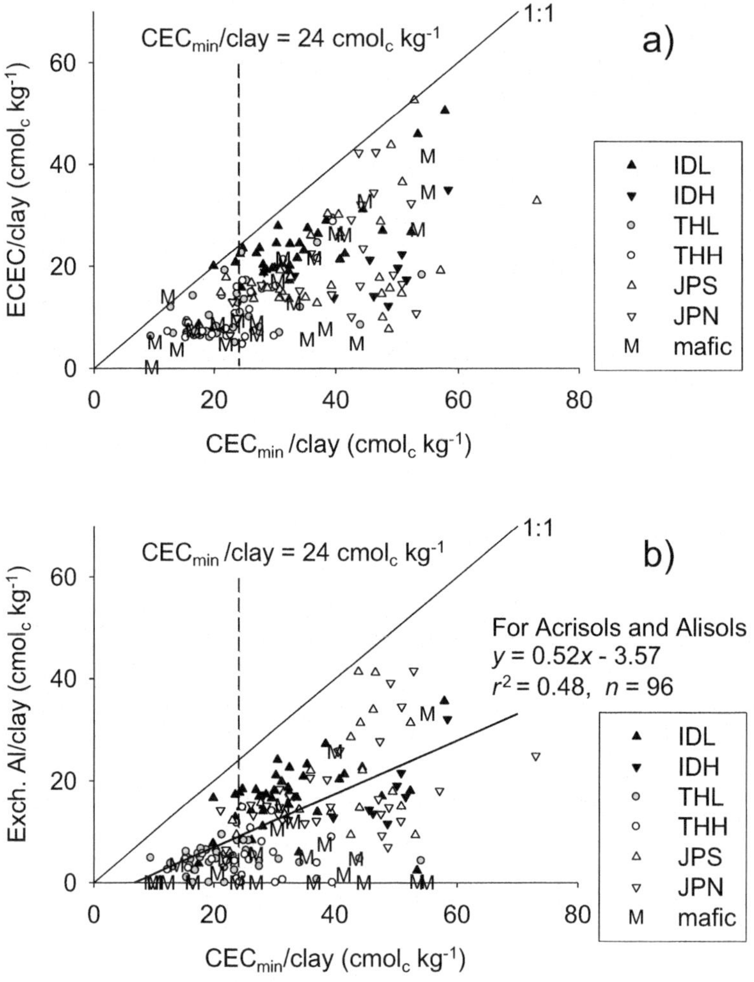

Fig. 3.10 Characteristics of the cation exchange capacity (CEC) of the soils (CEC$_{min}$/clay, CEC value per unit clay content; ECEC/clay, effective CEC (sum of exchangeable bases and Al) divided by clay content)

of Al toxicity should be higher in the soils from the udic or the perudic soil moisture regimes of Indonesia and Japan than in those from the ustic soil moisture regime of Thailand. Similarly high CEC/clay values were reported for soils with an udic soil moisture regime, e.g., 34.0 and 26.3 cmol$_c$ kg^{-1} in subsoil layers of upland soils in northern Sarawak, Malaysia (Tanaka et al. 2005), or average values of 34.7 and 41.2 cmol$_c$ kg^{-1} for 25 and 17 upland soils, respectively, in southern Sarawak, Malaysia (Tanaka et al. 2007). In contrast, for soils with an ustic soil moisture regime, lower values of CEC/clay were often reported. Toriyama et al. (2007) reported the CEC/clay values for subsoils of well-drained Acrisols in central Cambodia (60 to 100 m a.s.l.) to be typically around 10 cmol$_c$ kg^{-1}. According to

Watanabe et al. (2004), the clay mineralogy and CEC/clay values in upland soils in northern Laos (600 to 1100 m a.s.l.) were essentially similar to our results in that illite and kaolinite were dominant in the clay fraction, and the values of CEC/clay of B-horizon soils were calculated to be almost equal to or lower than the threshold value, i.e., 24 $cmol_c$ kg^{-1}. All these reports indicated that our results obtained in some limited regions in humid Asia can be applied to different countries with similar geological or climatic conditions.

The relationship obtained between CEC_{min}/clay and (exchangeable Al)/clay for the Acrisols and the Alisols (i.e., Exch. Al /clay $= -3.57 + 0.52$ CEC_{min}/clay) was close to that introduced in Driessen et al. (2001) for those soils in Indonesia, the Caribbean region, Rwanda, Cameroon, Peru, and Colombia (i.e., Al/clay $= -1.28 + 0.64$ CEC/clay). The overall properties of Al retention and, perhaps, of the proportion of permanent and variable negative charges of the soils in the present study were considered to be similar to those in the Acrisols and the Alisols from other regions.

There was a unique characteristic of the soils derived from the mafic parent materials compared to the soils from the sedimentary rocks (excluding limestone) or the felsic volcanic materials. These soils were more variable in pH (Fig. 3.9) or CEC_{min}/clay (Fig. 3.10a). According to Fig. 3.10b, the majority of the soils did not retain appreciable amounts of exchangeable Al, even if the value of CEC_{min}/clay was high. One reason must have been the relatively higher pH and the base saturation for some soils (Fig. 3.9). Another possible explanation was that the variable negative charges derived from the free oxide surfaces contributed to the apparent increase of the CEC at a higher pH range (i.e., 7). The actual CEC at a lower pH range at which Al can dominate was small. Such a wide variety in the chemical properties of the soils derived from the mafic materials (including limestone) in Java Island was also reported by Supriyo et al. (1992).

3.4.4 Relationship Between Clay Mineral Composition and CEC and pH of the Soils

It is widely known that the clay mineralogical properties of soils strongly affect soil physicochemical properties. In fact, the CEC_{min}/clay in the present study was influenced by the relative abundance of the 1.4-nm minerals in the clay fraction (Fig. 3.11a). The following equation was obtained:

$$CEC_{min}/clay(cmol_c\ kg^{-1}) = 0.201 \times (1.4\ nm\ minerals\ in\ \%) + 25.5\ (r^2 = 0.24, p < 0.01) \tag{3.2}$$

This equation suggested that the 1.4-nm minerals contribute to the CEC increase by 20.1 $cmol_c$ kg^{-1}. Since most of the 2:1 minerals in the Japanese soils were modified by the hydroxy-interlayered materials (Hirai 1995; Kitagawa 2005), the CEC values of these soils could be estimated to be lower than the possible negative charge derived from the 2:1 lattice structure (Funakawa et al. 2003). On the contrary, among the soils having the lowest amounts of the expandable 2:1 minerals (i.e., the Thai soils), the relative contribution of other mineral components was increasing; and the CEC_{min} attributable to the 1.4 nm minerals should be appreciably lower than the value of the intercept of the equation, i.e., 25.5 $cmol_c$ kg^{-1}. As a result, the actual contribution of the expandable 2:1 minerals to the CEC may be considerably higher than 20.1 $cmol_c$ kg^{-1} clay. The CEC of the upland soils in humid Asia was primarily determined by the relative abundance of expandable 2:1 minerals, on which the exchangeable Al was predominant, as represented by the equation described earlier: Exch. Al /clay $= -3.57 + 0.52$ CEC_{min}/clay.

Soil pH is the one of most important factors affecting a wide range of agronomical and environmental functions of soils (Robson 1989). In our present study, a negative correlation was observed between the relative abundance of the 1.4-nm minerals and the soil pH (Fig. 3.11b), and the following equation was obtained:

$$pH(H_2O) = 5.06 - 0.00803 \times (1.4 \text{ nm minerals in } \%)$$
$$+0.00451 \times (Fe_d \text{ in g kg}^{-1})(r^2 = 0.23, p < 0.01) \tag{3.3}$$

As most of the soils studied were acidic and the base saturation was lower than 50 %, the dominant cation retained on the cation exchange sites of soils was Al^{3+}. The equation above indicated that the presence of the expandable 2:1 minerals contributed to pH reduction by 0.008 units per percentage point of clay. In solution, the Al^{3+} ion behaved as a weak acid through a hydrolytic reaction: i.e., $Al^{3+} + OH^- = Al(OH)^{2+}$ (pKa = 5.0). The soils that were rich in exchangeable Al showed a distinct buffer zone against OH^- addition during titration (Funakawa et al. 1993, 2008). The strong acidity far below pH 5.0 of some soils must have been affected by the presence of a stronger acid(s), e.g., H^+, though the quantitative analysis of exchangeable H^+ is difficult, and the selectivity coefficient between Al^{3+} and H^+ ions on the permanent negative sites of soils is rarely determined.

On the other hand, the oxide surfaces represented by the Fe_d fraction were considered to increase the soil pH by 0.00451 units per g Fe_d kg^{-1} soil. As is well known, the oxide surfaces act as weak acids that have a zero point of charge (ZPC) of around 6–9 (McBride 1989). In lower pH regions, it might mitigate H^+ in solution through the protonation reaction: $M–OH + H^+ = M–OH_2^+$, where M– represents the metal ions (Fe, Al) at the surface of the soil particles, contributing to increased soil pH. As was discussed earlier, the soils derived from mafic volcanic materials were often rich in the Fe_d fraction originally inherent to the parent materials, and the soil pH was relatively high (Table 3.2). In humid Asia, the major distribution of the mafic parent materials was limited generally to specific regions, such as the volcanic belts of Java and Sumatra Islands. In the humid tropics of other continents, e.g., eastern Latin America or equatorial Africa, wider regions are covered by mafic materials. It can be said that the acid mitigation by oxide surfaces is relatively limited in humid Asia, unlike in other continents. The general results obtained in the present study are, therefore, considered to be specific for some regions of humid Asia, in which the sedimentary rocks or the felsic volcanic materials dominate as the parent materials of the soils.

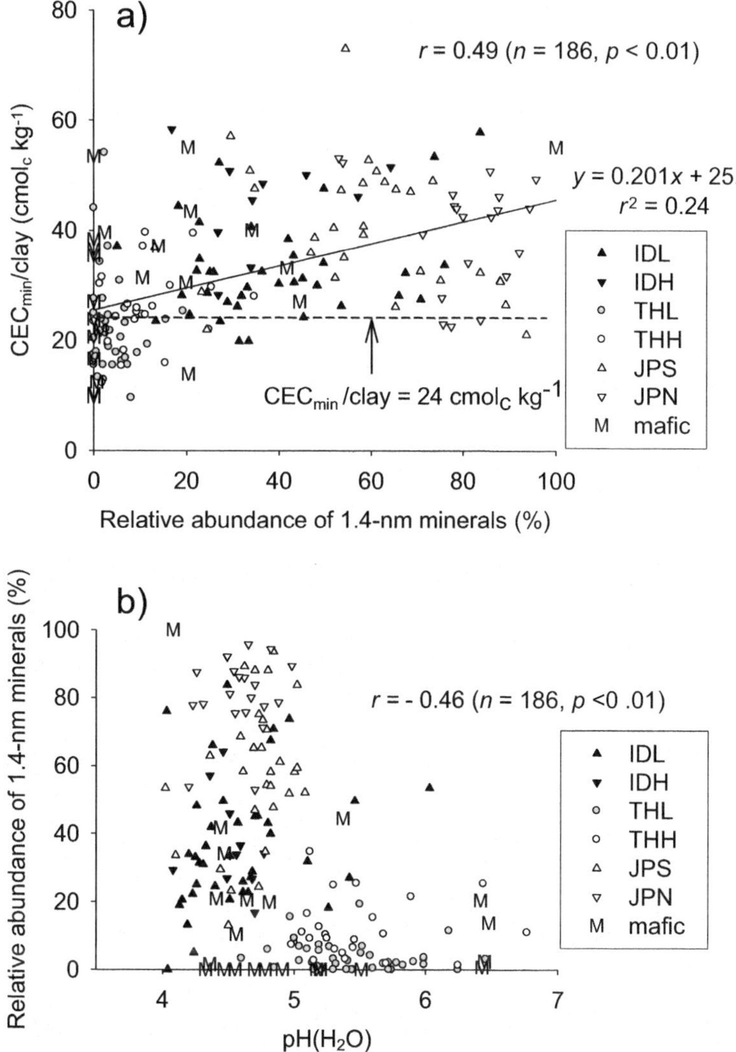

Fig. 3.11 Relationships between the relative abundances of the expandable 1.4-nm minerals in the clay fraction and (**a**) CEC_{min}/clay and (**b**) $pH(H_2O)$ of soils

3.4.5 Classification of the Soils Studied According to WRB (2006) and Its Relation to Mineral Weathering Conditions

Table 3.3 summarizes the classification of the soils studied according to IUSS Working Group WRB (2014). The RSGs to which "mafic" soils were classified are diverse, including the Alisols, the Acrisols, the Luvisols, and the Cambisols.

Table 3.3 Summary of the classification of the soils according to the World Reference Base

Sample group	Number of samples	Number of soils classified into different RSG						
		Andosols	Podzols	Alisols	Acrisols	Luvisols (or Lixisols)	Cambisols	
							(Dystric)	(Eutric)
IDL	39	0	0	34	2	0	2	1
IDH	10	0	0	5	0	0	5	0
THL	40	0	0	5	20	3(1)	9	2
THH	24	0	0	6	6	1	9	2
JPS	30	0	0	7	0	0	23	0
JPN	20	4	6	0	0	0	10	0
Mafic	23	0	0	7	4	4	5	3
Total	186	4	6	64	32	9	63	8

More detailed surveys focusing on the relationship between parent materials or primary minerals and the properties of the soils that developed are necessary.

Most of the samples of IDL and IDH belong to the Alisols according to WRB, except for those on relatively steep slopes in high mountains that are classified into Dystric Cambisols due to lack of a clear argic horizon. The higher soluble Al^{3+} in the Alisols may be a more serious constraint for agricultural production compared with the Acrisols. In contrast, the Acrisols were predominant in the THL samples, and the proportion of the Alisols and the Dystric Cambisols increased in the THH samples. In Japan, the majority of the soils were Dystric Cambisols, and some from the JPS were Alisols. In JPN, some soils were classified into Andosols or Podzols due to the relatively high influence of amorphous materials under cool temperate climates and/or the presence of nearby active volcanoes. It should be noted that no extremely weathered soils, such as Ferralsols or Plinthosols, were found in the present study, though there is still a necessity to extend future surveys to whole regions of humid Asia, especially to the relatively stable plains of northeast Thailand, central Cambodia, and southern Vietnam.

Thus, among the 186 soil samples, only nine and eight soils are classified into Luvisols (or Lixisols) and Eutric Cambisols, respectively, indicating that the upland soils in humid Asia are predominantly acidic. In the previous sections, dioctahedral mica inherent to the sedimentary rocks or the felsic igneous rocks was thought to weather to form expandable 2:1 minerals, i.e., vermiculitization, under the lower pH conditions associated with the udic or the perudic soil moisture regime. On the other hand, mica was relatively stable under the higher pH conditions associated with the ustic soil moisture regime, while other primary minerals, such as feldspars, were unstable and dissolve to form the kaolin minerals and gibbsite. These processes should be further analyzed critically considering the possibility of the selective dissolution of interlayered K^+ ion from mica under acidic conditions from a clay mineralogical viewpoint (as discussed in Fanning et al. 1989). Our finding in the present study suggested that the regional distribution patterns of

Acrisols and Alisols in humid Asia were strongly related to the pedogenetic conditions through clay mineral formation. In turn, such a close relationship suggested that the soils in humid Asia were dominantly on the course of pedogenesis under the present bioclimatic conditions, probably unlike tropical soils on plains in other continents.

3.5 Conclusion

A clear trend of clay mineral composition was observed for the soils in humid Asia in terms of geology and climate. In soils derived from mica-free parent materials, such as the mafic volcanic rocks, the clay mineral composition changed to the kaolin apex along the mineral axis of the 1.4-nm minerals. For other soils derived from the sedimentary rocks or the felsic parent materials, mica and kaolin minerals dominated with lesser amounts of 1.4-nm minerals in northern Thailand, while significant amounts of the 1.4-nm minerals formed in Indonesia and Japan. Based on these findings and the thermodynamic analysis using soil water extracts, the mineral weathering sequences in the soils from the sedimentary rocks or the felsic igneous rocks were postulated for each of the regions as follows. In Thailand, under higher pH conditions associated with the ustic moisture regime, mica was relatively stable, while other primary minerals, such as feldspars, were unstable and dissolved to form kaolinite and gibbsite. Under the lower pH conditions of the udic moisture regime in Japan and Indonesia, mica weathered to form the 1.4-nm minerals. The Japanese soils were young and had much Al–OH species in the soil solution, resulting in the formation of Al hydroxide between the 2:1 layers and gibbsite, while the Al hydroxides and gibbsite had not formed or were already removed from the more highly weathered Indonesian soils.

These distinct soil mineralogical properties were considered to affect the chemical properties of the soils, such as the CEC/clay and the pH, and also the taxonomic classification of the soils. The CEC/clay of the soils derived from the sedimentary rocks (excluding limestone) or the felsic parent materials showed a clear regional trend; that is, it was usually higher than 24 $\mathrm{cmol_c\ kg^{-1}}$ (corresponding to the Alisols if the argic horizon was recognized) under the udic and the perudic soil moisture regimes in Indonesia and Japan. However, it was predominantly lower than 24 $\mathrm{cmol_c\ kg^{-1}}$ (corresponding to the Acrisols) under the ustic soil moisture regime in Thailand. In contrast, soils derived from the mafic volcanic rocks or limestones were more variable in clay mineral composition, the CEC/clay or the pH, and were often high in Fe_d. The WRB classification was generally consistent with the regional trends for the chemical and the mineralogical properties of soils and successfully described the distribution patterns of acid soils in humid Asia using the criteria of CEC/clay $= 24\ \mathrm{cmol_c\ kg^{-1}}$.

Acknowledgements This chapter is derived in part from an article published in Geoderma, 19 April 2006, copyright Elsevier, available online: http://dx.doi.org/10.1016/j.geoderma.2006. 02.001 and in part from an article published in Soil Science and Plant Nutrition, October 2008, copyright Taylor & Francis, available online: http://www.tandfonline.com/doi/10.1111/j.1747-0765.2008.00294.x.

References

Adams F (1971) Ionic concentrations and activities in soil solutions. Soil Sci Soc Am Proc 35:420–426

Araki S, Kyuma K, Funakawa S, Lim S, Suh Y, Um K (1990) Comparative study of red yellow colored soils of Thailand, China, Korea and Japan. In: Transactions of 14th international congress of soil science, Kyoto, Japan, V:328–329

Araki S, Msanya BM, Magogo JP, Kimaro DN, Kitagawa Y (1998) Characterization of soils on various planation surfaces in Tanzania. In: 16th world congress of soil science, Montpellier, France

Borchardt G (1989) Smectites. In: Dixon JB, Weed SB (eds) Minerals in soil environments, Book Series No. 1, 2nd edn. Soil Science Society of America, Madison, pp 675–727

Bouma J, Hoeks J, van der Plas L, van Scherrenburg B (1969) Genesis and morphology of some Alpine podzol profiles. J Soil Sci 20:384–398

Driessen PM, Deckers JA, Spaargaren OC, Nachtergaele FO (eds) (2001) Lecture notes on the major soils of the world, World soil resources reports, vol 94. Food and Agriculture Organization of the United Nations, Rome

Fanning DS, Keramidas VZ, El-Desoky MA (1989) Micas. In: JB Dixon and SB Weed (eds) Minerals in soil environments, 2nd edn. Soil Science Society of America, Inc., Madison, Wisconsin, pp 551–634

Funakawa S, Nambu K, Hirai H, Kyuma K (1993) Pedogenetic acidification of forest soils in northern Kyoto. Soil Sci Plant Nutr 39:677–690

Funakawa S, Ashida M, Yonebayashi K (2003) Charge characteristics of forest soils derived from sedimentary rocks in Kinki District, Japan, in relation to pedogenetic acidification process. Soil Sci Plant Nutr 49:387–396

Funakawa S, Hirooka K, Yonebayashi K (2008) Temporary storage of soil organic matter and acid neutralizing capacity during the process of pedogenetic acidification of forest soils in Kinki District, Japan. Soil Sci Plant Nutr 54:434–438

Gjems O (1970) Mineralogical composition and pedogenic weathering of the clay fraction in Podzol soil profiles in Zalesine, Yugoslavia. Soil Sci 110:237–243

Goenadi DH, Tan KH (1988) Differences in clay mineralogy and oxidic ratio of selected LAC soils in temperate and tropical regions. Soil Sci 146:151–159

Hirai H (1995) Studies on the genesis of Brown Forest soils and their related soils in Japan. Doctoral thesis. Kyoto University, Kyoto, p 146

Hirai H, Araki S, Kyuma K (1989) Clay mineralogical properties of brown forest soils in northern Kyoto with special reference to their pedogenetic process. Soil Sci Plant Nutr 35:585–596

Hodges SC (1987) Aluminum speciation, a comparison of five methods. Soil Sci Soc Am J 51:57–64

Huang PM, Wang MK, Kampf N, Schulze DG (2002) Aluminum hydroxides. In: Dixon JB, Schulze DG (eds) Soil mineralogy with environmental applications, Book Series No. 7, Soil Science Society of America, Madison, pp 261–289

ISSS-ISRIC-FAO (1998a) World reference base for soil resources. World soil resources reports 84. Food and Agriculture Organization of the United Nations, Rome

ISSS-ISRIC-FAO (1998b) World Reference Base for Soil Resources: Atlas, Eds EM Bridges, NH Batjes and FO Nachtergaele. Food and Agriculture Organization of the United Nations, Rome

IUSS Working Group WRB (2014) World Reference Base for Soil Resources 2014. International soil classification system for naming soils and creating legends for soil maps. World Soil Resources Reports No. 106. FAO, Rome

Kanket W, Suddhiprakarn A, Kheoruenromne I, Gilkes RJ (2002) Clay mineralogy of some Thai Ultisols. In: 17th world congress of soil science, Bangkok, Thailand

Karathanasis AD (2002) Mineral equilibria in environmental soil systems. In: Dixon JB, Schulze DG (eds) Soil mineralogy with environmental applications, book series No. 7, Soil Science Society of America, Madison, pp 109–151

Karathanasis AD, Adams F, Hajek BJ (1983) Stability relationships in kaolinite, gibbsite, and Al-interlayered vermiculite soil systems. Soil Sci Soc Am J 47:1247–1251

Kitagawa Y (2005) Characteristics of clay minerals in Podzols and podzolic soil. Soil Sci Plant Nutr 51:151–158

Kittrick JA (1973) Mica-derived vermiculites as unstable intermediates. Clay Clay Miner 21:479–488

Koch CB, Bentzon MD, Larsen EW, Borggaard OK (1992) Clay mineralogy of two Ultisols from Central Kalimantan, Indonesia. Soil Sci 154:158–167

Lasaga AC (1998) Rate laws of chemical reactions. In: Kinetic theory in the earth sciences, Princeton University Press, Princeton, pp 3–151

Lindsay WL (1979) Aluminum, Silica, Aluminosilicate. In: Chemical Equilibria in Soils, Wiley, New York, pp 34–77

Malla PB (2002) Vermiculites. In: Dixon JB, Schulze DG (eds) Soil mineralogy with environmental applications, Book series No. 7, Soil Science Society of America, Madison, pp 501–529

Matsue N, Wada K (1989) Source minerals and formation of partially interlayered vermiculites in Dystrochrepts derived from Tertiary sediments. J Soil Sci 40:1–7

McBride MB (1989) Surface chemistry of soil minerals. In: Dixon JB, Weed SB (eds) Minerals in soil environments, 2nd edn. Soil Science Society of America, Inc., Madison, pp 35–88

Nagy KL (1995) Dissolution and precipitation kinetics of sheet silicates. In: White AF, Brantley SL (eds) Chemical weathering rates of silicate minerals, reviews in mineralogy, 31, Mineralogical Society of America, Washington, DC, pp 173–233

Norfleet ML, Karathanasis AD, Smith BR (1993) Soil solution composition relative to mineral distribution in Blue Ridge mountain soils. Soil Sci Soc Am J 57:1375–1380

Prasetyo BH, Suharta N, Subagyo H, Hikmatullah (2001) Chemical and mineralogical properties of Ultisols of Sasamba area, East Kalimantan. Indones J Agric Sci 2:37–47

Rai D, Kittrick JA (1989) Mineral equilibria and the soil system. In Minerals in Soil Environments, Book Series No. 1, 2nd ed., Dixon JB, Weed SB (eds.), Soil Science Society of America, Madison, pp 161–198

Rausell-Colom JA, Sweatman TR, Wells CB, Norrish K (1965) Studies in the artificial weathering of mica. In: Crawford EG (eds) Experimental pedology, Hallsworth, Butterworths, London, p 40–72

Reid-Soukup DA Ulery AL (2002) Smectites. In: Soil mineralogy with environmental applications, Book Series No. 7, Dixon JB, Schulze DG (eds), Soil Science Society of America, Madison, pp 467–499

Robson AD (ed) (1989) Soil acidity and plant growth. Academic Press Australia, Sydney, p 306

Ross RJ, Mortland MM (1966) A soil beidellite. Soil Sci Soc Am Proc 30:337–343

Soil Survey Staff (2014) Keys to soil taxonomy, Twelfth edn. United States Department of Agriculture, Natural Resources Conservation Service

SPSS (1998) SYSTAT 8.0. Statistics. SPSS Inc, Chicago, p 1086

Supriyo H, Matsue N, Yoshinaga N (1992) Chemistry and mineralogy of some soils from Indonesia. Soil Sci Plant Nutr 38:217–225

Tanaka S, Kendawang JJ, Yoshida N, Shibata K, Jee A, Tanaka K, Ninomiya I, Sakurai K (2005) Effect of shifting cultivation on soil ecosystems in Sarawak, Malaysia. IV. Chemical properties of the soils and runoff water at Niah and Bakam experimental sites. Soil Sci Plant Nutr 51:525–533

Tanaka S, Wasli ME, Kotegawa T, Seman L, Sabang J, Kendawang JJ, Sakurai K, Morooka Y (2007) Soil properties of secondary forests under shifting cultivation by the Iban of Sarawak, Malaysia in relation to vegetation condition. Tropics 16:385–398

Thornthwaite CW (1948) An approach toward a rational classification of climate. Geogr Rev 38:55–94

Thornthwaite CW, Mather JR (1957) Instructions and tables for computing potential evapotranspiration and the water balance. Climatology 10:3

Toriyama J, Ohta S, Araki M, Ito E, Kanzaki M, Khorn S, Pith P, Lim S, Pol S (2007) Acrisols and adjacent soils under four different forest type in central Cambodia. Pedologist 51:35–49

van Breemen, N., Brinkman, R. (1976) Chemical equilibria and soil formation. In Soil Chemistry. A. Basic Elements, Bolt GH, Wruggenwert MGM (eds), Elsevier, Amsterdam, pp 141–170

Watanabe E, Sakurai K, Okabayashi Y, Lasay N, Alounsawat C (2004) Soil fertility and farming systems in a slash and burn cultivation area of northern Laos. Southeast Asian Stud 41:519–537

Yoothong K, Moncharoen L, Vijarnson P, Eswaran H (1997) Clay mineralogy of Thai soils. Appl Clay Sci 11:357–371

Yoshinaga N, Matsue N, Virakornphanich P, Tantiseri B, Supa-Udomleark B (1995) Clay mineralogy of soils from upper north Thailand. Soil Sci Plant Nutr 41:337–343

Chapter 4
Soil-Forming Factors Determining the Distribution Patterns of Different Soils in Tanzania with Special Reference to Clay Mineralogy

Shinya Funakawa and Method Kilasara

Abstract Soil-forming conditions in Tanzania, which is located near the Great Rift Valley, vary with geological and climatic conditions. In the present study, 95 surface soil samples were collected from croplands, forests, and savannas in different regions across Tanzania and the physicochemical and mineralogical properties of the samples were analyzed. Reflecting the wide variation of climatic conditions and parent materials of soils, soil physicochemical properties varied widely. Clay mineral composition was basically similar to the soils developed under ustic moisture regime in Southeast Asia, in which mica and kaolin minerals were usually dominated. Based on a principal component analysis of the collected soil samples, five individual factors were determined. From the clay mineralogical composition and the relation between the geological conditions (or parent materials) and the annual precipitation and the scores of the five factors, the following observations were made: (1) The maximum scores of "SOM and amorphous compounds" were found at the volcanic center of the southern mountain ranges from the east of Mbeya to Lake Malawi. (2) The scores of the "available P and K" were high in the volcanic regions around Mt. Kilimanjaro and in the southern volcanic mountain ranges, presumably due to the intensive agricultural management with fertilizer application. (3) The 1.4-nm minerals were probably formed under conditions of high sodicity and were often observed in the soils near Lake Victoria. (4) In Tanzania, the volcanic regions and the Great Rift Valley region, where soil is generally more fertile than in other regions, are conducive to modernized agriculture. The semiarid regions in Tanzania suffer from water shortage, while the relatively humid areas have less fertile soil that predominantly contains kaolin minerals. These conditions are not favorable for agricultural production and should be taken into consideration

S. Funakawa (✉)
Graduate School of Global Environmental Studies, Kyoto University, Yoshida-Honmachi, Sakyo-ku, Kyoto 606-8501, Japan
e-mail: funakawa@kais.kyoto-u.ac.jp

M. Kilasara
Faculty of Agriculture, Sokoine University of Agriculture, Morogoro, Tanzania

© Springer Japan KK 2017

S. Funakawa (ed.), *Soils, Ecosystem Processes, and Agricultural Development*, DOI 10.1007/978-4-431-56484-3_4

when studying the feasibility of agricultural development in different areas in the future.

Keywords Clay mineralogy • Great Rift Valley • Soil fertility • Tanzania

4.1 Introduction

The pedogenetic conditions in Tanzania vary widely. In particular, the country has a wide variety of parent materials of soils because of the presence of volcanic mountains, the Great Rift Valley, and several plains and mountains with different elevations (hence, different temperatures). In addition, the amount and seasonal distribution pattern of the annual precipitation vary, from less than 500 mm to more than 2500 mm. The potential land use and agricultural production differ greatly among regions, owing to the presence of different soils.

There have been several reports on the distribution patterns of soils and their physicochemical and mineralogical properties. According to a review of the history of soil surveys in Tanzania by Msanya et al. (2002), the major soils observed in the country are ferric, chromic, and eutric Cambisols (39.7 %); these are followed by rhodic and haplic Ferralsols (13.4 %) and humic and ferric Acrisols (9.6 %). To obtain basic information on soil mineralogy, Araki et al. (1998) investigated soil samples collected from regions at different altitudes in the Southern Highland and reported that the cation exchange capacity (CEC) per unit amount of clay content showed a negative correlation with elevation, which was accompanied by clay mineralogical transformation from mica to kaolinite. The authors suggested that soil formation on different planation surfaces is controlled mainly by the geological time factor whereby the lower surfaces are formed at the expense of the higher surfaces. Szilas et al. (2005) analyzed the mineralogy of well-drained upland soil samples collected from important agricultural areas in different ecological zones in the subhumid and humid parts of Tanzania. They concluded that all soils were severely weathered and had limited but variable capacities to hold and release nutrients in plant-available form and to sustain low-input subsistence agriculture. Generally, there seems to be a consensus that the soils in Tanzania and its neighboring countries are not very fertile. According to the Global Soil Regions (Soil Survey Staff, 1999), distribution pattern of soils in East Africa may summarized by the dominance of Ultisols and Alfisols with accessorized distribution of Aridisols and Oxisols.

In the present study, the regional trend in soil fertility was investigated with special reference to the clay mineralogy and its forming conditions such as geology and climate. An understanding of why the distribution of some soil properties is influenced by soil-forming factors would help in planning an appropriate land-use strategy, which should make it possible to not only sustain and develop agricultural production but also maintain natural resources such as forest and woodland ecosystems.

4.2 Soil Samples and Analytical Procedure

Ninety-five samples of surface soils were collected from different regions of Tanzania (Figs. 4.1 and 4.2). All the sampling points were located on slopes or plains, covering regions with different parent materials and with a wide variety of annual precipitation (less than 250 to more than 1500 mm) (Fig. 4.2; prepared based on Atlas of Tanzania [1967]). Apparent lowland soils were excluded from the analysis. The parent materials of the soils were broadly classified according to the following categories: (1) volcanic rocks (mostly basic), (2) granite and other plutonic rocks, (3) sedimentary and metamorphic rocks, and (4) Cenozoic rocks and recent deposits. The sampling plots were used as croplands or were covered by

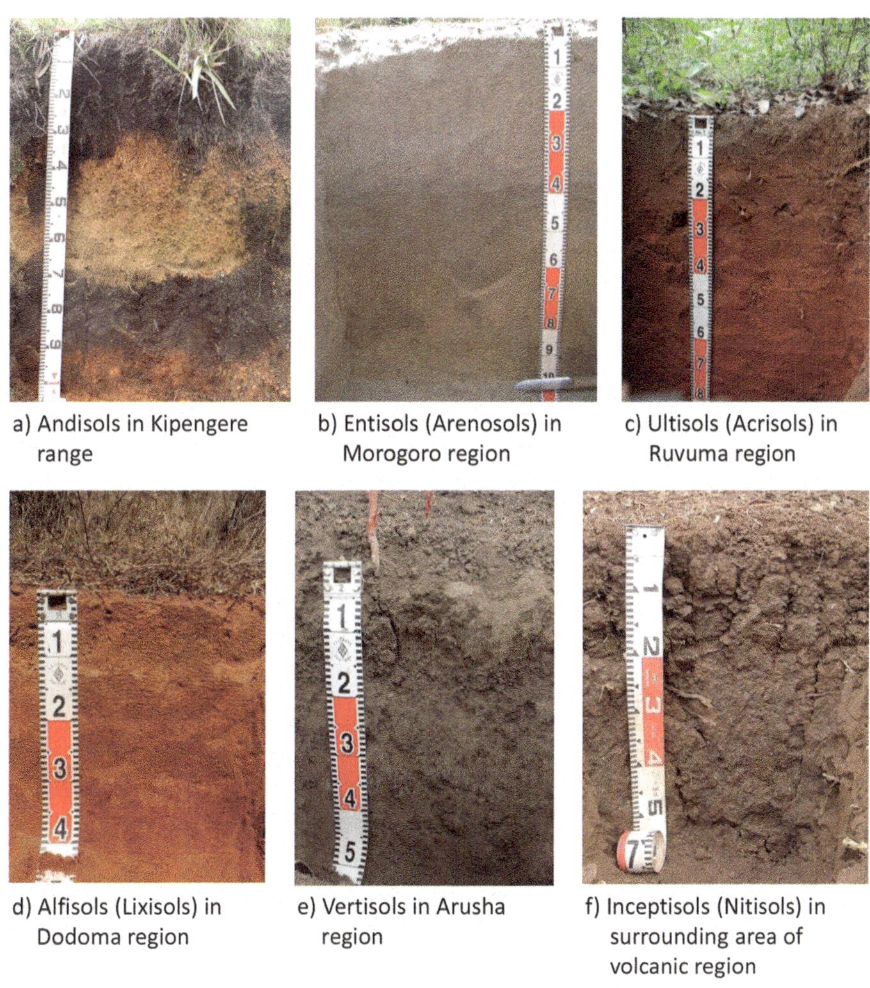

a) Andisols in Kipengere range

b) Entisols (Arenosols) in Morogoro region

c) Ultisols (Acrisols) in Ruvuma region

d) Alfisols (Lixisols) in Dodoma region

e) Vertisols in Arusha region

f) Inceptisols (Nitisols) in surrounding area of volcanic region

Fig. 4.1 Representative soil profiles observed in Tanzania

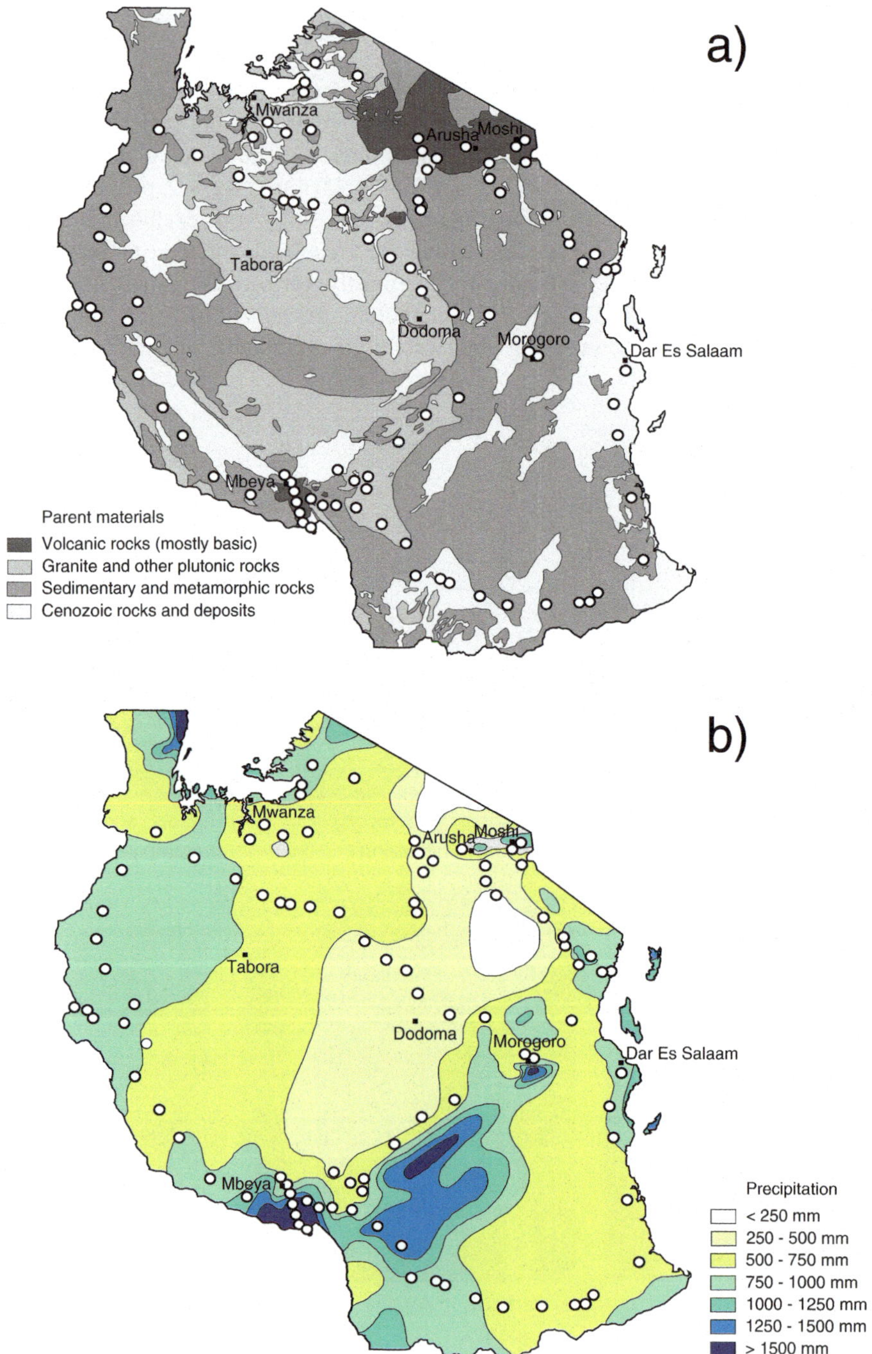

Fig. 4.2 Geological and climatic conditions of the sampling plots

seminatural vegetation (forest or woodland) or secondary vegetation that had grown after human disturbance.

The chemical and mineralogical properties of the samples were analyzed; we determined the $pH(H_2O)$, $pH(KCl)$, electrical conductivity (EC), exchangeable cations (bases and Al), cation exchange capacity (CEC), total C and total N content, available P (Bray-II method), particle size distribution, acid ammonium oxalate- and DCB-extractable oxides (Fe_o, Al_o, Si_o, Fe_d, and Al_d), and clay mineral composition (by X-ray diffraction). The data obtained were statistically analyzed with the software SYSTAT version 8.0 (SPSS 1998).

4.3 General Physicochemical and Mineralogical Properties of the Soils

Selected statistical values of the analytical data are listed in Table 4.1. The surface soils studied were usually slightly acidic, with the average values of $pH(H_2O)$ and $pH(KCl)$ being 6.17 and 5.37, respectively. The exchangeable Al content was low and the base saturation was high, exceeding 95 % on average; hence, soil acidity was not considered a serious constraint for agricultural production. Although the average soil texture was sandy clay loam to clay loam, the particle size distribution varied widely. The average C content was 20.7 g kg^{-1}. However, the obtained values of most of the aforementioned variables varied significantly over the regions under study, with their coefficients of variation exceeding 100 %; these results indicated a marked difference in the soil characteristics in various regions across Tanzania.

According to the X-ray analysis, several patterns of diffractograms were obtained (Fig. 4.3a–d); the dominant clay mineral was kaolinite, followed by clay mica. In the volcanic centers in the northern and southern mountains, X-ray amorphous clays from volcanic origin were observed. The 1.4-nm clay minerals such as smectite were occasionally dominated among the soils from northwestern region of the country. Figure 4.3e shows a summary of semiquantitative clay mineral composition of all the soils studied, except for the volcanic soils with X-ray amorphous. Although some soils developed on Cenozoic rocks and recent deposits contained fairly high amounts of expandable 1.4-nm minerals, the majority of the soils were dominated by mica and kaolin minerals with few amounts of 1.4-nm minerals. This composition was basically similar to that of the Asian soils developed under ustic soil moisture regime (Chap. 3 in this volume; Funakawa et al. 2008), indicating that the mineral weathering pathways postulated for the Asian soils could be applicable to Tanzanian soils under ustic soil moisture regime.

Table 4.2 lists the obtained data categorized according to the parent materials and land use. In terms of soil parent materials, the physicochemical and mineralogical properties of the volcanic-derived soils ($n = 12$) were unique; they were significantly different from the other soil groups in terms of CEC, total C content,

Table 4.1 Descriptive statistics for the physicochemical and mineralogical properties of the soils studied

Variable	Number of samples	Average	Minimum	Maximum	Standard deviation	Coefficient of variation (%)
pH(H$_2$O)	95	6.17	4.36	8.66	0.80	13.0
pH(KCl)	95	5.37	3.71	7.96	0.89	16.5
pH(NaF)	95	8.15	7.12	11.01	0.66	8.0
EC(μS dm^{-1})	95	74.3	10.0	325	59.4	79.9
CEC (cmol$_c$ kg^{-1})	95	14.0	1.61	59.5	11.0	78.6
Exch. Na (cmol$_c$ kg^{-1})	95	0.18	0.00	1.92	0.29	161
Exch. K (cmol$_c$ kg^{-1})	95	1.10	0.10	5.62	1.12	102
Exch. Mg (cmol$_c$ kg^{-1})	95	2.78	0.18	11.4	2.14	77.0
Exch. Ca (cmol$_c$ kg^{-1})	95	6.98	0.00	49.5	8.76	126
Exch. Al (cmol$_c$ kg^{-1})	95	0.21	0.00	2.99	0.51	241
Exch. bases (cmol$_c$ kg^{-1})	95	11.0	0.43	60.7	11.4	103
Base satur. (%)	95	95.4	49.0	101	10.5	11.0
Sand (%)	95	63.6	3.4	96.7	23.3	36.7
Silt (%)	95	11.2	0.2	48.1	11.3	101
Clay (%)	95	25.2	1.5	81.4	17.6	69.8
Total C (g kg^{-1})	95	20.7	2.13	152	24.4	124
Total N (g kg^{-1})	95	1.49	0.21	13.7	1.84	129
Available P (gP$_2$O$_5$ kg^{-1})	95	0.15	0.01	1.0	0.24	161
Fe$_o$ (g kg^{-1})	95	2.46	0.02	14.7	3.28	133
Al$_o$ (g kg^{-1})	95	3.61	0.08	64.3	9.34	259
Si$_o$ (g kg^{-1})	95	1.10	0.00	21.7	3.32	303
Fe$_d$ (g kg^{-1})	95	23.7	0.19	159	25.8	109
Al$_d$ (g kg^{-1})	95	4.55	0.01	50.9	7.48	164
0.7 nm minerals (%)	90	72.5	5.4	100	27.0	37.3
1.0 nm minerals (%)	90	19.6	0.0	91.2	21.4	109
1.4 nm minerals (%)	90	7.9	0.0	94.6	17.9	227

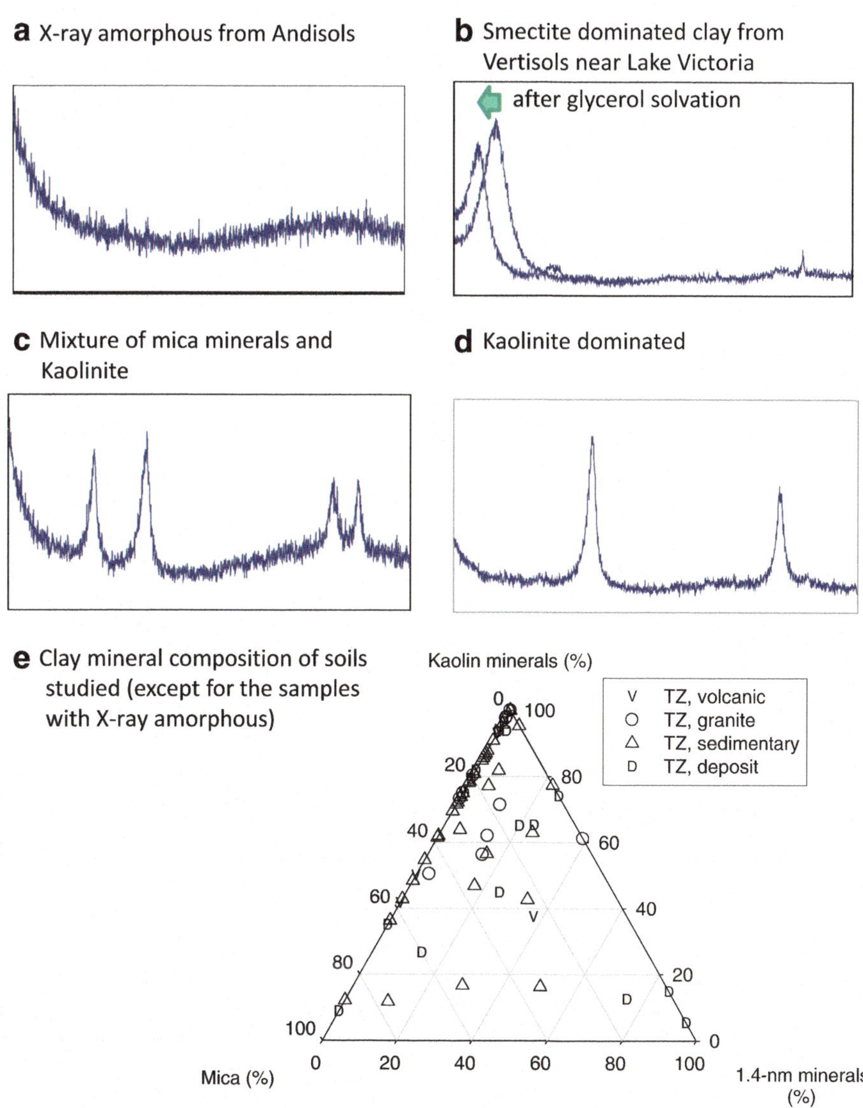

Fig. 4.3 Representative patterns of X-ray diffractograms of clay specimen (**a–d**) and summary of semi-quantitive clay mineral composition of all the soils studied (**e**)

available P, and free oxide-related properties. On the other hand, these soil properties usually did not differ significantly for different land uses.

A principal component analysis was conducted to summarize several soil parameters related to soil fertility. The studied variables included pH(H$_2$O); pH (KCl); pH(NaF); CEC; amounts of exchangeable Na$^+$, K$^+$, Mg^{2+}, Ca^{2+}, and Al^{3+};

Table 4.2 Average values of measured soil variables in terms of parent materials or land uses

Variable	Averages for soils from different parent materials				Averages for soils under different land uses		
	Volcanic rocks	Granite and other plutonic rocks	Sedimentary and metamorphic rocks	Cenozoic rocks and deposits	Natural and matured secondary vegetation	Incipient fallow vegetation	Cropland
Number of samples	12(9)[a]	14	50(48)[a]	19	37(35)[a]	16	42(39)[a]
pH(H$_2$O)	5.91 ab	5.70 a	6.28 ab	6.43 b	6.34 a	6.14 a	6.04 a
pH(KCl)	5.18 a	4.86 a	5.49 a	5.56 a	5.53 a	5.32 a	5.26 a
pH(NaF)	9.04 b	7.95 a	8.04 a	8.05 a	8.12 a	7.99 a	8.22 a
EC (μS dm^{-1})	103.2 b	48.1 a	78.7 ab	63.9 a	87.2 b	35.9 a	77.6 b
CEC (cmol$_c$ kg^{-1})	29.5 b	6.93 a	12.5 a	13.3 a	14.3 a	8.6 a	15.7 a
Exch. Na (cmolc kg^{-1})	0.26 ab	0.07 a	0.11 a	0.39 b	0.11 a	0.12 ab	0.27 b
Exch. K (cmolc kg^{-1})	2.49 b	0.45 a	1.11 a	0.68 a	1.17 a	0.70 a	1.19 a
Exch. Mg (cmolc kg^{-1})	4.34 b	1.25 a	2.70 ab	3.12 b	2.95 a	2.53 a	2.72 a
Exch. Ca (cmolc kg^{-1})	11.60 b	1.68 a	5.87 ab	10.85 b	8.02 a	4.17 a	7.13 a
Exch. Al (cmolc kg^{-1})	0.18 a	0.28 a	0.25 a	0.07 a	0.22 a	0.28 a	0.18 a
Exch. bases (cmol$_c$ kg^{-1})	18.7 b	3.4 a	9.8 ab	15.0 b	12.2 a	7.5 a	11.3 a
Base satur. (%)	97.4 ab	88.5 a	95.4 ab	99.0 b	96.5 a	92.3 a	95.5 a
Sand (%)	36.2 a	73.9 b	64.7 b	70.2 b	66.2 a	69.0 a	59.1 a
Silt (%)	28.7 b	6.6 a	9.2 a	9.0 a	9.5 a	7.0 a	14.3 a
Clay (%)	35.1 a	19.5 a	26.1 a	20.8 a	24.3 a	23.9 a	26.5 a
Total C (g kg^{-1})	43.3 b	12.5 a	20.1 a	13.9 a	28.4 a	11.8 a	17.2 a
Total N (g kg^{-1})	3.40 b	0.98 a	1.42 a	0.87 a	1.98 a	0.80 a	1.33 a
Available P (gP$_2$O$_5$ kg^{-1})	0.431 b	0.044 a	0.128 a	0.112 a	0.154 a	0.067 a	0.180 a
Fe$_o$ (g kg^{-1})	8.35 b	0.73 a	2.04 a	1.12 a	2.26 a	1.11 a	3.16 a
Al$_o$ (g kg^{-1})	13.89 b	1.50 a	2.76 a	0.91 a	4.22 a	1.11 a	4.02 a
Si$_o$ (g kg^{-1})	4.83 b	0.16 a	0.75 a	0.34 a	1.18 a	0.29 a	1.33 a
Fe$_d$ (g kg^{-1})	40.2 b	13.2 a	26.5 ab	11.4 a	23.2 a	26.2 a	23.1 a
Al$_d$ (g kg^{-1})	11.34 b	3.50 a	4.37 a	1.50 a	5.58 a	2.98 a	4.24 a
0.7 nm minerals (%)	75.8 a	80.4 a	73.3 a	63.0 a	73.9 a	79.6 a	68.3 a
1.0 nm minerals (%)	20.1 a	13.8 a	22.0 a	17.8 a	20.9 a	16.1 a	19.9 a
1.4 nm minerals (%)	4.2 a	5.8 a	4.7 a	19.2 b	5.2 a	4.3 a	11.8 a

[a]Parenthesis denotes the number of samples for XRD analysis (i.e., the percentage of 0.7, 1.0, and 1.4 nm minerals)

Table 4.3 Factor pattern for the first four principal components ($n = 95$)

Variable	PC1	PC2	PC3	PC4	PC5
pH(H$_2$O)	−0.19	−0.07	−0.94	0.10	0.04
pH(KCl)	−0.05	−0.11	−0.95	0.10	−0.02
pH(NaF)	0.84	0.00	−0.05	0.29	0.05
CEC	0.57	0.51	−0.09	0.39	0.43
Exch. Na	−0.01	−0.01	0.04	0.08	0.93
Exch. K	0.01	0.32	−0.29	0.84	0.07
Exch. Mg	−0.03	0.62	−0.42	0.36	0.39
Exch. Ca	0.05	0.28	−0.64	0.38	0.47
Exch. Al	0.20	0.12	0.64	−0.08	0.07
Sand	−0.30	−0.83	−0.11	−0.35	−0.17
Silt	0.45	0.29	0.05	0.65	0.21
Clay	0.11	0.91	0.11	0.04	0.09
Total C	0.89	0.20	0.08	0.02	0.01
Total N	0.88	0.19	0.14	0.03	0.01
Avail. P	0.06	0.02	−0.26	0.87	0.04
Fe$_o$	0.62	0.32	0.18	0.57	−0.01
Al$_o$	0.97	0.00	0.13	0.05	0.01
Si$_o$	0.94	−0.07	0.07	0.10	0.04
Fe$_d$	0.09	0.86	0.15	0.13	−0.27
Al$_d$	0.87	0.24	0.26	−0.06	−0.08
Eigenvalue	5.98	3.43	3.14	2.92	1.60
Proportion (%)	29.9	17.2	15.7	14.6	8.0
	"SOM and amorphous compounds" factor	"Texture" factor	"Acidity" factor	"Available P and K" factor	"Sodicity" factor

sand, silt, and clay content; total C and total N content; available P content; and Fe$_o$, Al$_o$, Si$_o$, Fe$_d$, and Al$_d$ content. Table 4.3 lists the factor pattern for the first five principal components after varimax rotation.

As seen from the list in Table 4.3, high positive coefficients were obtained for pH(NaF), total C, and total N, Fe$_o$, Al$_o$, Si$_o$, and Al$_d$ for the first component. These variables correspond to the properties derived from organic materials that are bound to amorphous compounds, which might have originated from recent volcanic activity. Hence, the first component is referred to as the "soil organic matter (SOM) and amorphous compounds" factor. The second component has high negative coefficients for sand content and high positive coefficients for clay content, exchangeable Mg, and Fe$_d$. These soil characteristics could be associated with parent materials and clay formation, that is, soils derived from mafic and/or clayey parent materials tend to exhibit fine-textured properties with high concentrations of exchangeable Mg and Fe$_d$ through rapid mineral weathering and clay formation. Hence, the second component is referred to as the "texture" factor. The coefficients corresponding to the third component have high positive or negative values for pH(H_2O), pH(KCl), and exchangeable Ca and Al, indicating that a close relationship exists between this component and soil acidity. This relationship can be referred to as the "acidity" factor. The fourth and fifth components are referred to as the "available P and K" and the "sodicity" factors, respectively, on the basis of the coefficients correlating each of the components and the soil variables.

4.4 Principal Component Analysis for Summarizing Soil Properties

From this analysis, a wide variety of soil parameters were categorized into five principal components, which accounted for 85.4 % of the total variance.

4.5 Pedogenetic Conditions Determining the Distribution Patterns of Factor Scores for Each of the Principal Components

Figure 4.4 shows a scattergram of the factor scores of SOM and amorphous compounds and those of available P and K. Both factor scores were significantly higher in soils derived from volcanic rocks than in other soils, but they were not statistically correlated. The factor scores are plotted on the geological map, as shown in Fig. 4.5. There are two representative volcanic areas in Tanzania, namely, Mt. Kilimanjaro and its surrounding region and Kipengere range between the east of Mbeya and Lake Malawi. Generally, the scores of the factor for SOM and amorphous compounds were highest in the region of Kipengere range, followed by some plots around Mt. Kilimanjaro (Fig. 4.5a), whereas the scores of the factor for available P and K tended to be high in both volcanic regions (Fig. 4.5b). Since these regions were already cultivated intensively, influence of chemical fertilizer might affect the high scores in this factor. Furthermore, as Msanya et al. (2007) indicated, the volcanic soils in the southern mountain ranges were rich in K, compared to several Japanese volcanic soils, likely reflecting lithological differences among the parent materials. The predominantly high scores of the factor for SOM and amorphous compounds in Kipengere range indicate a relatively incipient feature of soils after recent active volcanic events and potentially high soil fertility relating to SOM in these regions.

Figure 4.5c represents the distribution pattern of the factor scores of texture in terms of the geological conditions. There is a certain regional trend in these factor scores, though no statistical difference was observed in terms of the geological condition as a whole. Among the soils of volcanic origin, those in the northern

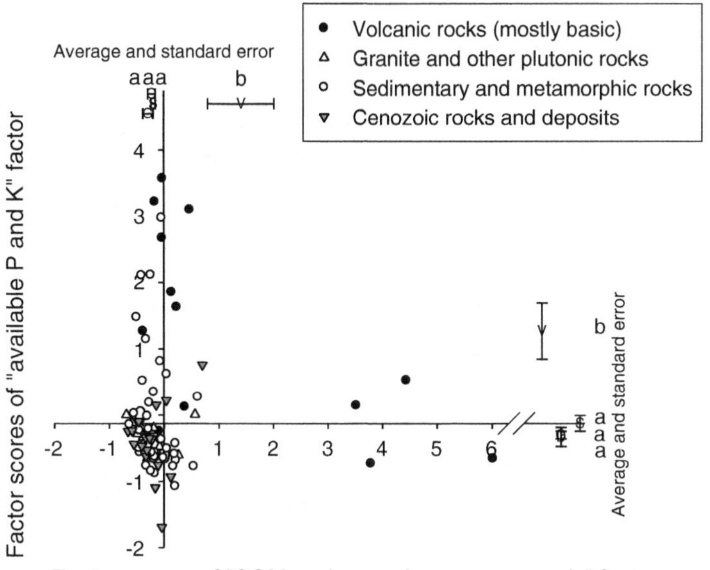

Fig. 4.4 Relationship between the scores of the "SOM and amorphous" and "available P and K" factors

volcanic regions exhibited higher scores in the texture factor, consistent with a previous report by Mizota et al. (1988), in which they postulated that these soils were in the advanced stages of weathering of volcanic materials. The scores were high in some soils from sedimentary and metamorphic rocks, which are mostly distributed in the western region around Kigoma and the hillslopes near Tanga, and, in contrast, were usually low in soils from granite, except for those of the southern highland.

Figure 4.6 shows the influence of the amount of precipitation on the scores of selected factors. There was no clear relationship between the amount of precipitation and the factor scores of acidity or texture. Although it was expected that a positive contribution of precipitation on mineral weathering might accompany soil acidification or the formation of clays and secondary Fe oxides, there was no correlation between them; this indirectly suggested that the influence of parent materials on soil properties was stronger than that of climatic factors among the soils studied.

                                                  S. Funakawa and M. Kilasara

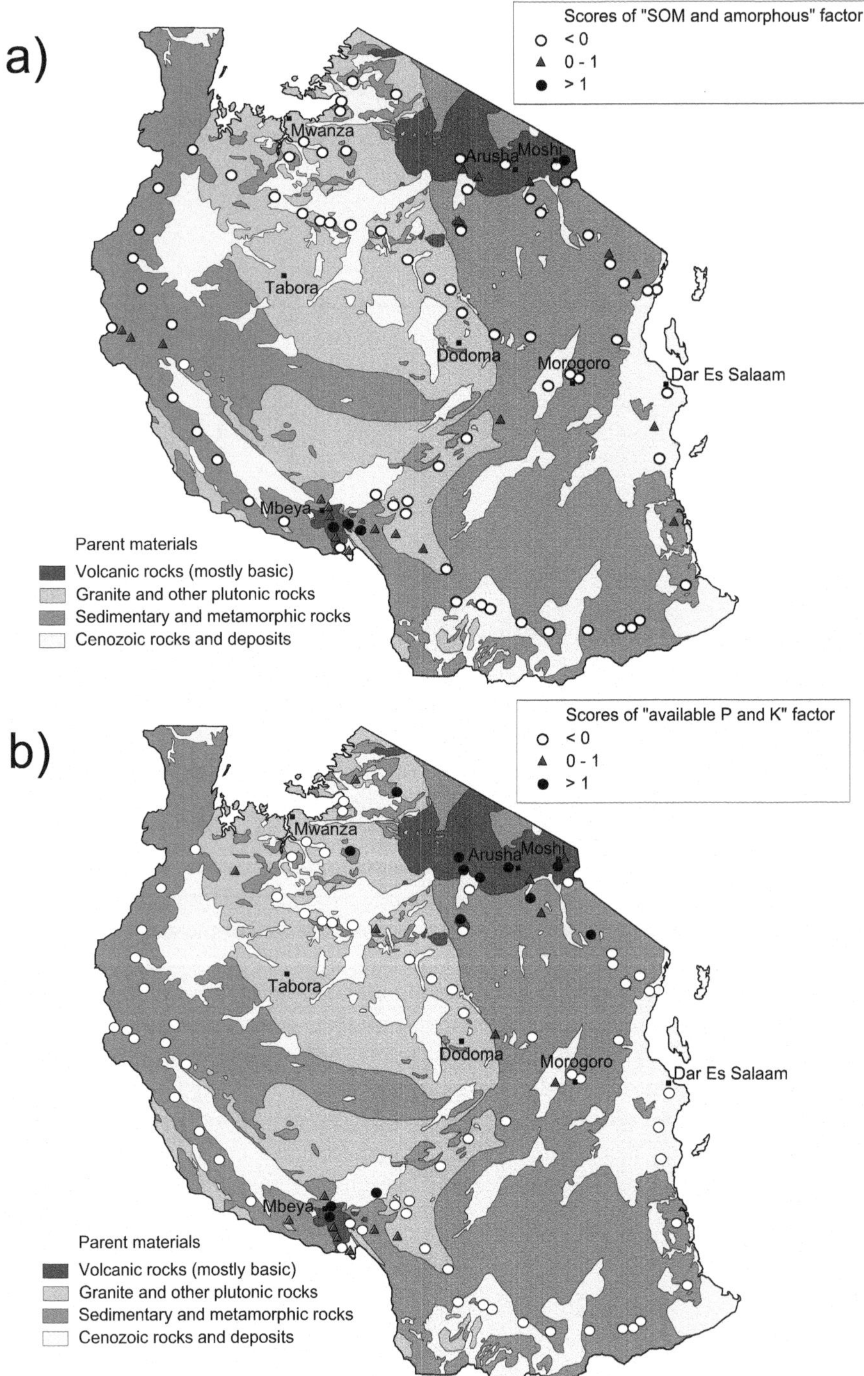

Fig. 4.5 Distribution patterns of scores of (**a**) "SOM and amorphous," (**b**) "available P and K," and (**c**) "texture" factors in relation to geological conditions

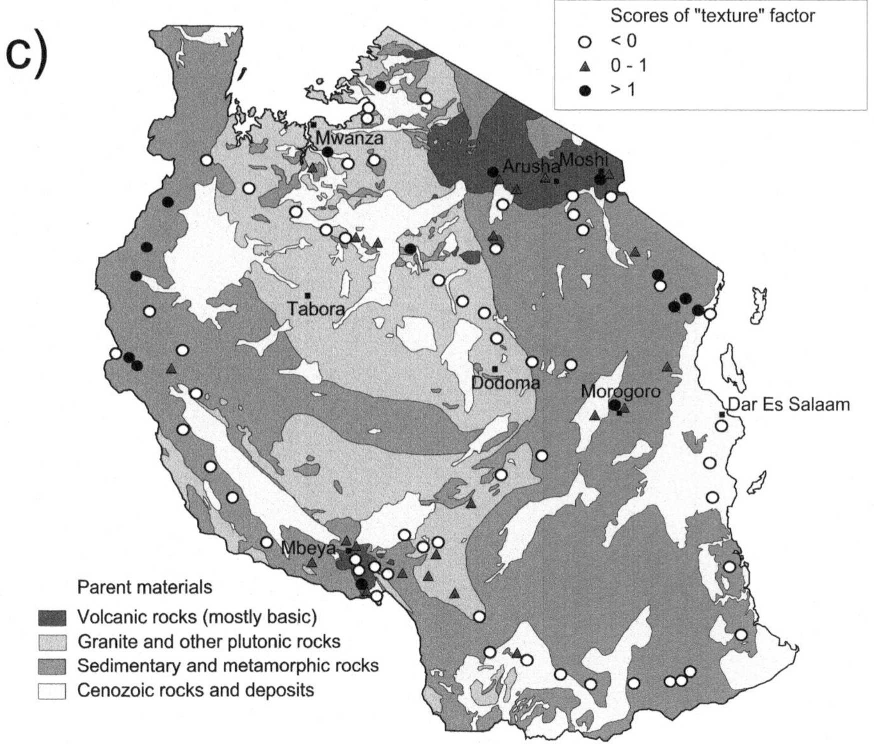

Fig. 4.5 (continued)

4.6 Pedogenetic Conditions Determining the Clay Mineralogy of the Soils

Figure 4.7 shows the distribution patterns of the clay mineralogy in relation to the geological and climatic conditions. The relative abundance of 1.4-nm minerals was often higher in the northern region of the Great Rift Valley and around Lake Victoria. On the other hand, the abundance of 0.7-nm minerals tended to be lower in the central steppe, which has lower precipitation than other regions. These relationships are more clearly presented in Fig. 4.8. Stepwise multiple regression indicated that the abundances of 1.4-nm minerals (mostly smectite) could be expressed by the following equation:

$$1.4 - \text{nm minerals} (\%) = 6.38 + 13.4 \text{ (sodicity factor)}$$
$$-9.78 \text{ (SOM/amorphous factor)} \quad (4.1)$$
$$+3.17 \text{ (P/K factor)}; r^2 = 0.58 \text{ } (p < 0.01, n = 90)$$

 S. Funakawa and M. Kilasara

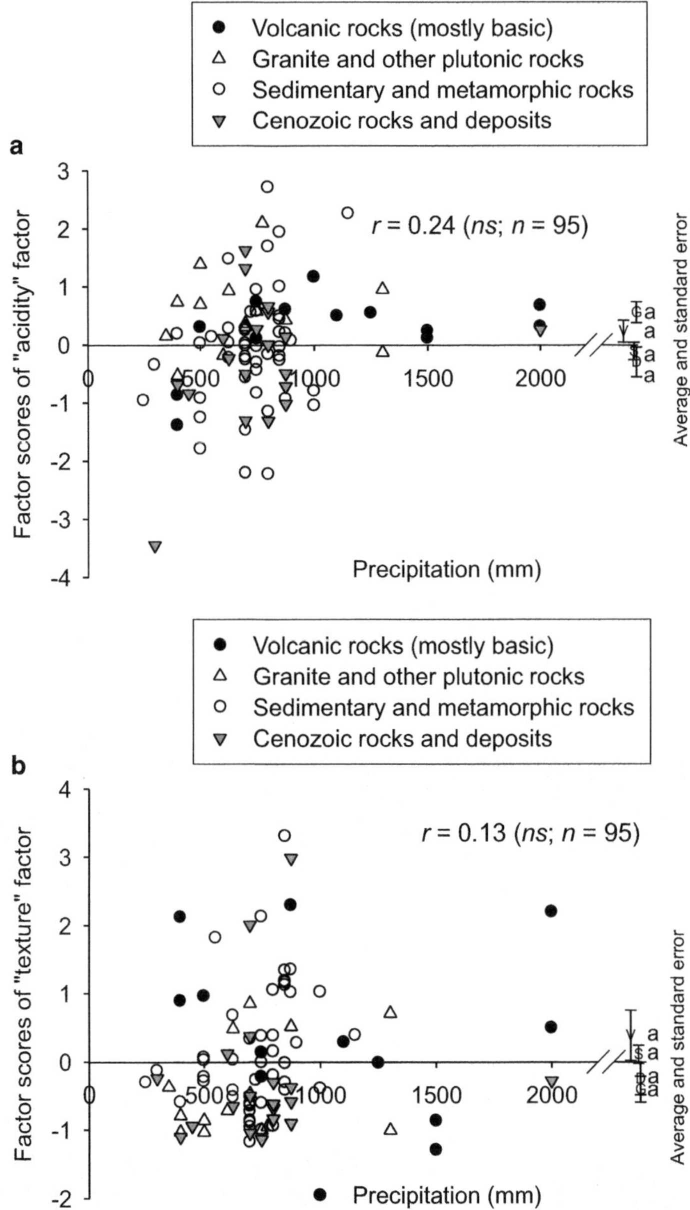

Fig. 4.6 Relationships between precipitation and scores of (**a**) "acidity" and (**b**) "texture" factor

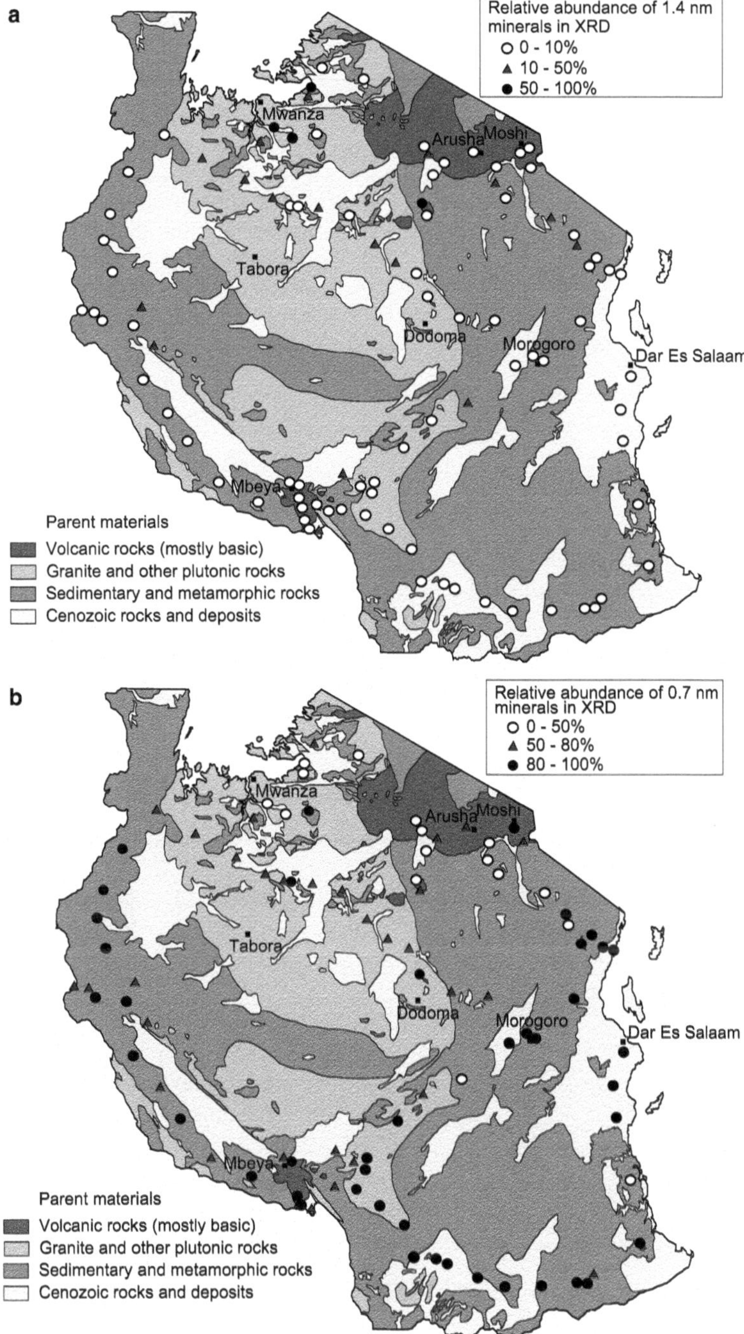

Fig. 4.7 Distribution patterns of clay mineralogy in relation to geological or climatic conditions. Abundances of (**a**) 1.4 nm and (**b**) 0.7 nm minerals

a)

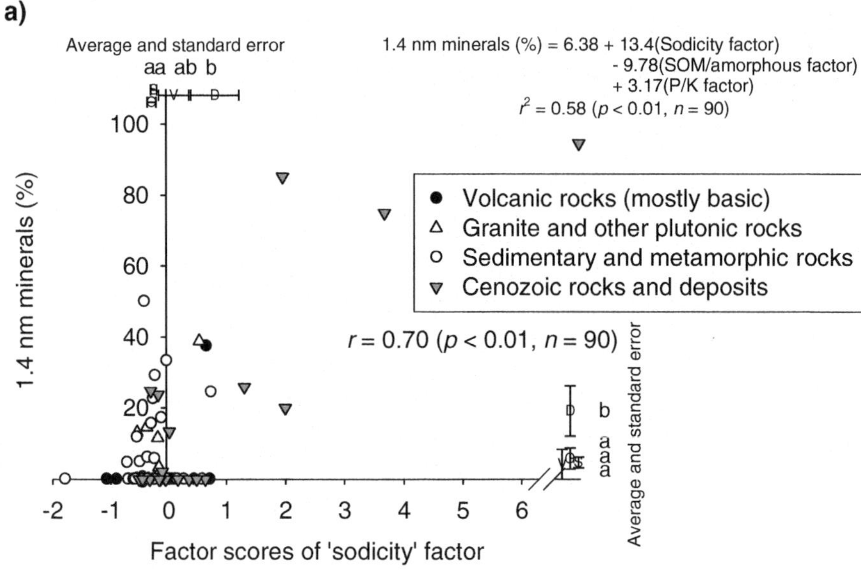

b)

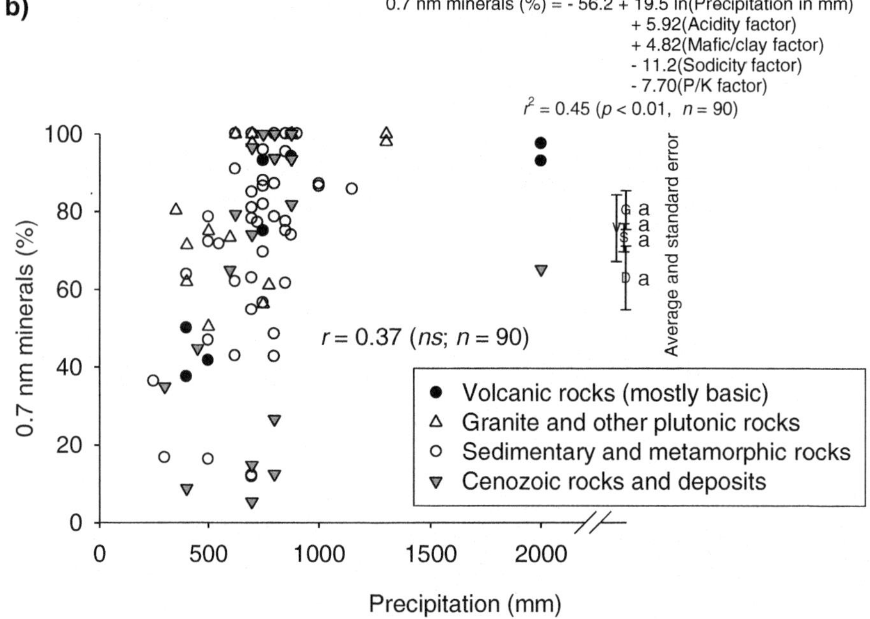

Fig. 4.8 Relationships between clay mineralogy and soil and climatic factors. Abundances of (**a**) 1.4 nm and (**b**) 0.7 nm minerals

The 1.4-nm minerals were probably formed under the strong influence of the high sodicity of the parent materials around the Great Rift Valley and were often observed in the soils in the flat plains near Lake Victoria.

On the other hand, the abundances of the 0.7-nm minerals (kaolin minerals) can be expressed by the following equation:

$$0.7 - \text{nm minerals } (\%) = -56.2 + 19.5\ln (\text{precipitation in mm})$$
$$+5.92 \text{ (acidity factor)} + 4.82 \text{ (texture factor)} - 11.2 \text{ (sodicity factor)}$$
$$-7.70 \text{ (P/K factor)}; \qquad r^2 = 0.45 \ (p < 0.01, n = 90)$$

$$(4.2)$$

From this equation, it can be stated that the kaolin formation is promoted under highly humid conditions with the positive influence of soil acidity and texture (or clayey parent materials) as well as the negative influence of sodicity. Hence, it can be inferred that the clay mineralogical properties of the soils studied herein were formed under the strong influence of the present climatic conditions as well as the parent materials on a countrywide scale in Tanzania.

4.7 General Discussion on the Soil Conditions in Tanzania with Specific Reference to Potential Agricultural Development

As stated earlier, soil fertility is considered high in and around the volcanic regions owing to the high SOM-related fertility and the high P and K nutrient status. In addition, the soils around Lake Victoria are fertile due to the strong influence of the 1.4-nm minerals, which also contribute to the high nutrient retaining potential of the soils. Both the regions, namely, the volcanic regions and the regions around Lake Victoria, are broadly included in the Great Rift Valley, which is the center of the intensive agricultural activities of the country. However, in other areas of Tanzania, soils are generally low in SOM-related parameters and the 1.4-nm minerals are virtually absent. The proportion of kaolin minerals increases with the precipitation; hence, soil fertility decreases in regions of high humidity. Soil fertility in terms of clay mineralogy is comparatively higher in dry regions than in humid regions because of the greater abundance of mica minerals; however, water availability decreases in such dry regions. Thus, the semiarid regions in Tanzania suffer from water shortage, while the relatively humid areas have less fertile soil that predominantly contains kaolin minerals; hence, these inherently adverse conditions are not favorable for agricultural production and should be taken into consideration when studying the feasibility of agricultural development in different areas in the future.

4.8 Conclusion

Reflecting the wide variation of climatic conditions and parent materials of soils, soil physicochemical properties varied widely in this study. Clay mineral composition was, however, basically similar to the soils developed under ustic moisture regime in Southeast Asia. From the principal component analysis of the collected soil samples, five individual factors—SOM and amorphous compounds, texture, acidity, available P and K, and sodicity—were determined. From the clay mineralogical composition and the relation between the geological conditions (or parent materials) and the annual precipitation and the scores of the four factors, the following observations were made:

1. The maximum scores of "SOM and amorphous compounds" were found at the volcanic center of the southern mountain ranges from the east of Mbeya to Lake Malawi.
2. The scores of the "available P and K" were high in the volcanic regions around Mt. Kilimanjaro and in the southern volcanic mountain ranges, presumably due to the intensive agricultural management with fertilizer application.
3. The abundance of 1.4-nm minerals (mostly smectite) can be expressed by the following equation:

$$1.4 - \text{nm minerals } (\%) = 6.38 + 13.4 \text{ (sodicity factor)}$$
$$-9.78 \text{ (SOM/amorphous factor)}$$
$$+3.17(\text{P/K factor}); r^2 = 0.58 \ (p < 0.01, n = 90)$$

The 1.4-nm minerals were probably formed under conditions of high sodicity and were often observed in the soils near Lake Victoria.

4. The abundance of 0.7-nm minerals (kaolin minerals) can be expressed by the following equation:

$$0.7 - \text{nm minerals } (\%) = -56.2 + 19.5\ln(\text{precipitation in mm})$$
$$+5.92 \text{ (acidity factor)} + 4.82 \text{ (texture factor)} - 11.2 \text{ (sodicity factor)}$$
$$-7.70 \text{ (P/K factor)}; \qquad r^2 = 0.45(p < 0.01, n = 90)$$

From this equation, it was found that kaolin formation is promoted by highly humid conditions and that it is influenced by the acidity and texture of the soil (or parent materials). Hence, it was inferred that the formation of the soils studied in the present study was strongly influenced by climatic conditions and parent materials.

5. In Tanzania, the volcanic regions and the Great Rift Valley region, where soil is generally more fertile than in other regions, are conducive to modernized agriculture. The semiarid regions in Tanzania suffer from water shortage, while the relatively humid areas have less fertile soil that predominantly contains kaolin minerals. These conditions are not favorable for agricultural

production and should be taken into consideration when studying the feasibility of agricultural development in different areas in the future.

Acknowledgments This chapter is derived in part from an article published in Soil Fertility Status and Its Determining Factors in Tanzania. By Shinya Funakawa, Hiroshi Yoshida, Tetsuhiro Watanabe, Soh Sugihara, Method Kilasara and Takashi Kosaki, 2012. InTech, available online: http://doi/10.5772/29199.

References

Araki S, Msanya BM, Magoggo JP, Kimaro DN, Kitagawa Y (1998) Characterization of soils on various planation surfaces in Tanzania. In: Summaries of 16th world congress of soil science, Montpellier, France, Vol I, p 310

Funakawa S, Watanabe T, Kosaki T (2008) Regional trends in the chemical and mineralogical properties of upland soils in humid Asia: with special reference to the WRB classification scheme. Soil Sci Plant Nutr 54:751–760

Mizota C, Kawasaki I, Wakatsuki T (1988) Clay mineralogy and chemistry of seven pedons formed in volcanic ash, Tanzania. Geoderma 43:131–141

Msanya BM, Magoggo JP, Otsuka H (2002) Development of soil surveys in Tanzania (Review). Pedologist 46:79–88

Msanya BM, Otsuka H, Araki S, Fujitake N (2007) Characterization of volcanic ash soils in southwestern Tanzania: morphology, physicochemical properties, and classification. Afr Stud Monogr (Supplementary Issue) 34:39–55

Soil Survey Staff (1999) Soil taxonomy: a basic system of soil classification for making and interpreting soil surveys. 2nd edn. Natural Resources Conservation Service. U.S. Department of Agriculture Handbook 436

SPSS (1998) SYSTAT 8.0. Statistics. SPSS, Chicago

Surveys and Mapping Division, Tanzania (1967) Atlas of Tanzania. Dar es Salaam, Tanzania

Szilas C, Møberg JP, Borggaard OK, Semoka JMR (2005) Mineralogy of characteristic well-drained soils of sub-humid to humid Tanzania. Acta Agric Scand Sect B – Plant Soil Sci 55:241–251

Chapter 5
Soil Fertility Status in Equatorial Africa: A Comparison of the Great Rift Valley Regions and Central/Western Africa

Shinya Funakawa and Takashi Kosaki

Abstract Soil fertility was measured in contrasting regions of sub-Saharan Africa, i.e., Tanzania, Rwanda, western D. R. Congo, Cameroon, Nigeria, Burkina Faso, Ivory Coast, and others, with special reference to geological and climatic conditions. The general properties of the soils in each region of equatorial Africa could be summarized as follows: when comparing the soils in central and western regions of equatorial Africa, the soils in Tanzania are affected more or less by the Great Rift Valley movement, including volcanic activity, and are considered relatively fertile. The base level is generally high, and soils are therefore not intensely acidic, and soil texture is intermediate. The soil organic matter (SOM) level is moderately high, partially affected by volcanic activity and relatively high elevation. Clay mineral composition also suggests that the soil in this region is somewhat less weathered and possibly supplies more mineral nutrients than soils in the other regions. A similar, but more definitive, advantage of volcanic soil could be found for the soils in the volcanic regions of the highlands of Rwanda and eastern D. R. Congo. The soils in this region are characterized by high base levels, high cation exchange capacity (CEC) values (which are influenced by the presence of both SOM and 2:1 clay minerals), intermediate to clayey soil texture, and relatively high SOM levels; the latter two are affected by parent materials and cool temperatures. In contrast to the Great Rift Valley regions, a large part of Cameroon is situated on the Cameroonian plateau, which is composed of Precambrian basement rocks under humid climates. The soils in this region are characterized by a strong acidic nature, high levels of exchangeable Al, fewer base components, moderately low SOM level, and clayey soil texture, dominated by inactive kaolin minerals. The soils in the western regions of equatorial Africa, such as the Nigeria/Benin and Burkina Faso/Ivory Coast/ Liberia regions, are commonly characterized by the presence of sandy soils. The sand

S. Funakawa (✉)
Graduate School of Global Environmental Studies, Kyoto University, Yoshida-Honmachi, Sakyo-ku, Kyoto 606-8501, Japan
e-mail: funakawa@kais.kyoto-u.ac.jp

T. Kosaki
Graduate School of Urban Environmental Sciences, 1-1, Minami-Osawa, Hachioji, Tokyo 192-0397, Japan

© Springer Japan KK 2017
S. Funakawa (ed.), *Soils, Ecosystem Processes, and Agricultural Development*,
DOI 10.1007/978-4-431-56484-3_5

content usually exceeds 70 %, while clay content is less than 20 %. As a result, the base reserve is typically low, and the SOM level is less than 10–15 g C kg^{-1} soil.

Keywords Central/Western Africa • Great Rift Valley • Soil fertility

5.1 Introduction

Africa has several distinctive geological features—(1) most of Africa is made up of very old rock, often exposed to the surface; (2) the continent is slowly being divided along the great East African Rift Valley, and volcanic activity is common; (3) large parts of African land have been geologically stable for millions of years; and (4) large parts of Africa are covered by recent sediments (Jones et al. 2013). From a pedogenetic aspect, such geological features provide varying types of parent materials for soil development in each of the regions, in part relating to topography and time factors.

However, climatic conditions, or soil moisture regimes, in the African continent distribute rather regularly with regard to latitude—a humid climate with the udic soil moisture regime is found in the equatorial regions, followed by sub-humid and/or semiarid climates with the ustic soil moisture regime, which are located approximately 5–10° in the north and south latitudes. Lastly, arid regions with the aridic soil moisture regime are found at 30° latitude to the north and south. Among them, the eastern part of the continent featured rather dry climates due to the dominating effects of the uplift highland plateau surrounding the Great Rift Valley (Jones et al. 2013).

Owing to the influences of the soil-forming factors introduced above, pedogenetic conditions are characterized by extensive volcanic activity and highland climates, with rather cooler and drier conditions (relative to the same latitude) in East Africa, typically around the Great Rift Valleys. Central and western regions of equatorial Africa are near rather stable landforms with relatively low elevation. Soils distributed in these contrasting regions are typically Andisols/Ultisols/Alfisols in the Great Rift Valley region and Oxisols/Ultisols/Alfisols in the central to western parts of equatorial Africa (Soil Survey Staff 1999).

In this chapter, we analyzed general soil properties, including the mineralogical characteristics that are definitively important in determining soil physical and chemical properties, with special reference to geological and climatic factors relevant to soil development, as a basis for further discussion and understanding of the ecological processes and agricultural practices in the following chapters.

5.2 Soil Sampling and Analytical Procedures

In total, 240 samples of surface soils were collected from different regions of equatorial Africa, including Tanzania, Rwanda, the Democratic Republic of Congo (D. R. Congo), Cameroon, Sao Tome, Nigeria, Benin, Burkina Faso, Ivory Coast,

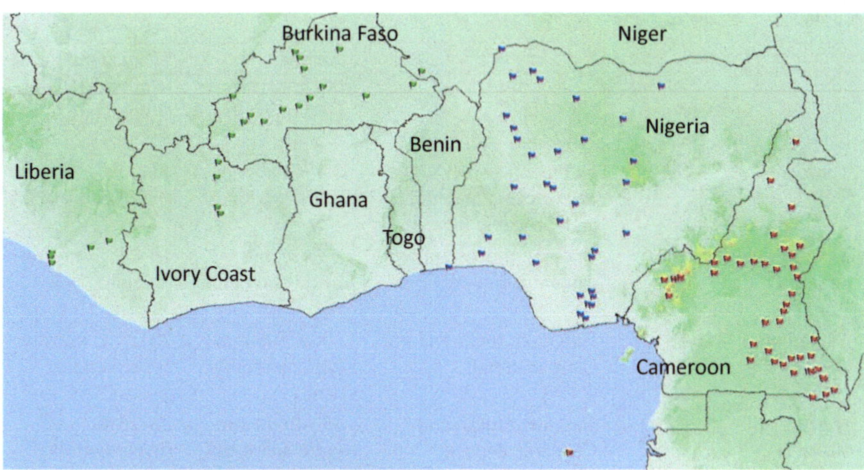

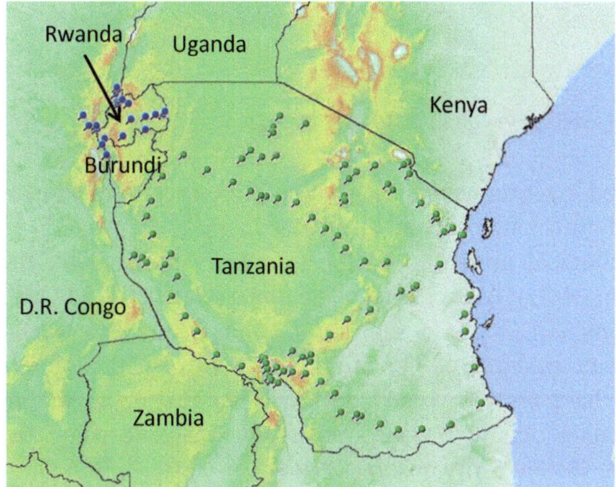

Fig. 5.1 Study soils

and Liberia (Fig. 5.1 and Table 5.1). All the sampling points were located on slopes or plains, covering regions with different parent materials and a wide variation in annual precipitation (250–3400 mm). Apparent lowland soils were excluded from the analysis. The climatic conditions, i.e., annual precipitation (AP) and mean annual air temperature (MAT) of the sampling plots, were estimated by the values recorded at a nearby meteorological station (World Climate 1996) after correction by elevation for MAT. Parent materials were identified according to field observation and geological maps. The parent materials of the soils were broadly classified according to the following categories: (1) basement rocks formed on Precambrian shields, mostly derived from migmatite, gneiss, schist, or other metamorphic rocks, (2) granite and other plutonic rocks, (3) sedimentary and metamorphic rocks, (4) Cenozoic rocks and recent deposits, and (5) volcanic rocks and ejecta. The sampling plots were once used

Table 5.1 Outline of sample soils

Sample group	Number of samples[a]	Major parent materials	Annual precipitation (soil moisture regimes[a])	Mean annual temperature (soil temperature regimes[a])
Tanzania	94	Sedimentary rocks, granite, volcanic materials, and Cenozoic deposit	250–2000 mm (mostly ustic)	13–26°C (thermic and hyperthermic)
Rwanda and eastern D. R. Congo	24	Basement rocks and volcanic materials	>1000 mm (ustic and udic)	16–23°C (mostly thermic)
Cameroon and Sao Tome	56	Basement rocks and volcanic materials	800–2500 mm (udic, partially ustic)	20–25°C (mostly hyperthermic)
Nigeria and Benin	41	Basement rocks and Cenozoic deposit	550–2400 mm (udic and ustic)	22–28°C (hyperthermic)
Burkina Faso, Ivory Coast, and Liberia	25	Basement rocks	550–3400 mm (udic and ustic)	25–29°C (hyperthermic)
Total	240			

[a]According to US Soil Taxonomy (Soil Survey Staff 2014)

as croplands or were covered by seminatural vegetation (forest or savanna), or secondary vegetation that had grown after human disturbance.

The chemical and mineralogical properties of the samples were used for the following statistical analyses: pH (H_2O), exchangeable cations (Na, K, Ca, Mg, and Al), cation exchange capacity (CEC), base saturation, total C and N content, particle size distribution, clay mineral composition (by X-ray diffraction for selected samples), and total base reserve (based on total chemical composition; for selected samples). The data obtained were statistically analyzed using the SYSTAT version 13 software (Systat Software, Inc. 2014).

5.3 General Physicochemical and Mineralogical Properties of the Soils

Selected average values of the analytical data, in terms of sampling regions and parent materials, respectively, are listed in Table 5.2. Soil pH was apparently lower in the Cameroonian soils compared to other soil types, with a significantly lower base saturation ($p < 0.05$), partially due to the high precipitation of this region. Both the exchangeable bases and CEC contents were higher in the Great Rift Valley in Rwanda and eastern D. R. Congo than in the central and western countries ($p < 0.05$). Both the higher content of soil organic matter (SOM), due to high elevation, and also the higher content of 2:1 minerals on relatively new land surfaces, might contribute to the high values of exchangeable bases and CEC in Rwanda and the

Table 5.2 Average values of selected chemical and mineralogical properties of soils from different regions

Sample group	Number of samples					pH(H$_2$O)		Exch. bases (cmol$_c$ kg^{-1})		CEC (cmol$_c$ kg^{-1})		Base saturation (%)	
	All	Natural forest	Secondary forest	Savanna	Cropland	AVE	SE	AVE	SE	AVE	SE	AVE	SE
Tanzania	94 (89)	22 (20)	14 (14)	16 (16)	42 (39)	6.15 b	0.08	10.53 b	1.06	14.0 c	1.1	95.3 c	1.1
Rwanda/eastern D. R. Congo	24 (21)	1 (0)	2 (2)	5 (4)	16 (15)	5.77 b	0.22	16.57 c	2.99	23.3 d	3.1	87.5 bc	4.2
Cameroon/Sao Tome	56 (23)	15 (6)	8 (5)	13 (8)	20 (4)	4.85 a	0.10	3.20 a	0.54	11.5 bc	0.7	53.9 a	3.8
Nigeria/Benin	41 (0)	0 (0)	0 (0)	0 (0)	41 (0)	5.91 b	0.13	2.75 a	0.31	4.7 a	0.5	84.2 bc	3.9
Burkina Faso/Ivory Coast/Liberia	25(0)	0 (0)	0 (0)	1 (0)	24 (0)	5.78 b	0.18	2.20 a	0.33	6.2 ab	1.0	79.2 b	4.9

Sample group	Total C (g kg^{-1})		Particle size distribution						Clay mineral composition					
			Sand (%)		Silt (%)		Clay (%)		1.4 nm (%)		1.0 nm (%)		0.7 nm (%)	
	AVE	SE	AVE	SE	AVE	SE	AVE	SE	AVE	SE	AVE	SE	AVE	SE
Tanzania	20.2 b	2.5	63.5 b	2.4	11.3 a	1.2	25.2 b	1.8	8 a	2	19 b	2	73 a	3
Rwanda/eastern D.P. Congo	35.5 c	4.9	38.7 a	3.9	25.9 c	2.8	35.4 c	3.6	13 a	5	17 ab	3	69 a	6
Cameroon/Sao Tome	15.2 ab	3.1	35.8 a	2.5	16.5 b	1.7	47.1 d	2.4	3 a	0	5 a	2	92 b	2
Nigeria/Benin	7.7 a	0.9	82.0 c	2.4	5.9 a	0.8	12.1 a	1.9	n.d.	n.d.	n.d	n.d	n.d	n.d
Burkina Faso,/Ivory Coast/Liberia	8.8 a	1.1	72.1 bc	3.6	12.0 ab	2.4	15.9 ab	2.0	n.d.	n.d.	n.d	n.d	n.d	n.d

(continued)

Table 5.2 (continued)

Parent materials	Number of samples						pH(H₂O)		Exch. bases (cmol_c kg⁻¹)		CEC (cmol_c kg⁻¹)		Base saturation (%)	
	All	Tanzania	Rwanda/ eastern D. R. Congo	Cameroon/ Sao Tome	Nigeria/ Benin	Burkina Faso/ Ivory Coast/ Liberia	AVE	SE	AVE	SE	AVE	SE	AVE	SE
Basement rocks	80 (22)	0(0)	3 (3)	37 (19)	19 (0)	21 (0)	5.33 a	0.11	2.67 a	0.43	7.8 a	0.6	66.1 a	3.5
Granite	15 (15)	14 (14)	1 (1)	0 (0)	0 (0)	0 (0)	5.61 ab	0.16	3.37 ab	0.72	7.2 ab	1.4	86.7 b	4.6
Sedimentary rocks	60 (52)	49 (49)	1 (1)	2 (2)	4 (0)	4 (0)	6.15 b	0.11	8.88 bc	0.95	11.6 ab	0.8	91.5 b	2.2
Cenozoic deposit	54 (33)	18 (18)	15 (15)	3 (0)	18 (0)	0 (0)	6.04 b	0.13	10.11 bc	1.65	12.0 b	1.6	90.0 b	2.6
Volcanic origin	31 (11)	13(8)	4 (1)	14 (2)	0 (0)	0 (0)	5.46 a	0.13	12.63 c	2.50	25.4 c	2.5	83.2 b	4.2

Sample group	Total C (g kg⁻¹)		Particle size distribution						Clay mineral composition					
			Sand (%)		Silt (%)		Clay (%)		1.4 nm (%)		1.0 nm (%)		0.7 nm (%)	
	AVE	SE	AVE	SE	AVE	SE	AVE	SE	AVE	SE	AVE	SE	AVE	SE
Basement rocks	8.7 a	1.0	57.0 b	3.0	10.6 a	1.1	32.0 bc	2.5	4 a	1	5 a	2	92 b	3
Granite	12.8 a	3.2	71.1 b	5.7	9.0 a	2.9	19.9 ab	3.8	5 ab	3	13 ab	4	81 ab	5
Sedimantary rocks	16.6 a	1.8	65.0 b	2.8	9.2 a	1.1	25.7 ab	2.1	4 a	1	22 b	3	74 ab	3
Cenozoic deposit	14.9 a	1.8	66.5 b	3.3	11.8 a	1.3	21.8 a	2.6	19 b	5	17 ab	4	64 a	6
Volcanic origin	47.0 b	7.3	30.9 a	3.2	31.4 b	2.7	37.7 c	3.7	5 a	3	17 ab	6	78 a	7

AVE average, SE standard error

The values with the same letters are not significantly different by Tukey test ($p < 0.05$)

Clay mineralogy was determined for soil samples numbered in the parenthesis

eastern D. R. Congo regions. The soils of the western countries, Nigeria, Benin, Burkina Faso, Ivory Coast, and Liberia, had a more sandy texture than the others. Such an extensive distribution of sandy soils was repeatedly reported for these regions.

When comparing the soils from different parent materials, volcanic soils exhibited unique characteristics in that (1) the level of exchangeable bases was high in spite of a low pH, presumably due to a high CEC with increased organic matter, and (2) fine-textured soils, especially with high silt contents, were the dominant soil type. Besides the volcanic soils, those from the Cenozoic deposit were less acidic with higher levels of exchangeable bases and CEC, often accompanying an increased level of expandable 2:1 minerals, all indicating a less weathered nature of the mineral components of the soils. It is possible to conclude that the soils derived from volcanic materials and Cenozoic deposits were, relatively speaking, more fertile than the others. In contrast, soils derived from basement rocks were usually acidic, low in CEC and exchangeable bases, and higher in clay and kaolin minerals. All these characteristics were expected for highly weathered soils like Oxisols and were found mainly in the soils of central and western countries. This is consistent with the observations discussed in Chap. 4 for Tanzanian soils, though the clay mineralogical composition of the central and western African countries exhibited more weathered features than the nonvolcanic soils in Tanzania, adjacent to the Great Rift Valley.

5.4 Principal Component Analysis for Summarizing Soil Properties

A principal component analysis was conducted to summarize several soil parameters related to soil fertility. The studied variables included pH (H_2O); amounts of exchangeable K^+, Mg^{2+}, Ca^{2+}, and Al^{3+}; sum of exchangeable bases, CEC; effective CEC; base saturation; total C and total N contents; as well as sand, silt, and clay content. Table 5.3 lists the pattern for the first three principal components, which had eigenvalues of higher than unity after varimax rotation.

As seen from the list in Table 5.3, high positive coefficients were obtained for exchangeable K, Ca, and Mg, sum of exchangeable bases, CEC, and ECEC for the first component. All these variables correspond to the soil properties relating to the retention of exchangeable bases. The score of this component highly correlated with the total base reserve as well, indicating that the first component could be termed an "base reserve" factor (Fig. 5.2). The second component had high negative coefficients for pH (H_2O), base saturation, and sand content, but high positive coefficients for exchangeable Al and clay content. These soil characteristics could be associated with soil acidity and fine texture, that is, strong acidity was considered to preferentially develop on relatively fine-textured soils under humid climatic conditions. Hence, the second component is referred to as the "acidity and

Table 5.3 Factor pattern for the first four principal components ($n = 240$)

Variable	PC1	PC2	PC3
pH(H_2O)	0.49	−0.75	−0.14
Exch. K	0.74	−0.06	0.15
Exch. Ca	0.92	−0.13	0.19
Exch. Mg	0.91	−0.09	0.20
Exch. Al	−0.28	0.84	0.01
Exch. Bases	0.96	−0.12	0.20
CEC	0.69	0.20	0.65
ECEC	0.95	−0.02	0.20
Base saturation	0.37	−0.80	0.08
Total C	0.14	0.01	0.96
Total N	0.16	0.01	0.96
Sand	−0.37	−0.76	−0.38
Silt	0.29	0.29	0.62
Clay	0.30	0.80	0.11
Eigenvalue	5.34	3.28	2.54
Proportion (%)	37.2	23.4	18.1
	"Base reserve" factor	"Acidity and clay" factor	"SOM" factor

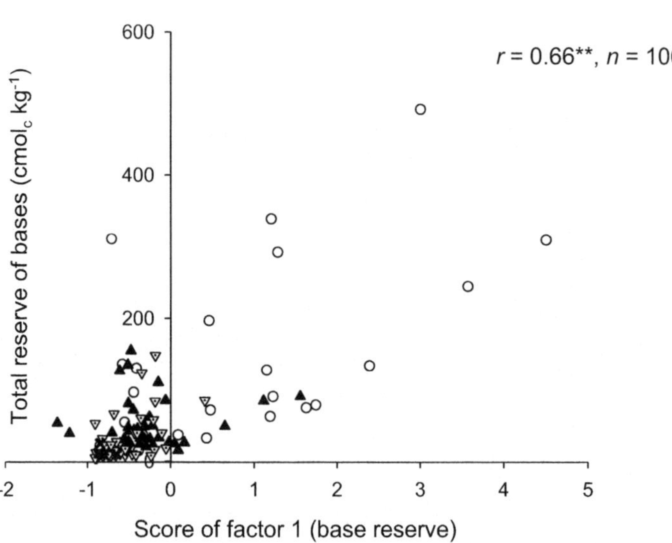

Fig. 5.2 Relationship between the score of factor 1 and total base reserve

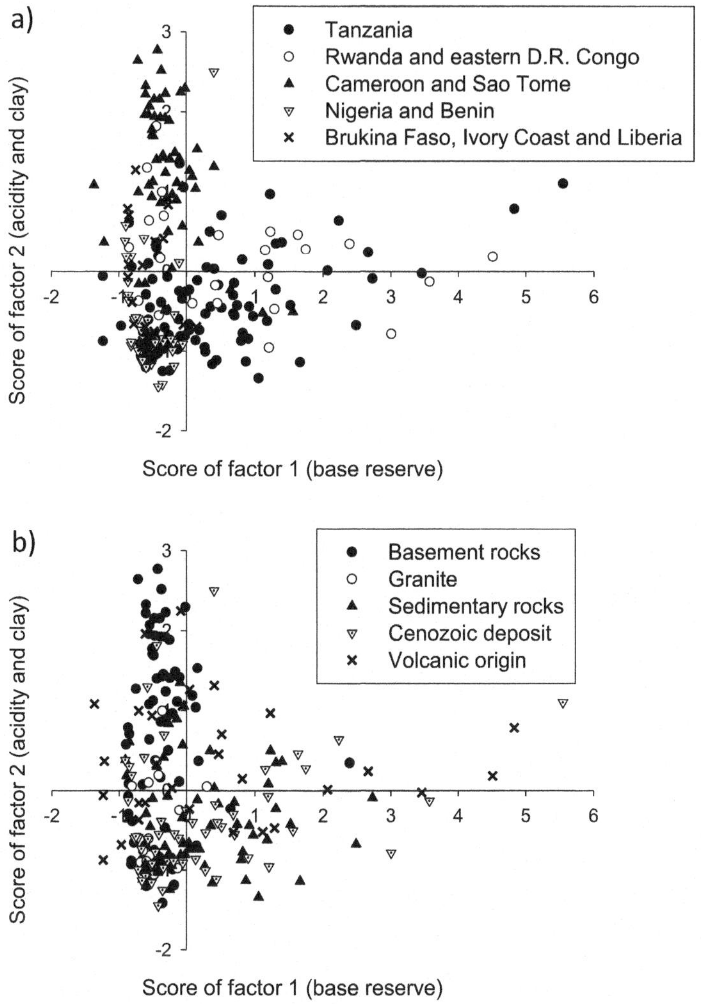

Fig. 5.3 Relationship between the scores of factors 1 and 2

clay" factor. It is perhaps a slightly unusual situation that "base reserve" and "acidity and clay" factors, which were usually considered to be negatively related, were actually independently recognized in this analysis. Figure 5.3 plots the relationship between the scores of factors 1 and 2. We can say that the samples distributed in quadrants II and IV followed the commonly held belief that, if acidity proceeds, the base reserve would be depleted. Many Tanzanian soils derived from sedimentary rocks, which were rich in base reserves under relatively dry climates, and most Cameroonian soils derived from basement rocks, which have been acidified under humid climate and are hence low in base levels, are typically distributed in these quadrants. However, some soils in Tanzania and Rwanda/

eastern D. R. Congo were rich in bases due to the relatively less weathered features of Cenozoic deposits or volcanic soils, while were simultaneously acidic due to high precipitation. They are distributed in quadrant I. In contrast, many of the sandy soils that developed on some Cenozoic deposit and/or sedimentary rocks, with low scores in factor 2, were found in quadrant III. The soils distributed in quadrants I and III in Fig. 5.3 were unique, but somewhat commonly observed in the Great Rift Valley and western Africa, respectively.

The coefficients corresponding to the third component have high positive values for CEC, total C, total N, and silt content, indicating that a close relationship exists between this component and SOM accumulation. As discussed earlier, the soils of volcanic origin accumulated higher SOM and unweathered features, such as high silt content. Therefore, this component is referred to as the "SOM" factor.

From this analysis, a wide variety of soil parameters were categorized into three principal components, which accounted for 78.7 % of the total variance.

5.5 Climatic Conditions Affecting Factor Scores for Each of the Principal Components

Figure 5.4 shows scattergrams of climatic parameters, such as AP and MAT, and the scores of three factors, i.e., base reserve, acidity and clay, and SOM. Table 5.4 indicates the average values of the factor scores by each of the regions or by a difference in parent materials. According to Fig. 5.3a and b, AP could contribute to a decrease in the base reserve and an increase in soil acidity, through intense leaching and mineral weathering. However, such a direct influence may be limited judging from the rather low correlation coefficients. The soils with low scores in factor 1, which distributed in regions of low AP in spite of their overall negative correlation, suggested two possibilities: that is, (1) cumulative mineral weathering might have consecutively occurred under more humid climatic conditions in the past, and consequently decreased the base reserve during that period, and (2) the domination of sandy soils (mostly composed of quartz) could directly decrease the total reserve as well as the amount of exchangeable bases. Similarly, in Fig. 5.3b and Table 5.4, soils in western Africa, including those in the most humid regions, showed the lowest scores in factor 2 ("acidity and clay" factor) compared with the highest scores in the clayey soils of the Cameroonian plateau. Judging from the close relationship between clay content and soil acidity, i.e., a clear negative correlation between clay content and soil pH (H_2O) ($r = -0.41**$) or exchangeable Al ($r = 0.46**$), the presence of a certain amount of clay may be a prerequisite to develop soil acidity.

Figure 5.4c shows that the influence of lower temperatures may enhance the accumulation of SOM, together with a possible positive influence of volcanic materials to SOM accumulation in highland regions of the Great Rift Valley. Lower scores in this factor were commonly observed for the soils in Nigeria/ Benin and the Burkina, Faso/Ivory, Coast/Liberia regions (Fig. 5.4c and

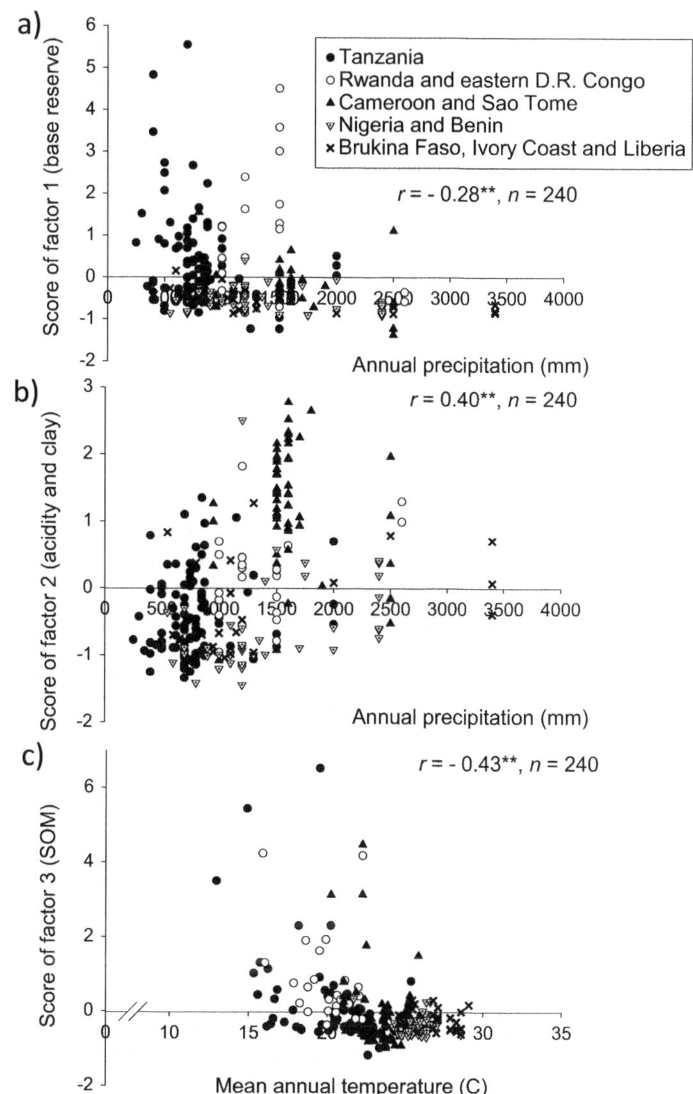

Fig. 5.4 Relationship between the climatic conditions and scores of different factors

Table 5.4). Both a hot climate at a low elevation and the absence of volcanic-derived materials, as well as the dominance of sandy soils, clearly characterized the low SOM status in these regions; therefore, the utilization of SOM-related resources for agriculture would also be difficult.

These results suggest that soil chemical properties, in terms of base status, soil acidity, and SOM, would not effectively characterize the soils within the Nigeria/Benin and Burkina Faso/Ivory Coast/Liberia regions, which were similarly dominated by soils with a sandy texture and low CEC (Table 5.2).

Table 5.4 Average values of factor scores of soils from different regions or parent materials

Sample group	Factor scores					
	Base reserve factor		Acidity and clay factor		SOM factor	
	AVE	SE	AVE	SE	AVE	SE
Tanzania	0.34 b	0.12	−0.42 a	0.06	0.05 a	0.11
Rwanda/eastern Zaire	0.82 b	0.29	0.17 b	0.13	0.84 b	0.25
Cameroon and Sao Tome	−0.28 a	0.06	1.22 c	0.12	−0.07 a	0.14
Nigeria and Benin	−0.55 a	0.05	−0.59 a	0.11	−0.37 a	0.04
Burkina Faso, Ivory Coast, Liberia	−0.55 a	0.05	−0.33 ab	0.14	−0.22 a	0.06

Parent materials	Factor scores					
	Base reserve factor		Acidity and clay factor		SOM factor	
	AVE	SE	AVE	SE	AVE	SE
Basement rocks	−0.40 a	0.05	0.45 b	0.14	−0.35 a	0.04
Granite	−0.42 ab	0.07	−0.36 ab	0.16	−0.13 a	0.15
Sedimentary rocks	0.22 bc	0.11	−0.36 a	0.08	−0.14 a	0.07
Cenozoic deposit	0.22 bc	0.17	−0.35 a	0.11	−0.14 a	0.07
Volcanic origin	0.42 c	0.29	0.33 b	0.14	1.48 b	0.35

AVE average; SE standard error
The values with the same letters are not significantly different by Tukey test ($p < 0.05$)

5.6 Pedogenetic Conditions Determining the Clay Mineralogy of Soils

Figure 5.5 shows the relationship between the scores of factors 1 or 2 and the relative abundance of 0.7-nm kaolin minerals. There are moderate but significant negative and positive correlations ($p < 0.05$), respectively. The combination of two factors in the multiple regression analysis successfully improved the simulation as described by the following equation:

0.7-nm minerals (%) = 78.1–11.7 ("base reserve" factor) + 8.69 ("acidity and clay" factor)–4.75 ("SOM" factor); $r^2 = 0.34$ ($p < 0.01$, $n = 136$)

In this equation, the negative contribution of the "base reserve" factor might indicate the result of mineral weathering, i.e., as a result of mineral weathering, the base reserve would decrease and 0.7-nm minerals would increase. However, the positive contribution of the "acidity and clay" factor may suggest that soil acidification is a driving force of mineral weathering. The factor relating to SOM accumulation may indicate an inhibitory effect of SOM on mineral weathering, combined with the low temperature of highlands, as well as the presence of recently supplied volcanic materials rich in weatherable minerals. It is notable, however, that the value of intercept of this equation was already high as 78.1 %, and most soils were already subjected to considerable mineral weathering, not like in the soils of humid Asia (Chap. 3 in this volume; Funakawa et al. 2008),

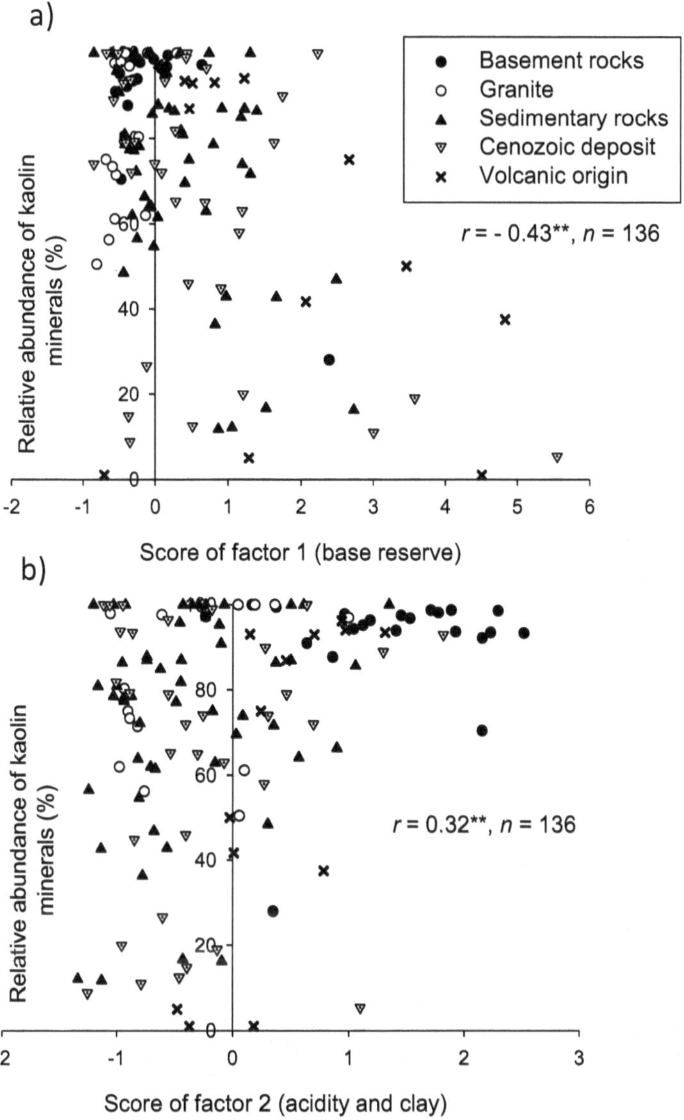

Fig. 5.5 Relationship between the scores of factors 1 and 2 and relative abundance of kaolin minerals

where the relative abundance of kaolin minerals was approximately 60 % on average.

Figure 5.6 shows the contribution of factors 1 and 2 on the relative abundance of mica minerals and expandable 2:1 minerals, respectively. For both relationships, the relative abundances of mica and expandable 2:1 minerals were apparently affected by the scores of factor 1 or 2, i.e., an increase in the "acidity and clay"

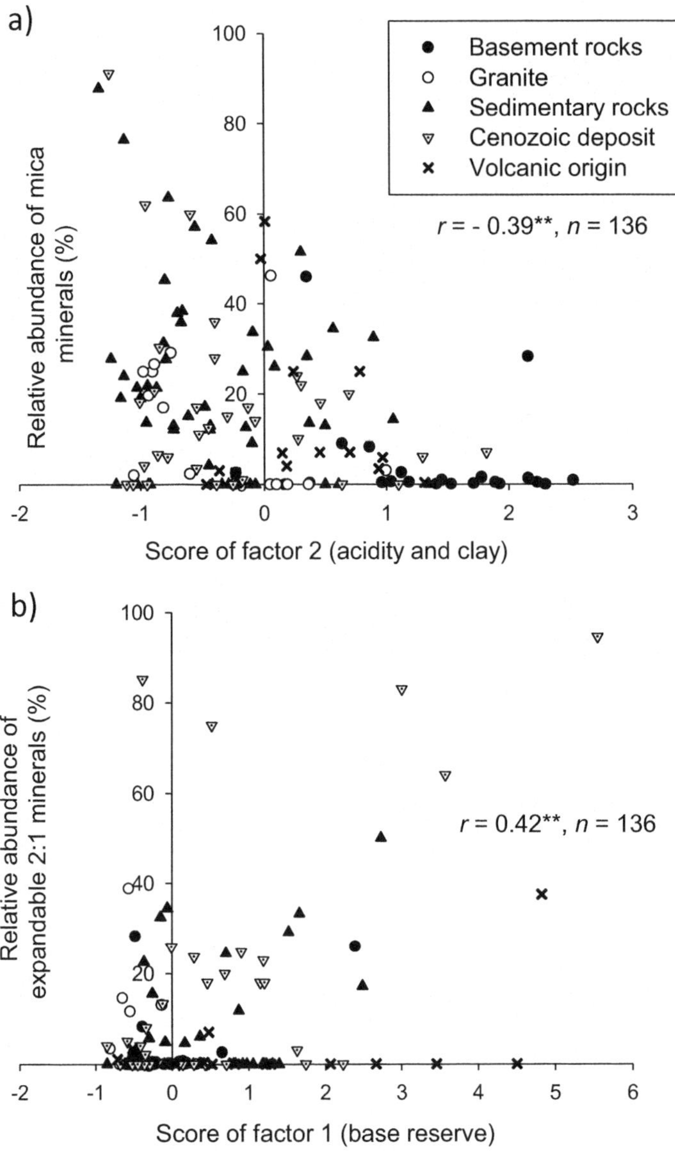

Fig. 5.6 Relationship between the scores of factors 1 and 2 and relative abundances of mica minerals and expandable 2:1 minerals

factor decreased the relative abundance of mica minerals (Fig. 5.6a), while an increase in the "base reserve" factor resulted in greater amounts of expandable 2:1 minerals (Fig. 5.6b). However, at the same time, there were many exceptions to such simple relationships. For example, in Fig. 5.6a, several soils of volcanic origin, Cenozoic deposit, or sedimentary rocks exhibited a very low mica mineral content, in the range of low scores for factor 2. Similarly, an increase in the factor 1 scores did not always accompany an increase in the expandable 2:1 minerals for the soils derived from the Cenozoic deposit and/or sedimentary rocks (Fig. 5.6b). There could be two possibilities for such a discrepancy: (1) one is that the parent materials of some soils actually did not contain mica minerals or expandable 2:1 minerals as primary minerals, and these minerals have never been formed under the pedogenetic conditions of these soils, and (2) the other is that the past climatic conditions, which might have developed clay mineralogy and the "base reserve" and/or soil "acidity and clay" factors of the soils, does not actually correlate with the present climates.

Thus the overall limited correlations among present climatic factors, soil physicochemical properties, and clay mineral composition suggest that the soils in equatorial Africa, especially those in central and western Africa, have developed under fluctuating pedogenetic conditions for a long period of time. During such a long history, climatic conditions have considerably fluctuated and vegetation cover has also changed. As a result, the mass movement of surface soil materials through wind and/or water erosion, involving both removal and deposition, has considerably modified soil properties (Driessen et al. 2001). This is not the case for upland soils in the humid areas of Asia, where the respective contributions of geology (or parent materials) and climate could be more easily distinguished (Chaps. 2 and 3 in this volume; Funakawa et al. 2008).

5.7 General Discussion on the Soil Conditions for Different Regions of Equatorial Africa, with Specific Reference to Potential Agricultural Development

Based on the preceding sections, the general properties of the soils in the respective regions of equatorial Africa could be summarized as follows:

For soils in Tanzania, we generally concluded in Chap. 4 that clay mineral composition was characterized by the dominance of mica and kaolin minerals. From the clay mineralogical composition and the relationship between the geological conditions (or parent materials) and climatic factors (MAT and AP) and the scores of the five factors determined by the principal component analysis, the following observations were made: the scores of both "SOM and amorphous compounds" and "available P and K" were high in/around the volcanic regions of Mt. Kilimanjaro and in the southern volcanic mountain ranges, and the 1.4-nm minerals were often observed in the soils near Lake Victoria. Hence, the volcanic regions, where soil is generally more fertile than in other regions, are conducive to modernized agriculture. The semiarid regions in Tanzania suffer from water shortages, while the relatively humid areas have less fertile soil that predominantly contains kaolin minerals.

In this chapter, when comparing the Tanzanian soils with those of the central and western regions of equatorial Africa, a similar conclusion could be made: the soils in Tanzania are affected more or less by the Great Rift Valley movement, including volcanic activity, and are considered to be relatively fertile in that (1) base reserve is generally high, (2) soil acidity is not intense and soil texture is intermediate, and (3) SOM level is moderately high, partially affected by volcanic activity and relatively high elevation. The chemical and mineralogical characteristics of the volcanic soils are generally similar to those observed in the previous study (Mizota et al. 1988). Clay mineral composition also suggests that the soils in Tanzania are for the most part less weathered and possibly supply more mineral nutrients than soils found in other regions.

A similar but more definitive advantage of the volcanic soils could be found for the soils in volcanic region in the highlands of Rwanda and eastern D. R. Congo. The soils in this region are characterized by (1) a high base reserve; (2) high CEC values, contributed both by the presence of SOM and 2:1 clay minerals; (3) intermediate to clayey soil texture; and (4) high SOM levels, affected both by parent materials (volcanic origin) and cool temperatures (high elevation). The high base reserve and CEC were considered to be closely related to the greater SOM found in this region, as discovered by Asadu et al. (1997). In fact, this region was already extensively cultivated and, consequently, was densely populated, sometimes reaching up to 1000 people per square kilometer (Jones et al., 2013). This is probably because the soils here are exceptionally fertile among the regions of equatorial Africa. The detailed chemical properties of the soils in this region are extensively reported by Vander Zaag et al. (1984) and Yamoah et al. (1990).

In contrast to the Great Rift Valley regions mentioned above, a large part of Cameroon is situated on the Cameroonian plateau, composed of Precambrian basement rocks with humid climates. The soils in this region are characterized by (1) a strong acidic nature, high in exchangeable Al; (2) low base levels; (3) moderately low SOM level; and (4) clayey soil texture, dominated by inactive kaolin minerals. Compared with the remaining western regions of equatorial Africa, the soils in Cameroon are typically clayey, presumably because this territory has been continuously maintaining forest vegetation, even during the driest era, and fine soil components have not been considerably eroded. As a result, clayey acidic soils are still extensively covering this region. Although the soils are strongly acidic, the level of exchangeable Al, 2.05 cmol$_c$ kg^{-1} soil on average, is much lower than those for the rainforests of humid Asian regions (5–20 cmol$_c$ kg^{-1} for soils in Kalimantan, Indonesia), in which expandable 2:1 minerals with high CEC dominate; highly weathered Oxisols with low CEC values are the major components of the Cameroonian plateau. Low-input agricultural practices presently dominate this region. Agricultural development for the future, however, can advance after the introduction of modernized techniques, such as appropriate fertilization, when compared to the present agricultural development of Oxisols in other continents, such as Cerrado, Brazil.

The soils in the western regions of equatorial Africa, such as the Nigeria/Benin, Burkina Faso/Ivory, and Coast/Liberia regions, are commonly characterized by the presence of sandy soils. The sand content usually exceeds 70 %, while clay content is less than 20 %. As a result, the base reserve is typically small, and the SOM level is also low, less than 15 g C kg^{-1} soil. Concerning soil fertility, the agricultural development of these regions would be difficult unless the continuous application of chemical fertilizers is guaranteed. However, it should be noted that the population density of southern Nigeria is sometimes already as high as > 500 people per square kilometer (Jones et al. 2013). Establishing a stable agricultural system is urgently needed in these regions.

References

Asadu CLA, Diels J, Vanlauwe B (1997) A comparison of the contributions of clay, silt, and organic matter to the effective CEC of soils of subSaharan Africa. Soil Sci 162:785–794

Driessen PM, Deckers JA, Spaargaren OC, Nachtergaele FO (eds) (2001) Lecture notes on the major soils of the world, World Soil Resources Reports, vol 94. Food and Agriculture Organization of the United Nations, Rome

Funakawa S, Watanabe T, Kosaki T (2008) Regional trends in the chemical and mineralogical properties of upland soils in humid Asia: with special reference to the WRB classification scheme. Soil Sci Plant Nutr 54:751–760

Jones A, Breuning-Madsen H, Brossard M, Dampha A, Deckers J, Dewitte O, Gallali T, Hallett S, Jones R, Kilasara M, Le Roux P, Micheli E, Montanarella L, Spaargaren O, Thiombiano L, Van Ranst E, Yemefack M, Zougmoré R (eds) (2013) Soil Atlas of Africa. European Commission, Publications Office of the European Union, Luxembourg

Mizota C, Kawasaki I, Wakatsuki T (1988) Clay mineralogy and chemistry of seven pedons formed in volcanic ash, Tanzania. Geoderma 43:131–141

Soil Survey Staff (1999) Soil taxonomy: a basic system of soil classification for making and interpreting soil surveys. 2nd edn. Natural Resources Conservation Service. U.S. Department of Agriculture Handbook 436

Soil Survey Staff (2014) Keys to soil taxonomy. 12th edn. United States Department of Agriculture, Natural Resources Conservation Service

Systat Software, Inc (2014) SYSTAT 13 Getting Started. IL USA, Chicago

Vander Zaag P, Yost RS, Trangmar BB, Hayashi K, Fox RL (1984) An assessment of chemical properties for soils of Rwanda with the use of geostatistical techniques. Geoderma 34:293–314

World Climate (1996) http://www.worldclimate.com. Accessed 5 Jan 2016

Yamoah CF, Burleigh JR, Malcolm MR (1990) Application of expert systems to study of acid soils in Rwanda. Agric Ecosyst Environ 30:203–218

Chapter 6
Significance of Active Aluminum and Iron on Organic Carbon Preservation and Phosphate Sorption/Release in Tropical Soils

Tetsuhiro Watanabe

Abstract The effects of active Al and Fe (acid ammonium oxalate-extractable Al and Fe: Al_o and Fe_o) on the preservation of organic carbon (OC) and sorption/release of phosphate in tropical soils are presented. For OC preservation, Al_o and Fe_o, Fe oxides, and clay fraction have been assumed to be important components stabilizing the organic matter. The $Al_o + Fe_o$ content explained more than 60 % of variation in OC for the all soil groups classified by the degree of soil weathering (weak, moderate, and strong). The quantitative relationships between OC and $Al_o + Fe_o$ were of similar order for all the soil groups. In contrast, Fe oxides and clay contents are less correlated with OC. Effects of Al_o and Fe_o on phosphate sorption capacity and its extractability after sorption were also examined. Furthermore, the retardation effects of phosphate sorption into micro- and meso-pores of Al_o and Fe_o were assessed. The phosphate sorption capacity was strongly correlated with $Al_o + Fe_o$. The proportion of labile phosphate relative to added phosphate and the rate of phosphate released in sequential extractions were negatively correlated with $Al_o + Fe_o$ and porosity of the soils (0.7–4-nm pore specific surface area (SSA) relative to the total SSA). These results showed that Al_o and Fe_o are important components of tropical soils that influence OC preservation and phosphate sorption/release. Carbon dynamics and availability of fertilized phosphate will be affected by the amount of Al_o and Fe_o in the tropical soils.

Keywords Acid ammonium oxalate-extractable aluminum and iron • Organic matter • Phosphorus • Clay content • Iron oxides

T. Watanabe (✉)
Graduate School of Agriculture, Kyoto University, Kitashirakawa-Oiwakecho, Sakyo-ku, Kyoto 606-8502, Japan
e-mail: nabe14@kais.kyoto-u.ac.jp

© Springer Japan KK 2017
S. Funakawa (ed.), *Soils, Ecosystem Processes, and Agricultural Development*,
DOI 10.1007/978-4-431-56484-3_6

6.1 Overview

Active Al and Fe are the fractions that are extracted by acid ammonium oxalate (Al_o and Fe_o) (Shang and Zelazny 2008). Their levels are used for classifying soils as Andisols and Spodosols, and the Andic and Spodic subgroups in other orders in USDA Soil Taxonomy (Soil Survey Staff 1999); and Andosols, Nitisols, and Podosols in WRB (IUSS 2014). Al_o and Fe_o are extracted from amorphous Al and Fe, poor crystalline aluminosilicates, ferrihydrite, and Al- and Fe-humus complexes, but little or none from gibbsite, goethite, hematite, and layered silicates (Shang and Zelazny 2008). The importance of Al_o and Fe_o in the preservation of organic carbon (OC) and sorption and release of phosphate in soils is well known for soils derived from volcanic materials and those formed in temperate or cooler regions. However, the importance has not been thoroughly examined for highly weathered tropical soils where contents of Al_o and Fe_o are very low. Moreover, the contributions of Al_o and Fe_o were not compared with Fe oxides and clay fractions, which are also important to OC preservation and phosphate sorption/release.

Soil OC is critically important to the physical and chemical properties of soil and, therefore, its agricultural productivity. OC improves soil structure, cation exchange capacity (CEC), and biological activity in soil. OC is also related to climate change because it is a large terrestrial carbon pool. Al_o and Fe_o, Fe oxides, and fine particle fractions (clay) contribute to the preservation of OC in soils (von Lützow et al. 2006; Sollins et al. 1996; Rumpel and Koegel-Knabner 2011). Wiseman and Püttmann (2006) recognized that Al_o and Fe_o are strongly correlated to OC content, but poorly correlated to clay content. Kleber et al. (2005) discovered that stable OC, which is isolated by chemical treatment using NaOCl, is strongly related to Al_o and Fe_o but unrelated to free Fe oxides or to clay. In contrast, a positive correlation between OC and clay contents (Ohta and Effendi 1992; Konen et al. 2003) and a correlation between OC and free Fe oxides have been identified (Eusterhues et al. 2003; Lorenz et al. 2009). Although the importance of Al_o and Fe_o is suggested in eight Cambodian soils (Acrisols and Podosols) (Toriyama et al. 2015) and four Tanzanian soils (Acrisols and Alisols) (Kirsten et al. 2016), few studies have examined the preservation of OC by Al_o and Fe_o in tropical soils with low in Al_o and Fe_o contents due to strong weathering. Moreover, the contributions of Al_o and Fe_o, Fe oxides, and clay fractions to OC preservation in tropical soils have not been compared, and it is not clear which component has the largest effect on the preservation of OC in tropical soils.

The sorption and release of added phosphate in soils is important for both agriculture and the environment because the availability of added phosphate affects agricultural production and phosphate leaching causes eutrophication of water bodies. Compared to carbon preservation, the effects of Al and Fe oxides/hydroxides on phosphate sorption and release have been studied in detail (e.g., Sharpley 1983; Freese et al. 1992; Colombo et al. 1994; Hartono et al. 2006). Phosphate sorption processes in soils include fast ligand exchange reactions onto surface sites of Fe and Al oxides/hydroxides and subsequent slow reactions that deposit phosphate within soil particles (McGechan and Lewis 2002; Parfitt 1978; Addiscott and Thomas 2000). Generally, less phosphate is released than the amount added because of deposition of phosphate within soil particles (McGechan and Lewis 2002) in addition to inner-sphere complexation caused by ligand exchange reactions. The slow reactions include the physical process of diffusion into micro- and meso-pores (intra- and interparticle) and the subsequent chemical process of adsorption onto the inside surface of the pores (Arai and Sparks 2007), which will be referred to as sorption into micro- and meso-pores. Micro- and meso-pores are defined as pores of < 2 nm and 2–50 nm, respectively (Everett 1972). Phosphate trapped in micro- and meso-pores is difficult to be extracted because desorbed phosphate from surface sites in micro- and meso-pores has to diffuse into bulk solution before extraction or release from soil particles. Very few studies have studied the adsorbent porosity or the effects of micro- and meso-pores on phosphate sorption and extraction from soils (Torrent et al. 1992). Soil pore structures are likely to affect phosphate extraction, and inhibitory effects on phosphate extraction by micro- and meso-pores of non-soil materials have been reported (Cabrera et al. 1981; Strauss et al. 1997; Colombo et al. 1994; Makris et al. 2004). Clarifying the characteristics of phosphate release in soils with different porosities would be useful for predicting phosphate availability to plants and the risk of phosphate leaching after fertilization.

In this chapter, the importance of Al_o and Fe_o on OC preservation and phosphate sorption and release processes is emphasized on tropical soils where information on Al_o and Fe_o is limited. For OC preservation, the key component(s) associated with OC preservation include: Al_o and Fe_o, Fe oxides, and clay contents. For phosphate, the effects of Al_o and Fe_o on the sorption and release are clarified. Furthermore, the inhibitory effect on the extractability of phosphate sorption into micro- and meso-pores, which results from Al_o and Fe_o, is examined.

6.2 Effects of Active Al and Fe on Organic Carbon Preservation

6.2.1 Soil Samples

Soils were collected from various tropical countries (Cameroon, Tanzania, Indonesia, and Thailand) with tropical rain forest, monsoon, and savanna climates. Soils were also collected from Japan with a humid temperate climate to compare soils formed under tropical climate conditions with those under temperate climatic conditions. Subsurface horizons, most of which were identified as B horizons, were used. Soils strongly affected by volcanic ash were not included. Although carbon content is generally higher in surface horizons, the mass of OC stored in subsoil is generally larger in the subsurface horizons because the total volume of soil is much larger (Jobbagy and Jackson 2000; Kaiser et al. 2002; Rumpel and

Koegel-Knabner 2011). Thus, it is essential to study the soil components that contribute to soil OC preservation in subsurface soils to obtain a proper understanding of carbon preservation in the whole profile.

In Cameroon, 53 samples from subsurface horizons were collected at elevations ranging from 310 to 1650 m. The climate type is Aw, according to Köppen's classification, and the mean annual temperature (MAT) and mean annual precipitation (MAP) at the sites range from 20 to 27 °C and from 1070 to 2450 mm, respectively. The soil moisture and temperature regimes are udic and isohyperthermic, respectively. The primary soil types identified were Oxisols and Ultisols, and the primary parent materials were mica schist, granite, and granodiorite.

In Tanzania, 33 samples from subsurface horizons were collected at elevations ranging from 20 to 1940 m. There is a distinct dry season, and the climate type is Aw. MAT and MAP at the sites range from 18 to 27 °C and from 620 to 1500 mm, respectively. The soil moisture and temperature regimes are ustic and isohyperthermic, respectively. The primary soils types were Inceptisols and Ultisols, and the primary parent materials were granite, schist, gneiss, mudstone, and shale.

In Indonesia, 38 samples from subsurface horizons were collected at elevations ranging from 20 to 1450 m. The climate type is Af and Am, and the MAT and MAP at the sites range from 19 to 27 °C and from 1970 to 4300 mm, respectively. The soil moisture and temperature regimes are perudic/udic and isohyperthermic, respectively. The principal soil types were Ultisols and Inceptisols, and the main parent materials were mudstone, sandstone, and andesite.

In Thailand, 54 samples from subsurface horizons were collected at elevations ranging from 250 to 2000 m. The climate type is Aw and Cw. The MAT and MAP at the sites range from 17 to 26 °C and from 960 to 1800 mm, respectively. The soil moisture and temperature regimes are ustic and hyperthermic/thermic, respectively. The primary soils were Ultisols and Cambisols and the principal parent materials were granite, gneiss, schist, mudstone, and sandstone.

In Japan, 34 samples from subsurface horizons were collected at elevations ranging from 90 to 1600 m. The climate type was Cfa, and MAT and MAP at the sites ranged from 7 to 16 °C and from 1450 to 3920 mm, respectively. The soil moisture and temperature regimes are mostly udic/perudic and mesic/thermic, respectively. The primary soil type was Inceptisol, and the primary parent materials were shale and granite.

6.2.2 Analytical Procedures

The chemical and mineralogical properties for the samples were analyzed: pH (H_2O); pH(KCl); total C (which is assumed as OC in acidic soils without carbonates), Fe (Fe_t), Ca, Mg, K, and Na contents; clay content; sodium dithionite- and citrate-extractable Fe (Fe_d); and acid ammonium oxalate-extractable Al and Fe (Al_o

and Fe_o (McKeague and Day 1966)). Fe_d is a measurement of Fe in crystalline and poorly crystalline Fe oxides (Blakemore et al. 1981). Soil samples were grouped based on degree of weathering because of a wide range of pedogenic development and Al_o and Fe_o levels. Hierarchical cluster analysis was applied using standardized values of weathering indices: $(Fe_d - Fe_o)/Fe_t$, Fe_o/Fe_d, and total reserve in bases (TRB). $(Fe_d - Fe_o)/Fe_t$ represents the relative content of crystalline Fe oxides in relation to total Fe. The Fe activity ratio (Fe_o/Fe_d) reflects the poorly crystalline fraction of total Fe oxides. TRB is the sum of basic cations $(Ca + Mg + K + Na)$ $(cmol_c\ kg^{-1})$ and gives a chemical estimation of weatherable minerals (IUSS Working Group WRB 2015). TRB mainly reflects the bases in primary minerals although it also includes those located on exchange sites and possibly in the secondary clay minerals.

MAT and MAP of each site were taken from WorldClim (http://www.worldclim.org/).

The Shapiro–Wilk test was used to determine if a group of data was normally distributed. The climatic data and soil properties of the groups determined by cluster analysis were subjected to analyses of variance and then multiple comparisons analyses. Each multiple comparisons analysis was performed using the Tukey test if the data was normally distributed and using Dunn's method if the data was not normally distributed. The contributions of the soil components to OC preservation were analyzed by correlation and single variable linear regression analyses. In all analyses, $P < 0.05$ indicated statistical significance.

6.2.3 Grouping of Soils Based on Degree of Weathering

The soil samples were divided into three groups: weakly weathered (Ww), moderately weathered (Mw), and strongly weathered (Sw) using weathering indices. More than half of the Indonesian, Tanzanian, and Cameroonian soils were classified into the Sw group. Thai soils were mainly classified into Sw and Mw groups. The Ww group consisted mainly of Japanese soils. The Sw group had low Fe_o/Fe_d value (indicating low activity of free Fe oxides), a low TRB value (indicating few primary minerals), and a high $(Fe_d - Fe_o)/Fe_t$ value (indicating a high ratio of crystalline Fe oxides) (Fig. 6.1). The Mw group had similar values of Fe_o/Fe_d and $(Fe_d - Fe_o)/Fe_t$ to the Sw group but higher TRB values than Sw. In contrast, the Ww group had high activity of free Fe oxides, high primary mineral content, and a low ratio of crystalline Fe oxides to total Fe.

The climate of sampling sites was characterized for each group. MAT in Sw, Mw, and Ww groups were $23 \pm 2\ °C$ (mean $\pm$ standard deviation), $23 \pm 4\ °C$, and $16 \pm 6\ °C$, respectively. The MAT was lower in the Ww group than Sw and Mw groups. MAP was generally high and $1460 \pm 230\ mm$, $1800 \pm 870\ mm$, and $2180 \pm 1050\ mm$, respectively. MAP of Ww group was higher than that of the Sw group, but not significantly different from Mw group. Most of the samples in Sw and Mw were collected from a high temperature climate, while Ww were from low

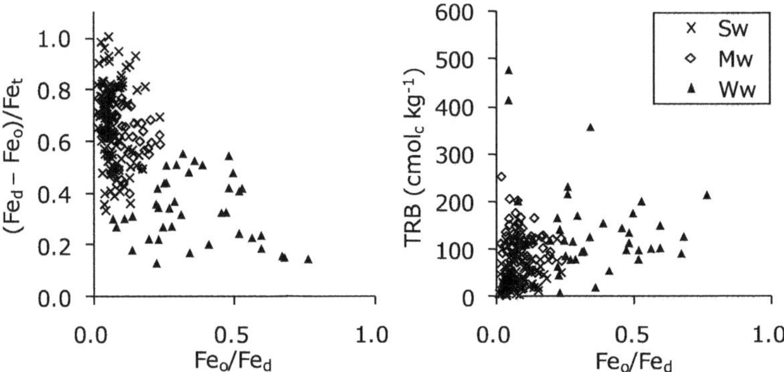

Fig. 6.1 Relationship between weathering indices: crystallinity of free iron oxides (Fe_o/Fe_d), relative amount of crystalline iron oxides ((Fe_d-Fe_o)/Fe_t), and total reserve in bases (TRB). *Sw, Mw, Ww* strongly, moderately, weakly weathered soil groups

temperature climate, and all the samples collected from a humid climate exceeded 1000 mm of precipitation.

The differences between soil properties for each weathering groups are shown in Table 6.1. The soils were acidic with pH (H_2O) of approximately 5. The mean values of OC content were higher in the Ww group than in other groups. The Al_o and Fe_o values of the Ww group were higher than those of the Mw and Sw groups. The mean values of Al_o and Fe_o in the Sw and Mw group were less than half those of the Ww group. The mean values of $Al_o + Fe_o$ were much lower than the diagnostic criteria for andic soil properties (< 72–74 cmol kg^{-1} calculated from 20 g kg^{-1} of $Al_o + 1/2$ Fe_o as the criteria for Andisols (Soil Survey Staff 1999)). Fe_d and clay contents were higher in the Sw group than that in the Ww group. These results suggest higher contents of clay and crystalline Fe oxides and lower contents of OC, Al_o, and Fe_o in the Sw group soils, which formed under high temperature conditions, than in the Ww group soils, which formed under lower temperature conditions.

6.2.4 Soil Components Contributing to Organic Carbon Preservation

For all samples and for the samples in each group, OC was strongly correlated with Al_o, Fe_o, and $Al_o + Fe_o$, except for Fe_o in Mw and Sw groups, with which OC was moderately correlated (Fig. 6.2). OC was weakly or moderately correlated with Fe_d or clay. The correlation coefficient for OC with Al_o was significantly higher than OC and Fe_o in the Sw and Mw groups, and the coefficient for Fe_o was higher than Al_o in the Ww group. They indicate that relative importance of Al_o and Fe_o may be different; Al_o may be more important in Sw and Mw groups, and Fe_o may be more important in the Ww group.

Table 6.1 Mean and standard deviation (*SD*) of soil properties for all samples and each sample group

		All samples		Strongly weathered soil group			Moderately weathered soil group			Weakly weathered soil group		
		$n = 212$		$n = 111$			$n = 59$			$n = 42$		
		Mean	*SD*	Mean	*SD*		Mean	*SD*		Mean	*SD*	
Organic C	cmol kg^{-1}	110	± 90	88	± 55	b	82	± 49	b	190	± 150	a
Al_o	cmol kg^{-1}	11	± 11	8	± 6	b	9	± 6	b	20	± 17	a
Fe_o	cmol kg^{-1}	7.5	± 9.0	4	± 3.5	b	6	± 4.8	b	18	± 15	a
Fe_d	cmol kg^{-1}	71	± 54	80	± 61	a	66	± 43	ab	53	± 42	b
Clay	%	45	± 19	48	± 20	a	46	± 17	ab	37	± 15	b
$pH(H_2O)$		5.1	± 0.8	5.1	± 0.8	a	5.2	± 0.8	a	4.9	± 0.8	a
Fe_o/Fe_d	mol mol^{-1}	0.13	± 0.14	0.07	± 0.04	c	0.10	± 0.06	b	0.34	± 0.18	a
$(Fe_d - Fe_o)/Fe_t$	mol mol^{-1}	0.60	± 0.19	0.69	± 0.15	b	0.63	± 0.10	ab	0.34	± 0.14	a
TRB	$\text{cmol}_c\,\text{kg}^{-1}$	82	± 66	39	± 22	b	124	± 34	a	138	± 95	a

Al_o, Fe_o acid ammonium oxalate-extractable Al, Fe, Fe_d dithionite citrate-extractable Fe, Fe_o/Fe_d crystallinity of free iron oxides, $(Fe_d - Fe_o)/Fe_t$ relative amount of crystalline iron oxides in total Fe, *TRB* total reserve in bases
Different letters indicate the data were statistically different among the groups at $P < 0.05$

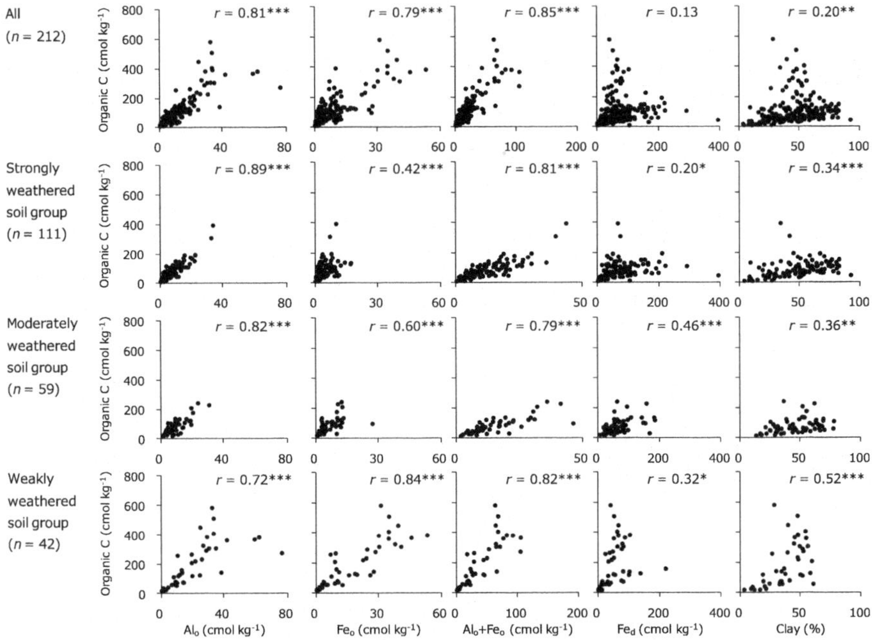

Fig. 6.2 Relationship between selected soil properties (Al_o and Fe_o acid ammonium oxalate-extractable Al and Fe, Fe_d dithionite citrate-extractable Fe, clay content) and soil organic carbon content. $* P < 0.05$; $** P < 0.01$; $*** P < 0.001$

$Al_o + Fe_o$ explained more than 60 % of the variation in OC content for all of the samples as well as for the samples in each group (Table 6.2). The regression coefficients were similar between the groups (3.9–5.3), suggesting quantitative relationships were similar in all the groups. Fe_d and clay contents only explain < 25 % of the variation of OC in any of the sample groups.

Because $Al_o + Fe_o$ explained a large part of the variation in OC in all of the groups (Table 6.2, Fig. 6.2), it indicates that this component contributes markedly to carbon preservation, not only in temperate soils but also in strongly weathered soils even though Al_o and Fe_o were very low (Table 6.1). Furthermore, the regression coefficients were similar in range in all groups, indicating quantitative relationships between $Al_o + Fe_o$ and OC were similar regardless of the degree of weathering or climatic conditions.

The correlations of Fe_d and clay contents with OC were also significant, but not strong (Fig. 6.2). Fe_d and clay contents explained < 25 % of the variations in OC for all sample groups (Table 6.2). The results in this study were contrast to strong correlations between OC and clay contents (Arrouays et al. 1995; Nichols 1984; Konen et al. 2003) and OC and Fe_d (Eusterhues et al. 2003; Kaiser and Guggenberger 2000) in previous studies. Correlations between OC and clay or Fe_d have been found for similar soil types in a region probably because the mineralogy in the clay fraction or activity of Fe_d was similar. In this study, a

Table 6.2 Regression equations for total carbon content by selected soil components

Group							R^2	n
All	Organic C (cmol kg^{-1})	=	30	+	4.3	×	Al$_o$ + Fe$_o$ (cmol kg^{-1})	0.72	212
Sw		=	22	+	5.3	×		0.65	111
Mw		=	24	+	3.9	×		0.62	59
Ww		=	34	+	4.1	×		0.66	42
All	Organic C (cmol kg^{-1})	=	91	+	0.2	×	Fe$_d$ (cmol kg^{-1})	0.01	212
Sw		=	74	+	0.2	×		0.03	111
Mw		=	46	+	0.5	×		0.20	59
Ww		=	131	+	1.1	×		0.08	42
All	Organic C (cmol kg^{-1})	=	62	+	0.99	×	Clay (%)	0.04	212
Sw		=	44	+	0.92	×		0.11	111
Mw		=	33	+	1.1	×		0.12	59
Ww		=	−2	+	5.2	×		0.25	42

Sw, Mw, Ww strongly, moderately, weakly weathered soil groups, *Al$_o$, Fe$_o$* acid ammonium oxalate-extractable Al, Fe, *Fe$_d$* dithionite citrate-extractable Fe

wide range of humid soils was analyzed and mineralogy of the clay fraction was variable, consisting of hydroxy-Al interlayered vermiculite in Japanese soils, kaolinite and mica in Thai and Tanzanian soils, kaolinite with/without vermiculite in Indonesian soils, and kaolinite in Cameroon soils (see Chaps. 2, 3, and 4). The Fe activity (Fe_o/Fe_d ratio) was also diverse in this study (Table 6.1), even in each group classified by the degree of soil weathering.

In the Sw group, Al_o was more strongly correlated with OC than with Fe_o. This may have been because Al is more soluble than Fe in soils under low pH conditions and had a greater opportunity to associate with OC (Oades 1989; Shang and Tiessen 1998). The Sw group had mean value of Fe_o that is less than half of Al_o, probably due to crystallization of Fe oxides, resulting in lower activity of Fe and lesser amounts of Fe_o. The lower contents of Fe_o compared with Al_o would result in poor correlation with OC in the Sw soils, whereas the Ww group had the similar contents of Al_o and Fe_o and both would contribute OC preservation in the soils.

6.2.5　Conclusion

Active Al and Fe (Al_o and Fe_o) are the most important components for OC preservation not only in the less weathered temperate soils but also in highly weathered tropical soils. In the more weathered soils, Al_o may be more important than Fe_o probably because Al is more soluble at low pH and has greater opportunity to associate with OC, whereas Fe tends to crystallize and lose its reactivity compared with Al.

6.3　Effects of Active Al and Fe on Sorption and Release of Phosphate

6.3.1　Soil Samples

To evaluate the effects of Al_o and Fe_o, porosity of soil, and phosphate sorption into micro- and meso-pores on the extractability of phosphate from various soil types, soils with a wide range of Al_o and Fe_o contents, which are likely related to soil porosity, were used. The values of Al_o and Fe_o were 0.88–74 g kg^{-1} and 0.79–26 g kg^{-1}, respectively, and the values of $Al_o + 1/2 Fe_o$, based on the criteria for Andisols (Soil Survey Staff 1999), were 1.7–87 g kg^{-1}. Thirty subsurface soil samples from Thailand, Indonesia, and Japan were used in this study.

The soils from Thailand and Indonesia developed under tropical climatic conditions and were mostly classified as Ultisols and Alfisols with high kaolinite contents. The soils from Japan developed under temperate climatic conditions and were mostly classified as Inceptisols with relatively high hydroxy-interlayered vermiculite contents. Four soils from Japan that developed on volcanic ash were also used, and they contained large amounts of short-range ordered minerals, identified from the high acid ammonium oxalate-extractable Si levels (Shang and Zelazny 2008), and were classified as Andisols.

6.3.2 Analytical Procedures

The chemical and mineralogical properties of the samples were analyzed: total P, pH (H_2O), clay content, Fe_d, Al_o, and Fe_o. The maximum phosphate sorption capacity was determined from the Langmuir sorption isotherm, which is commonly used to determine phosphate sorption capacity (McGechan and Lewis 2002). Extraction experiments could then be performed with every soil sample having equal phosphate loading conditions. Preparation of phosphate-sorbed samples was based on maximum phosphate sorption capacity.

Phosphate extractability was determined using a modified Hedley's fractionation method (Hedley et al. 1982; Tiessen and Moir 2008) after adding an amount of phosphate equal to the maximum phosphate sorption capacity and incubating for 24 h. The phosphate fractions are classified as most labile (Resin-P), less labile ($NaHCO_3$-P), more stable Fe- and Al-associated (NaOH-P), Fe- and Al-associated in soil aggregates (Sonicate/NaOH-P), Ca-associated (HCl-P), and most stable forms (Residual-P) (Tiessen and Moir 2008; Hedley et al. 1982). Although this fractionation method has limitations in the selectivity for different fractions (Pierzynski et al. 2005; Condron and Newman 2011), relative lability among Resin-P, $NaHCO_3$-P, and other fractions is maintained.

The rate of phosphate release was determined using sequential AEM extractions (Sato and Comerford 2006) for each soil after adding the maximum sorption capacity of phosphate and incubating for 24 h. AEM-extracted P (AEM-P) consists mainly of highly mobile phosphate ions, which have a rapid transfer rate between soil and solution. Applying the AEM extraction repeatedly allows the phosphate release rate to be determined from a regression curve using the exponential equation (Sharpley 1996; Roboredo and Coutinho 2006):

$$\sum_{t=1}^{8} \text{AEM-P}_t = a\left(1 - e^{-bt}\right)$$

where a is a constant that is closely related to the maximum amount of phosphate released using AEM, b is a constant that represents the rate of phosphate released using the AEM strips, and t is the number of extractions. The constant b was used to represent the phosphate extractability.

Specific surface area (SSA) characterization was performed on soil samples with and without phosphate addition using the N_2 adsorption method. The total surface areas (SSA_{total}) of the samples were calculated using the Brunauer–Emmett–Teller

(BET) equation (Brunauer et al. 1938). The SSAs of micro-pores (≤ 2 nm) were calculated using the t-plot procedure (Lippens and Deboer 1965), and the SSAs of meso-pores (2–50 nm) were determined using the Barrett–Joyner–Halenda (BJH) method (Barrett et al. 1951).

Normality of the data was tested using the Kolmogorov–Smirnov test. A Spearman rank correlation (rs) was applied because most of the data were not normally distributed. It was used to assess correlations between the maximum phosphate sorption capacity and soil components (e.g., Al_o, Fe_o, $Al_o + Fe_o$, Fe_d, and clay content) and between desorption values and soil components or porosity of the studied soils. The contributions of soil components to maximum phosphate sorption capacity were analyzed by single variable linear regression analyses.

6.3.3 Maximum Phosphate Sorption Capacity

The maximum phosphate sorption capacity (Table 6.3) was calculated using the Langmuir equation. The phosphate sorption data for all of the soils had a high correlation ($r^2 > 0.98$). The maximum phosphate sorption capacity had a high positive correlation with Al_o ($rs = 0.89, P < 0.001$) and $Al_o + Fe_o$ ($rs = 0.91, P < 0.001$; Fig. 6.3) and a weaker correlation with Fe_o ($rs = 0.62, P < 0.001$). $Al_o + Fe_o$ explained the maximum phosphate sorption capacity well:

$$\text{Maximum phosphate sorption capacity (cmol kg}^{-1})$$
$$= 0.11 \times (Al_o + Fe_o)(\text{cmol kg}^{-1}) + 2.0 \ (R^2 = 0.95, n = 30)$$

In contrast, clay content and Fe_d did not significantly correlate with the maximum phosphate sorption capacity. Thus, most of the following analyses are shown using $Al_o + Fe_o$ as the most important soil components for phosphate sorption.

6.3.4 Extractability of Added Phosphate

The modified Hedley's fractionation method showed that most of the added phosphate was extracted as Resin-P, $NaHCO_3$-P_i, and $NaOH$-P_i (Table 6.3). The percentage of phosphate extracted in each fraction correlated significantly with $Al_o + Fe_o$; the most labile phosphate (Resin-P) extract was negatively correlated with $Al_o + Fe_o$, and the less labile and more stable Fe and Al-associated phosphate ($NaHCO_3$-P_i and $NaOH$-P_i) extracts were positively correlated with $Al_o + Fe_o$ (Table 6.4). These results indicate that soils with higher $Al_o + Fe_o$ have lower proportions of labile Resin-P and higher proportions of less labile $NaHCO_3$-P_i and $NaOH$-P_i.

Sequential extraction by AEM showed the slow release of added phosphate from soils with high $Al_o + Fe_o$. The amounts released were highly fitted to the exponential equation well ($R^2 \geq 0.80$), and the calculated constant b had P-values <0.001. The constant b, representing the phosphate release rate, had a significant negative

Table 6.3 Soil properties and results of Hedley's fractionation and extraction by anion exchange membrane

Sample	Al_o+Fe_o	P max sorption capacity	Hedley's fractionation								Rate constant for desorption with AEM
			Resin	$NaHCO_3$	$NaHCO_3$	NaOH	NaOH	sonicNaOH	HCl	Residual	
			-P	$-P_i$	$-P_o$	$-P_i$	$-P_o$	-P	-P	-P	
	cmol kg^{-1}	cmol kg^{-1}	%	%	%	%	%	%	%	%	extraction No.$^{-1}$
Thailand											
TH1	15	3.1	67	14	0	17	0	1	0	0	1.50
TH2	9.5	1.6	76	5	4	13	1	1	0	0	1.62
TH3	7.7	0.9	70	10	1	12	1	0	0	6	1.73
TH4	30	5.2	65	14	0	15	4	1	0	0	0.97
Indonesia											
ID3	24	2.6	64	12	0	18	2	2	0	0	1.38
ID5	31	6.2	63	15	0	15	0	1	0	6	1.07
ID7	14	3.5	67	15	0	17	0	1	0	0	1.48
ID9	17	2.2	76	13	0	10	0	1	0	0	1.59
ID10	19	3.6	67	15	0	17	0	1	0	0	1.43
ID12	6.3	2.2	75	5	2	15	2	1	0	0	1.76
ID13	9.8	3.1	80	6	0	13	0	1	0	0	2.07
ID14	9.8	2.7	72	12	0	14	1	1	0	0	1.63
ID15	17	3.5	65	15	0	17	0	1	0	2	1.40
ID17	11	2.5	76	9	0	13	1	1	0	0	1.63
ID18	10	2.5	76	5	3	12	1	1	0	1	1.62
Japan											
JP1	17	4.6	60	15	1	18	2	1	0	4	1.36
JP4	16	3.2	58	18	0	17	0	1	0	7	1.11
JP6	28	5.9	65	15	0	14	0	1	0	5	0.87
JP9	38	8.8	52	16	0	22	8	1	0	1	0.87

(continued)

Table 6.3 (continued)

Sample	Al_o+Fe_o	P max sorption capacity	Hedley's fractionation								Rate constant for desorption with AEM
			Resin	NaHCO$_3$	NaHCO$_3$	NaOH	NaOH	sonicNaOH	HCl	Residual	
			-P	-P$_i$	-P$_o$	-P$_i$	-P$_o$	-P	-P	-P	
	cmol kg^{-1}	cmol kg^{-1}	%	%	%	%	%	%	%	%	extraction No.$^{-1}$
JP11	21	5.0	58	20	0	20	0	2	0	0	1.16
JP12	17	3.5	62	15	1	18	2	1	0	1	1.27
JP13	50	9.6	45	12	5	33	2	1	0	2	0.87
JP16	15	3.8	62	12	0	22	0	2	0	1	1.39
JP17	36	7.2	53	17	2	22	3	1	0	1	0.88
JP18	44	6.0	65	14	0	15	2	1	0	3	0.89
JP19	12	2.8	57	15	0	20	0	1	0	6	1.09
Volcanic, Japan											
JP15	320	34	40	12	0	17	7	3	0	21	0.69
JP20	180	22	45	17	3	31	2	2	0	0	0.83
JP21	110	17	56	17	1	24	1	1	0	0	1.06
JP22	110	18	54	14	0	14	8	1	0	8	0.89

Al_o, Fe_o acid ammonium oxalate-extractable Al, Fe, P phosphorus, P_i inorganic phosphorus, P_o organic phosphorus, *sonicNaOH* NaOH extracted fraction with sonication, *AEM* anion exchange membrane

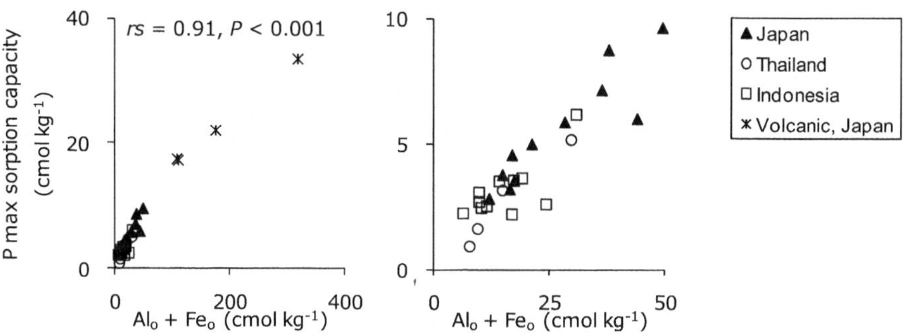

Fig. 6.3 Relationship between acid ammonium oxalate-extractable Al and Fe contents (Al$_o$ and Fe$_o$) and maximum phosphate sorption capacity (*left*: small scale; *right*: large scale)

Table 6.4 Spearman correlation coefficients between the value of extraction results and acid ammonium oxalate-extractable Al and Fe contents and the porosity of the studied soils

	Hedley's fractionation			
	Resin	NaHCO$_3$	NaOH	
	-P	-P$_i$	-P$_i$	Rate constant for desorption with AEM
Al$_o$ + Fe$_o$	−0 79***	0.55**	0.55**	−0 91***
Al$_o$	−0.81***	0.47**	0.56**	−0 90***
Fe$_o$	−0.44*	0.36*	0.23	−0.66***
SSA$_{0.7-4nm}$/ SSA$_{total}$	−0 72***	0.42*	0.46*	−0 70***
ΔSSA$_{0.7-4nm}$	−0.68***	0.53**	0.47*	−0 74***

P phosphorus, *P$_i$* inorganic phosphorus, *AEM* anion exchange resin, *Al$_o$, Fe$_o$* acid ammonium oxalate-extractable Al, Fe, *SSA$_{total}$* total specific surface area (SSA), *SSA$_{0.7-4nm}$* SSA of 0.7–4 nm pores, *ΔSSA$_{0.7-4nm}$* decrease in SSA of 0.7–4-nm pores after phosphate sorption treatment
*** $P < 0.001$, ** $P < 0.01$, * $P < 0.05$

correlation with Al$_o$ + Fe$_o$ (Table 6.4, Fig. 6.4), indicating that soils with higher Al$_o$ + Fe$_o$ have a lower phosphate release rate.

Japanese soils generally showed lower Resin-P proportions and lower release rates than soils from the other regions with the exception of soils TH4 and ID5 (Thai and Indonesian soils) and soil JP16 (Japanese soil) (Table 6.3, Fig. 6.4). The TH4 and ID5 were high-altitude (Thai) or volcanic (Indonesian) soils with high Al$_o$ + Fe$_o$ (30 and 31 cmol kg^{-1}, Table 6.3). JP16 was from Okinawa, an island in southwest Japan with a warmer climate than other parts of Japan. Thus, these soils would show intermediate values between the tropical countries and Japan.

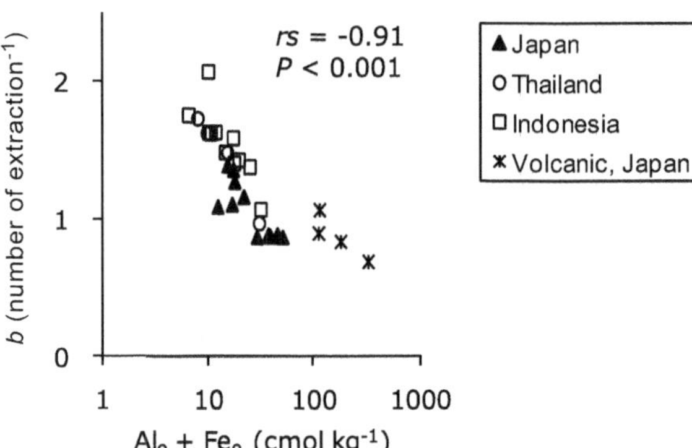

Fig. 6.4 Relationship between acid ammonium oxalate-extractable Al and Fe contents (Al_o and Fe_o) and the constant b, representing phosphate release rate using an anion exchange membrane

6.3.5 Effects of Micro- and Small Meso-Pores on Extractability of Sorbed Phosphate

The results indicate that Al_o and Fe_o reduced the extractability of added phosphate because of phosphate sorption into micro- and small meso-pores. Phosphate sorption treatment resulted in a decrease in SSA_{total} (Table 6.5). When the decrease in micro- and small meso-pore SSA ($\Delta SSA_{0.7-4-nm}$) was high, the proportions of labile Resin-P and less labile $NaHCO_3$-P_i and $NaOH$-P_i measured using Hedley's method were low and high, respectively, and the constant b for sequential extraction was low (Table 6.4, Fig. 6.5). $\Delta SSA_{0.7-4-nm}$ was positively correlated with $Al_o + Fe_o$ ($rs = 0.74$, $P < 0.001$) and the $SSA_{0.7-4-nm}/ SSA_{total}$ ratio ($rs = 0.80$, $P < 0.001$), suggesting that sorption into micro- and meso-pores occurred in soils with high Al_o and Fe_o contents and high porosities. The desorption of phosphate might be the same in all pore sizes, but the diffusion of the desorbed phosphate to bulk solution might be limited in the micro- and meso-pores.

The Japanese soil samples had high porosity (Table 6.5), and the decrease in the micro- and small meso-pore SSAs with phosphate treatment ($\Delta SSA_{0.7-4-nm}$) was greater in these samples than in the Thai and Indonesian samples (Fig. 6.6). When the amount of $Al_o + Fe_o$ was similar, the proportion of Resin-P measured using Hedley's method and the constant b for sequential extraction were lower for the Japanese soils than for the other soils (Table 6.3 and Fig. 6.4). For the Thai and Indonesian soil samples, the amounts of 0.7–4-nm micro- and meso-pores and the reduced amount in pore size after phosphate treatment were small (Fig. 6.6), indicating that this pore size range appeared to have less effect on the sorption and extraction of phosphate in these soils. Short-range ordered minerals, i.e., allophone and imogolite, contributed to the large micro- and meso-pore surface areas, their decrease with phosphate sorption, and the low phosphate extractability in the volcanic soils from Japan. The Si_o contents were similar in nonvolcanic Japanese, Thai, and Indonesian soils (mean ± standard deviation: 0.45 ± 0.33, 0.45 ±, 0.11, and 0.52 ± 0.13 g kg^{-1}, respectively), and therefore they appear to contain little, if any, allophone and imogolite. The crystallinity of the Al and Fe oxides/

Table 6.5 Acid ammonium oxalate-extractable Al and Fe contents (Al_o and Fe_o), specific surface area (SSA) of total and micro- and meso-pores, and decrease of SSA resulting from phosphate sorption

Region	Sample	$Al_o + Fe_o$ cmol kg^{-1}	SSA_{total} m^2 g^{-1}	$SSA_{0.7-4nm}$ m^2 g^{-1}	$SSA_{0.7-4nm}/SSA_{total}$	ΔSSA_{total} m^2 g^{-1}	$\Delta SSA_{0.7-4nm}$ m^2 g^{-1}
Thailand							
	TH1	14.8	25	4.5[a]	18	12	1.7
	TH2	9.5	24	4.4[a]	19	1.6	0.3
	TH3	7.7	15	3.4[a]	22	0.8	0.2
	TH4	29.8	38	7.1[a]	19	5.4	0.7
Indonesia							
	ID3	24.3	57	17	31	4.7	0.9
	ID5	30.9	54	31	58	28	25
	ID7	14.2	32	5.8[a]	18	6.5	0.8
	ID9	16.9	20	3.5[a]	18	2.6	0.2
	ID10	19.2	33	6.6[a]	20	4.1	0.9
	ID12	6.3	28	5.4[a]	19	2.9	0.4
	ID13	9.8	34	5.7[a]	17	7.0	0.7
	ID14	9.8	19	3.7[a]	20	2.3	0.4
	ID15	17.3	23	4.7[a]	20	1.4	0.5
	ID17	11.5	23	4.4[a]	19	5.2	0.4
	ID18	10.3	29	4.9[a]	17	6.8	0.6
Japan							
	JP1	17.0	33	13	40	9.3	9.0
	JP4	16.5	19	3.2[a]	17	5.2	0.5
	JP6	28.4	45	15	32	9.6	8.1
	JP9	38.0	43	11	25	6.2	5.2
	JP11	21.3	51	24	46	8.3	10

(continued)

Table 6.5 (continued)

Region	Sample	$Al_o + Fe_o$ cmol kg^{-1}	SSA_{total} m^2 g^{-1}	$SSA_{0.7-4nm}$ m^2 g^{-1}	$SSA_{0.7-4nm}/SSA_{total}$	ΔSSA_{total} m^2 g^{-1}	$\Delta SSA_{0.7-4nm}$ m^2 g^{-1}
	JP12	17.4	38	22	59	7.9	11
	JP13	49.6	33	20	60	4.6	5.0
	JP16	14.9	27	4.6[a]	17	5.0	0.3
	JP17	36.5	34	9.3	27	4.7	4.4
	JP18	44.2	26	12	46	8.3	9.1
	JP19	12.0	27	6.2	23	4.7	2.0
Volcanic, Japan							
	JP15	320	94	89	94	34	38
	JP20	176	61	49	81	5.1	8.2
	JP21	111	64	54	84	1.0	3.4
	JP22	109	74	64	86	18	19

Al_o, Fe_o acid ammonium oxalate-extractable Al, Fe, SSA_{total} total specific surface area (SSA), $SSA_{0.7-4nm}$ SSA of 0.7–4-nm pores, ΔSSA_{total}, $\Delta SSA_{0.7-4nm}$ decrease in SSA of total and 0.7–4-nm pores after phosphate sorption treatment

[a]Micropore (<2 nm) was not detected

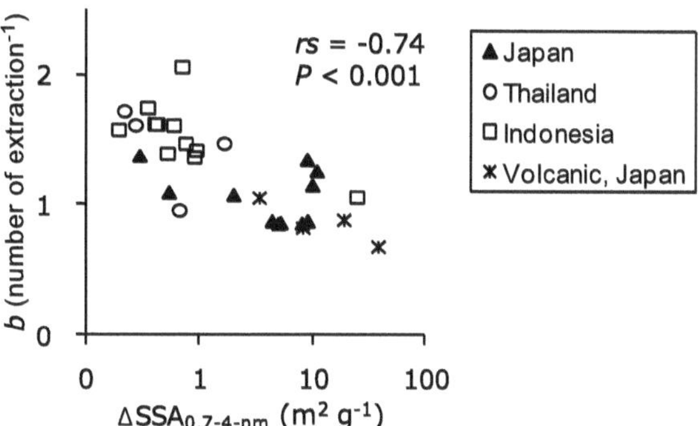

Fig. 6.5 Relationship between the decrease in specific surface area of micro- and meso-pores (0.7–4 nm) after phosphate sorption treatment ($\Delta SSA_{0.7-4\ nm}$) and the constant b representing phosphate release rate using an anion exchange membrane

hydroxides might be lower in the Japanese soils because they were younger soils such as Inceptisols and Andisols with interference in Al and Fe crystallization from organics and Si (Hsu 1989; Schwertmann and Taylor 1989), resulting in higher soil porosity, a decrease in micro- and meso-pores with phosphate sorption, and lower phosphate extractability.

The availability of phosphate to plants and the potential for phosphate leaching after application to agricultural land might differ depending on particle surface porosity derived from Al_o and Fe_o and phosphate sorption into micro- and small meso-pores. Lower soil loading with phosphate leads to lower phosphate extractability (Barros et al. 2005). If the same amount of phosphate was applied to the soils in the present study, soils with high Al_o and Fe_o (and therefore large maximum sorption capacities) would release sorbed phosphate more slowly. The mobility of applied phosphate differs between the Japanese and the Thai and Indonesian soils even when the amount of Al_o and Fe_o was similar. The Japanese soils were rich in micro- and meso-pores that sorbed phosphate inside the pores, whereas Thai and Indonesian soils had low porosity and larger pores, and the outer surfaces probably sorbed the most phosphate. This should result in higher phosphate mobility in Thai and Indonesian soils than in Japanese soils.

6.3.6 Other Soil Properties and Phosphate Extractability

Other than the parameters listed in Table 6.4, soil physicochemical properties that could contribute to phosphate extractability such as clay content, total carbon, Fe_d, $Fe_d - Fe_o$, and Fe_o/Fe_d showed weak or no correlation with the proportion of Resin-P from Hedley's method (rs values were -0.20, -0.55, -0.18, 0.08, and -0.46, respectively) or the constant b for AEM sequential extraction (rs values were -0.13, -0.44, 0.03, 0.27, and -0.48, respectively). This indicated that these properties did not affect phosphate extractability in the samples analyzed.

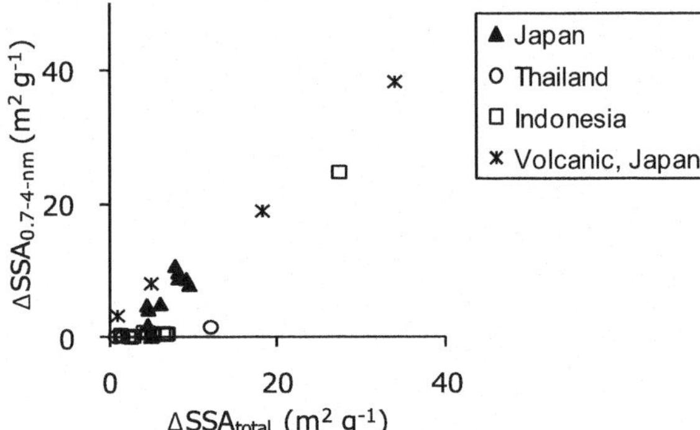

Fig. 6.6 Relationship between the decrease in total specific surface area (ΔSSA_{total}) and the decrease in SSAs of micro- and meso-pores (0.7–4 nm) after phosphate sorption treatment ($\Delta SSA_{0.7-4\ nm}$)

6.3.7 Conclusion

The importance of Al_o and Fe_o on the phosphate sorption capacity and the inhibitory effect of micro- and meso-pores on phosphate released using various types of soils were presented. Phosphate sorption capacity strongly depended on Al_o and Fe_o contents. Release of sorbed phosphate from soils with high Al_o and Fe_o contents and high porosity was more difficult than from soils with low porosity. The sorbed phosphate was more easily released from tropical soils with lower porosity. Estimations of phosphate release for plant nutrition or phosphate leaching in soils should consider the porosity of soils.

6.4 Significance of Active Al and Fe on Organic Carbon Preservation and Phosphate Sorption/Release in Tropical Soils

Active Al and Fe (Al_o and Fe_o) had a significant effect on both OC preservation and phosphorus sorption and release. The contents of $Al_o + Fe_o$ explained more than 60 % of variation in OC contents, and the quantitative relationships of OC with $Al_o + Fe_o$ were of similar order for all the groups classified by the degree of weathering, i.e., weakly, moderately, and strongly weathered soils. Maximum sorption capacity of phosphate was highly correlated with $Al_o + Fe_o$ ($rs = 0.91$, $P < 0.001$) and 95 % of variation was explained by $Al_o + Fe_o$. Release of sorbed phosphate was retarded by porosity of Al_o and Fe_o. These statistically significant effects of Al_o and Fe_o are notable especially when compared to poor correlations of clay and Fe oxide contents. It is important to highlight the utility of Al_o and Fe_o in humid regions for soils derived from volcanic materials and soils developing in temperate or cooler regions but also in highly weathered tropical soils.

Acknowledgement This chapter is derived in part from Watanabe et al. (2015) published in Soil Science, Wolters Kluwer Health, Inc., doi:10.1097/ss.0000000000000116.

References

Addiscott TM, Thomas D (2000) Tillage, mineralization and leaching: phosphate. Soil Tillage Res 53(3–4):255–273. doi:10.1016/s0167-1987(99)00110-5

Arai Y, Sparks DL (2007) Phosphate reaction dynamics in soils and soil components: a multiscale approach. In: Sparks DL (ed) Advances in agronomy, vol 94. Advances in agronomy. Elsevier, San Diego, pp 135–179. doi:10.1016/s0065-2113(06)94003-6

Arrouays D, Vion I, Kicin JL (1995) Spatial-analysis and modeling of topsoil carbon storage in temperate forest humic loamy soils of France. Soil Sci 159(3):191–198. doi:10.1097/00010694-199515930-00006

Barrett EP, Joyner LG, Halenda PP (1951) The determination of pore volume and area distributions in porous substances. I. Computations from nitrogen isotherms. J Am Chem Soc 73(1):373–380. doi:10.1021/ja01145a126

Barros NF, Comerford NB, Barros NF (2005) Phosphorus sorption, desorption and resorption by soils of the Brazilian Cerrado supporting eucalypt. Biomass Bioenergy 28(2):229–236. doi:10.1016/j.biombioe.2004.08.005

Blakemore LC, Searle PL, Daly BK (1981) Methods for chemical analysis of soils. New Zealand Soil Bureau Scientific Report 10A. DSIR, New Zealand

Brunauer S, Emmett PH, Teller E (1938) Adsorption of gases in multimolecular layers. J Am Chem Soc 60:309–319. doi:10.1021/ja01269a023

Cabrera F, Dearambarri P, Madrid L, Toca CG (1981) Desorption of phosphate from iron oxides in relation to equilibrium pH and porosity. Geoderma 26(3):203–216. doi:10.1016/0016-7061(81)90016-1

Colombo C, Barron V, Torrent J (1994) Phosphate adsorption and desorption in relation to morphology and crystal properties of synthetic hematites. Geochim Cosmochim Acta 58(4):1261–1269. doi:10.1016/0016-7037(94)90380-8

Condron LM, Newman S (2011) Revisiting the fundamentals of phosphorus fractionation of sediments and soils. J Soils Sediments 11(5):830–840. doi:10.1007/s11368-011-0363-2

Eusterhues K, Rumpel C, Kleber M, Kogel-Knabner I (2003) Stabilisation of soil organic matter by interactions with minerals as revealed by mineral dissolution and oxidative degradation. Org Geochem 34(12):1591–1600. doi:10.1016/j.orggeochem.2003.08.007

Everett DH (1972) Manual of symbols and terminology for physicochemical quantities and units, Appendix II: definitions, terminology and symbols in colloid and surface chemistry. Pure Appl Chem 31(4):577–638

Freese D, Vanderzee S, Vanriemsdijk WH (1992) Comparison of different models for phosphate sorption as a function of the iron and aluminum oxides of soils. J Soil Sci 43(4):729–738

Hartono A, Funakawa S, Kosaki T (2006) Transformation of added phosphorus to acid upland soils with different soil properties in Indonesia. Soil Sci Plant Nutr 52(6):734–744. doi:10.1111/j.1747-0765.2006.00087.x

Hedley MJ, Stewart JWB, Chauhan BS (1982) Changes in inorganic and organic soil phosphorus fractions induced by cultivation practices and by laboratory incubations. Soil Sci Soc Am J 46(5):970–976

Hsu PH (1989) Aluminum hydroxides and oxyhydroxides. In: Dixon JB, Weed SB (eds) Minerals in soil environments. SSSA Book Series, vol 1, 2nd edn. Soil Science Society of America, Madison, pp 331–378

IUSS (2014) World reference base for soil resources 2014. World Soil Resources Reports No. 106. FAO, Rome

IUSS Working Group WRB (2015) World reference base for soil resources 2014. Update 2015. World Soil Resources Reports No. 106. FAO, Rome

Jobbagy EG, Jackson RB (2000) The vertical distribution of soil organic carbon and its relation to climate and vegetation. Ecol Appl 10(2):423–436

Kaiser K, Guggenberger G (2000) The role of DOM sorption to mineral surfaces in the preservation of organic matter in soils. Org Geochem 31:711–725

Kaiser K, Eusterhues K, Rumpel C, Guggenberger G, Kögel-Knabner I (2002) Stabilization of organic matter by soil minerals – investigations of density and particle-size fractions from two acid forest soils. J Plant Nutr Soil Sci-Zeitschrift Fur Pflanzenernahrung Und Bodenkunde 165 (4):451–459

Kirsten M, Kaaya A, Klinger T, Feger K-H (2016) Stocks of soil organic carbon in forest ecosystems of the Eastern Usambara Mountains, Tanzania. Catena 137:651–659. doi:10. 1016/j.catena.2014.12.027

Kleber M, Mikutta R, Torn MS, Jahn R (2005) Poorly crystalline mineral phases protect organic matter in acid subsoil horizons. Eur J Soil Sci 56:717–725

Konen ME, Burras CL, Sandor JA (2003) Organic carbon, texture, and quantitative color measurement relationships for cultivated soils in north central Iowa. Soil Sci Soc Am J 67 (6):1823–1830

Lippens BC, Deboer JH (1965) Studies on pore systems in catalysts. V the t method. J Catal 4 (3):319–323. doi:10.1016/0021-9517(65)90307-6

Lorenz K, Lal R, Jimenez JJ (2009) Soil organic carbon stabilization in dry tropical forests of Costa Rica. Geoderma 152(1–2):95–103. doi:10.1016/j.geoderma.2009.05.025

Makris KC, Harris WG, O'connor GA, Obreza TA (2004) Phosphorus immobilization in micropores of drinking-water treatment residuals: implications of long-term stability. Commun Soil Sci Plant Anal Plant Soil Environ Sci Technol 38:6590–6596

McGechan MB, Lewis DR (2002) Sorption of phosphorus by soil, part 1: principles, equations and models. Biosyst Eng 82:1–24

McKeague JA, Day JH (1966) Dithionite- and oxalate-extractable Fe and Al as aids in differentiating various classes of soils. Can J Soil Sci 46:13–22

Nichols JD (1984) Relation of organic-carbon to soil properties and climate in the southern Great Plains. Soil Sci Soc Am J 48(6):1382–1384

Oades JM (1989) An introduction to organic matter in mineral soils. In: Dixon JB, Weed SB (eds) Minerals in soil environments. vol SSSA Book Series 1, 2nd edn. SSSA, Madison, WI, pp 89–159

Ohta S, Effendi S (1992) Ultisols of "lowland Dipterocarp forest" in East Kalimantan, Indonesia, II. Status of carbon, nitrogen, and phosphorus. Soil Sci Plant Nutr 38:207–216

Parfitt RL (1978) Anion adsorption by soils and soil materials. In: Brady NC (ed) Advances in agronomy, vol 30. Academic Press, Inc., pp 1–50

Pierzynski GM, McDowell RW, Sims JT (2005) Chemistry, cycling, and potential movement of inorganic phosphorus in soils. In: Sims JT, Sharpley AN (eds) Phoshprus: agriculture and the environment. vol 46. ASA, CSSA, SSSA, Madison, pp 53–86

Roboredo M, Coutinho J (2006) Chemical characterization of inorganic phosphorus desorbed by ion exchange membranes in short- and long-term extraction periods. Commun Soil Sci Plant Anal 37(11–12):1611–1626. doi:10.1080/00103620600710215

Rumpel C, Koegel-Knabner I (2011) Deep soil organic matter-a key but poorly understood component of terrestrial C cycle. Plant Soil 338(1–2):143–158. doi:10.1007/s11104-010-0391-5

Sato S, Comerford NB (2006) Assessing methods for developing phosphorus desorption isotherms from soils using anion exchange membranes. Plant Soil 279(1–2):107–117. doi:10.1007/s11104-005-0437-2

Schwertmann U, Taylor RM (1989) Iron oxides. In: Dixon JB, Weed SB (eds) Minerals in soil environments. SSSA Book Series, vol 1, 2nd edn. Soil Science Society of America, Madison, pp 379–438

Shang C, Tiessen H (1998) Organic matter stabilization in two semiarid tropical soils: Size, density, and magnetic separations. Soil Sci Soc Am J 62(5):1247–1257

Shang C, Zelazny L (2008) Selective dissolution techniques for mineral analysis of soils and sediments. In: Ulery AL, Drees LR (eds) Methods of soil analysis Part 5 mineralogical methods. Soil Science Society of America, Inc., Madison, pp 33–80

Sharpley AN (1983) Effect of soil properties on the kinetics of phosphorus desorption. Soil Sci Soc Am J 47:462–467

Sharpley AN (1996) Availability of residual phosphorus in manured soils. Soil Sci Soc Am J 60 (5):1459–1466

Soil Survey Staff (1999) Soil taxonomy, second ed. edn. USDA-NRCS, Washington, DC

Sollins P, Homann P, Caldwell BA (1996) Stabilization and destabilization of soil organic matter: Mechanisms and controls. Geoderma 74(1–2):65–105

Strauss R, Brummer GW, Barrow NJ (1997) Effects of crystallinity of goethite: II. Rates of sorption and desorption of phosphate. Eur J Soil Sci 48(1):101–114. doi:10.1111/j.1365-2389. 1997.tb00189.x

Tiessen H, Moir JO (2008) Characterization of available P by sequential extraction. In: Carter MR, Gregorich EG (eds) Soil sampling and methods of analysis. 2nd edn. CRC Press, Boca Raton, FL, pp 293–306

Toriyama J, Hak M, Imaya A, Hirai K, Kiyono Y (2015) Effects of forest type and environmental factors on the soil organic carbon pool and its density fractions in a seasonally dry tropical forest. For Ecol Manag 335:147–155. doi:10.1016/j.foreco.2014.09.037

Torrent J, Schwertmann U, Barron V (1992) Fast and slow phosphate sorption by goethite-rich natural materials. Clay Clay Miner 40(1):14–21. doi:10.1346/ccmn.1992.0400103

von Lützow M, Kögel-Knabner K, Ekschmitt K, Matzner E, Guggenberger G, Marshner B, Flessa H (2006) Stabilization of organic matter in temperate soils: mechanisms and their relevance under different soils conditions- a review. Eur J Soil Sci 57:426–445

Watanabe T, Hase E, Funakawa S, Kosaki T (2015) Inhibitory effect of soil micropores and small mesopores on phosphate extraction from soils. Soil Sci 180(3):97–106. doi:10.1097/ss. 0000000000000116

Wiseman CLS, Püttmann W (2006) Interactions between mineral phases in the preservation of soil organic matter. Geoderma 134:109–118

Part II
Ecosystem Processes in Forest-Soil Systems Under Different Geological, Climatic, and Soil Conditions

Chapter 7
Soil Acidification Patterns Under Different Geological and Climatic Conditions in Tropical Asia

Kazumichi Fujii and Arief Hartono

Abstract Tropical forests are characterized by highly weathered and acidified soils, whilst patterns and processes of soil acidification are diverse under different geological and climatic conditions. To identify the dominant processes of soil acidification in Southeast Asia, proton budgets were quantified for plant-soil systems in Indonesia and Thailand. The net proton generation by plant uptake was consistently high in the tropical forests. Acidification of soils can function as nutrient acquisition strategies of plants that promote cation mobilization through mineral weathering and cation exchange reaction. Soil solution composition indicated that organic acids are dominant anions that drive acidification in the highly acidic soils from sandy sedimentary rocks. Production of organic acids in the O horizons can be enhanced by the high activities of fungal enzymes (peroxidases) especially in the lignin-rich and P-poor litters on the highly acidic soils. On the other hand, bicarbonate also contributed to cation mobilization in the moderately acidic soils from clayey sedimentary rocks and ultramafic rocks (Indonesia) and under monsoon climate with distinct dry season (Thailand). The spatiotemporal variation in fine roots (plant uptake) and organic and carbonic acids can lead to different pathways of pedogenesis, i.e., incipient podzolization (Al eluviation/illuviation) and ferralitization (in situ weathering). The differences in acid-neutralizing capacities of parent materials and climatic patterns can generate the variability in soil acidity, and plant and microbial feedbacks can further reinforce the patterns of soil acidification.

Keywords Acid-neutralizing capacity • Dissolved organic matter • Proton budget • Soil acidity • Tropical forest

K. Fujii (✉)
Forestry and Forest Products Research Institute, 1 Matsunosato, Tsukuba, Ibaraki, Japan
e-mail: fkazumichi@affrc.go.jp

A. Hartono
Department of Soil Science and Land Resources, Bogor Agricultural University, Bogor 16680, Japan

© Springer Japan KK 2017
S. Funakawa (ed.), *Soils, Ecosystem Processes, and Agricultural Development*,
DOI 10.1007/978-4-431-56484-3_7

7.1 Introduction

Acidification is a common soil-forming process under humid climate where precipitation exceeds evapotranspiration (Krug and Frink 1983; Hallbäcken and Tamm 1986). In tropical regions, acidic soils (pH $<$ 5.5) are widespread; they cover 30 % of the world's total land area and 60 % of the total area in the tropics (Sanchez and Logan 1992). The acidic soils in tropical regions generally exhibit the low concentrations of exchangeable bases, low solubility of phosphorus (P), and high concentrations of exchangeable Al and solution Al ions (Al^{3+}). These factors related to soil fertility can limit productivity of plants (esp., crop species) in tropical regions.

The decline of plant productivity has also been reported in some of tropical forests on the highly acidified and weathered soils (Vitousek and Howarth 1991). On the other hand, the massive aboveground biomass of tropical forests can be developed even on highly acidic soils in Southeast Asia (Fujii 2014). There still exist several knowledge gaps between high productivity and soil acidity.

Soil acidification induced by acidic deposition or agriculture is well known as one of soil degradation, whilst acidification is an essential process for plant nutrient acquisition. In tropical forests, the large amounts of nutrient need to be supplied from the acidic soils for plant biomass production. Basic cations (K, Mg, and Ca) and P are released from mineral dissolution and cation (or anion) exchange reactions. Most of these reactions require acidification of soil. To assess the drivers and ecological significance of soil acidification, the dominant processes of soil-acidifying processes need to be identified.

Soil acidification is caused by nitrification, the dissociation of organic acids and carbonic acid, and the excess uptake of cations over anions by plants (Van Breemen et al. 1983; Binkley and Richter 1987). Soil solution studies have revealed the roles of biologically processed anions in soil solution, i.e., organic acids, carbonic acid, and nitric acid, as drivers of cation mobilization (Ugolini and Sletten 1991). Further, theory and calculation of proton budgets in a plant-soil system can quantify contribution of external and internal proton sources to soil acidification, as it can also include proton release associated with cation uptake by plants (Van Breemen et al. 1983; Johnson 1977; Fujii et al. 2008). These studies showed that soil-acidifying processes can vary from ecosystem to ecosystem. Organic acids [dissolved organic matter (DOM)] generally play roles in acidification of the Spodosol soils in coniferous forests, while carbonic and nitric acids contribute to acidification of the Inceptisol and Andisol soils in temperate broad-leaved forests (Ugolini and Sletten 1991; Fujii et al. 2008). Calculation of proton budgets in each of soil horizon provides a better understanding of proton generation and consumption within the soil profiles.

The soil-acidifying processes can also vary between temperate and tropical forests. In several tropical forests from America, the production and leaching of carbonic acids from intensive root and microbial respiration have been reported to be a major cause of soil acidification (Johnson 1977; McDowell 1998). The hypothesis has been proposed that carbonic acid leaching has developed in tropical forests to utilize high soil CO_2 pressure to acquire exchangeable bases and to minimize leaching losses of bases from base-poor soils (Johnson 1977).

Tropical forests exhibit wide variation in vegetation, geology, and climate. Regarding specific aspects in Southeast Asia, Ashton (1988) hypothesized that Dipterocarpaceae exhibit high species diversity, tall stature, and large biomass production through adaptation to acidic soils via ectomycorrhizal associations. The high host specificity of ectomycorrhizae can cause competitive advantage and family-level monodominance of Dipterocarpaceae in Southeast Asia, whereas most of the dominant trees in America and Africa associate with vesicular-arbuscular mycorrhizae (Connell and Lowman 1989). In Southeast Asian tropical forests, nutrient acquisition of trees in the Ultisol soils may be different from those in the Oxisol and Ultisol soils of America and Africa, in that Asian Ultisol soils are richer in weatherable minerals because of steep slopes or relatively young geological ages (Fujii et al. 2011a).

We applied theory of proton budgets to individual soil horizons to identify the dominant soil-acidifying processes and the site-specific and common aspects of acidification processes in Southeast Asia by comparing with the other tropical regions.

7.2 Materials and Methods

7.2.1 Descriptions of Experimental Plots

The experimental sites include five sites from Indonesia and one site from Thailand. The sites are selected to compare effects of geology (serpentine vs mudstone vs sandstone) and climate (tropical humid vs tropical monsoon) (Table 7.1).

Five sites of tropical forests in Indonesia consisted of natural secondary forests in Kuaro (KR1, KR2, and KR3) and Bukit Soeharto (BS) and pristine forest in Bukit Bankirai (BB), East Kalimantan Province, Indonesia. The parent materials are largely sedimentary rocks, but there occur patches of serpentine (ultramafic) intrusion. The KR1, KR2, and KR3 plots are located along a traverse across serpentine-sedimentary rocks (mudstone). The KR1 soil is located on the serpentine belt and is classified as a Rhodic Eutrudox (Soil Survey Staff 2006). The KR2 soil is located in the transitional zone between serpentine and mudstone, and the KR3 soil is located on mudstone (Fig. 7.1). The KR2 and KR3 soils are classified as Typic Paleudults. The BS and BB soils are derived from sedimentary rocks (sandstone) and are classified as Typic Paleudults. The mean annual air temperature is 26.8 °C at all plots. The annual precipitation is 2256 mm year^{-1} (KR1, KR2, and KR3), 2187 mm year^{-1} (BS), and 2427 mm year^{-1} (BB), respectively. At the KR sites, the vegetation is dominated by *Harpullia arborea* and *Bauhinia purpurea* at KR1, *Harpullia arborea* and *Durio* spp. at KR2, and *Harpullia arborea*, *Artocarpus lanceolata*, and *Durio* spp. at KR3. At the BS and BB sites, the vegetation is

Table 7.1 Description of experimental sites

Site	KR1	KR2	KR3	BS	BB	RP
Coordinates	S1°51′, E116°02′	S1°49′, E115°59′	S1°49′, E115°56′	S0°51′, E117°06′	S1°01′, E116°52′	N19°50′, E100°20′
Mean annual air temperature (°C)	27	27	27	27	27	25
Mean annual precipitation (mm)	2256	2256	2256	2187	2427	2084
Elevation (m)	92	204	167	99	80	697
Soil type[a]	Rhodic Eutrudox	Typic Paleudults	Typic Paleudults	Typic Paleudults	Typic Paleudults	Typic Haplustults
Parent material	Serpentine	Mudstone serpentine	Mudstone	Sandstone	Sandstone	Mudstone
Vegetation	*Harpullia arborea Bauhinia purpurea*	*Harpullia arborea Durio* spp.	*Harpullia arborea Artocarpus lanceolata Durio* spp.	*Shorea leavis Dipterocarpus cornutus*	*Shorea leavis Dipterocarpus cornutus*	*Lithocarpus* sp., *Eugenia* sp.

[a]Soils were classified according to Soil Taxonomy (Soil Survey Staff 2006)

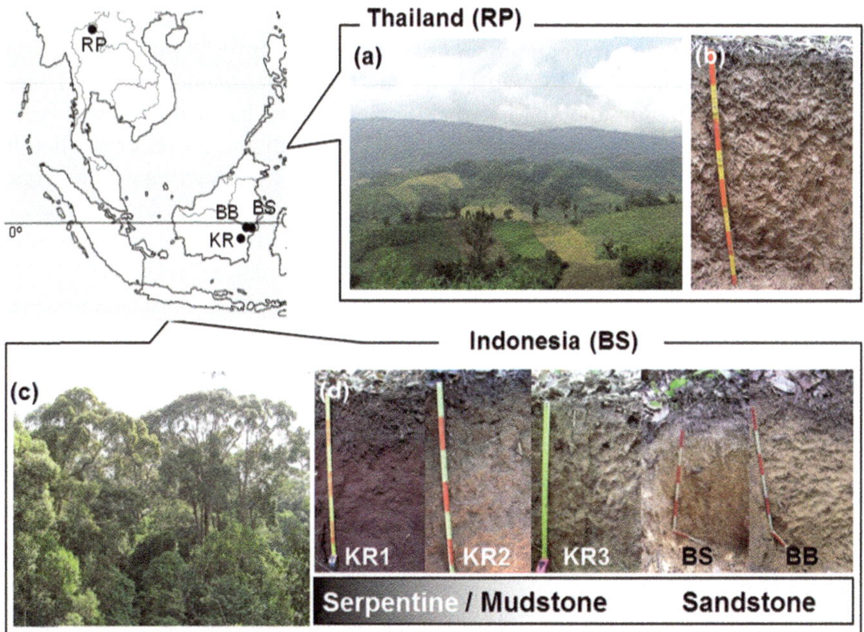

Fig. 7.1 Landscapes and soils of the experimental sites in Thailand (RP) and Indonesia (KR, BS, BB); (**a**) secondary forest/shifting cultivation landscape (RP), (**b**) Ultisol soil, (**c**) tropical lowland forest (BS), and (**d**) Oxisol and Ultisol soils in KR and Ultisol soils in BS and BB

dominated by *Shorea laevis* and *Dipterocarpus cornutus*. The detailed information of sites and soils is given in Fujii et al. (2009a, 2011c).

One site in Thailand was located in Ban Rakpaendin, Chiang Rai Province, Thailand, where the mean annual air temperature and annual precipitation are 25.0 °C and 2084 mm year^{-1}, respectively. There are distinct dry and wet seasons: the dry season is from November to March and the wet season with most of the rainfall is from April to October. Soils are derived from sedimentary rocks (mudstone) and classified as Typic Haplustults (Soil Survey Staff 2006).The vegetation is dominated by *Lithocarpus* sp. and *Eugenia* sp. in RP.

7.2.2 Soil Materials

Soil samples were air-dried and crushed to pass through a 2 mm sieve and then analyzed by the following methods. Soil pH was measured using a soil to solution (H_2O or 1 M KCl) ratio of 1:5 (w/v) after shaking for 1 h. Total carbon contents were determined using a CN analyzer (Vario Max CN, Elementar Analysensystem GmbH). Particle size distribution was determined by the pipette method. The exchangeable basic cation concentrations and CEC were determined using the ammonium acetate (1 M and pH 7.0) method. Ca and Mg in the extracts were measured using atomic absorption spectroscopy (A-A-640-01, Shimadzu), whereas Na and K were measured by flame photometry. Exchangeable Al and H were determined by titration after 1 M KCl extraction and exchangeable Al was determined using aluminon colorimetry. The contents of well and poorly crystalline Fe oxides and organic Fe compounds (Fe_d) were estimated by extraction with a citrate-bicarbonate mixed solution buffered at pH 7.3 with the addition of sodium dithionite (DCB) at 80°C (Mehra and Jackson 1960). Elemental analyses were made in hydrofluoric acid-sulfuric acid digests of soil samples (Jackson 1958). The Fe and Al contents were determined by inductively coupled plasma atomic emission spectrometry (ICP-AES, SPS1500, Seiko Instruments Inc.).

7.2.3 Plant Materials

Circular litter traps of 60 cm diameter were used to collect litterfall. Fine root biomass in the O horizon was estimated by collecting the roots from 30 cm $\times$ 30 cm quadrats. Fine root biomass in the mineral soil was estimated by collecting the roots in 5 cm depth intervals in cores of 0.1 L volume. Five replicates were used for all three of these measurements. Roots were rinsed in distilled water to remove soil materials. Aboveground biomass was estimated by applying the diameters of stems at breast height (DBH) to the regression equations obtained by Yamakura et al. (1986). DBH was measured by tree census. Wood increment was estimated by stem analysis (tree ring analysis) and using the regression equation for estimating tree biomass according to Johnson and Risser (1974). Wood samples were collected using an increment borer. Plant samples were oven-dried at 70°C for 48 h and then weighed and milled. The C and N concentrations in plant materials were determined using a CN analyzer. The Na and K concentrations in plant samples were determined by flame photometry, Mg and Ca concentrations by atomic absorption spectroscopy, Fe and Al contents by ICP-AES, and P concentrations colorimetrically (UV-VIS spectrophotometer UV-1200, Shimadzu) after nitric-sulfuric acid wet digestion. The Cl and S concentrations were determined by high performance liquid chromatography (HPLC, Shimadzu) after combustion according to Busman et al. (1983).

7.2.4 Throughfall and Soil Solution

Soil solutions were collected in five replicates using tension-free lysimeters, each draining a surface area of 200 cm^2, beneath the O (A for RP), A (BA for RP) and B (Bt1 for RP) horizons (0, 5 and 30 cm depths) at all plots. Throughfall was collected using a precipitation collector, also using five replicates. Sample solutions were filtered through 0.45 μm membrane filters before each analysis. The concentrations of H^+ in solution were determined with a glass electrode. The concentrations of Na^+, K^+, NH_4^+, Mg^{2+}, Ca^{2+}, Cl^-, NO_3^-, and SO_4^{2-} in solution were determined by HPLC. The concentrations of total Fe and Al in solution were determined by ICP-AES. The total charge equivalent of Al ions (Al^{n+}) was calculated as the equivalent sum of Al^{3+}, $AlOH^{2+}$, and $Al(OH)_2^+$. The concentrations of DOC, inorganic carbon (IC), and total dissolved N (TDN) were determined using a total organic carbon analyzer (TOC-V_{CSH}, Shimadzu). The concentrations of HCO_3^- in solution were determined from the solution pH and IC concentration, based on pKa = 6.3. The anion deficit, if any, was assigned to the negative charge of organic acids. The concentrations of dissolved organic N (DON) were calculated by subtracting NH_4^+ and NO_3^- from TDN concentrations (DON = TDN $- NH_4^+$ $- NO_3^-$).

7.2.5 *Monitoring Temperature and Volumetric Water Content in Soils*

The soil temperature at 5 cm depth was measured with a thermistor probe (107 Temperature Probe, Campbell Scientific, Inc.) using two replicates, while volumetric water contents in soils at depths of 5, 15, and 30 cm were measured with time domain reflectometer probes (CS615 Water Content Reflectometer, Campbell Scientific, Inc.) in three replicates. Data were recorded using data loggers at 30 min intervals (Campbell Scientific, Inc., CR-10X) during the study.

7.2.6 *Ion Fluxes*

Ion fluxes leached from each horizon were calculated by multiplying the water fluxes by ion concentrations in throughfall and soil solutions. Throughfall water fluxes were measured using a precipitation collector, while the half-hourly fluxes of soil water percolating at depths of 5, 15, and 30 cm were estimated by applying Darcy's law to the unsaturated hydraulic conductivity and the gradient of the hydraulic heads at each depth. The one-dimensional, vertical flow equation (Richards' equation) in the unsaturated soil zone is written as

$$C(h)\frac{\delta h}{\delta t} = \frac{\delta}{\delta z}\left[K(h)\left(\frac{\delta h}{\delta z}+1\right)\right] - S(h) \tag{7.1}$$

where C (m^{-1}) is differential water capacity, t (day) is time, Z (m) is height, h (m) is soil water pressure head, K (m day^{-1}) is unsaturated hydraulic conductivity, and S (day^{-1}) is a sink term accounting for water uptake by vegetation and lateral water flow. The unsaturated hydraulic conductivity and soil water pressure heads at depths of 5, 15, and 30 cm were estimated using the saturated hydraulic conductivity, water retention curves of soil, and volumetric water content monitored at 30 min intervals at each depth (Mualem and Dagan 1978; Van Genuchten 1980). In this process, the Mualem-Van Genuchten parameters describing the unsaturated hydraulic conductivity and soil water pressure heads were adjusted to meet the soil water budget according to Klinge (2001). A detailed description of the calculation of soil water fluxes was given in our previous report (Fujii et al. 2008). The water fluxes for each month were calculated by summing the half-hourly water fluxes thus estimated. Ion fluxes for each month were calculated by multiplying the water fluxes by ionic concentrations in throughfall and soil solution during each month. The annual ion fluxes were each calculated by summation.

7.2.7 Proton Budgets

Net proton generation (NPG) resulting from excess cation uptake by vegetation, nitrification, dissociation of organic acids, dissociation of carbonic acid, and net proton influx from the overlying horizon can be calculated based on the input-output budget of ions in soil vegetation systems and including solute leaching and vegetation uptake (Table 7.2; Van Breemen et al. 1983, 1984). In the present study, proton budgets were quantified for each soil horizon compartment (O, A, and B horizons) using throughfall and soil solution to describe proton generation and consumption in the soil profiles (Fujii et al. 2008). Net proton generation associated with excess cation uptake by vegetation (NPG_{Bio}), which consists of wood increment and litterfall, is calculated as

$$NPG_{Bio} = (Cation)_{bio} - (Anion)_{bio} \tag{7.2}$$

where (Cation) and (Anion) represent the equivalent sum of cations and anions, respectively. The ionic species considered in the present study were Na^+, K^+, Mg^{2+}, Ca^{2+}, Fe^{3+}, and Al^{n+} for cations and Cl^-, $H_2PO_4^-$, and SO_4^{2-} for anions. The suffix "bio" represents ion fluxes caused by plant uptake ($kmol_c\ ha^{-1}\ year^{-1}$).

Net proton generation associated with the transformation of nitrogen (NPG_{Ntr}) is

$$NPG_{Ntr} = \left\{ (NH_4^+)_{in} - (NH_4^+)_{out} \right\} + \left\{ (NO_3^-)_{out} - (NO_3^-)_{in} \right\} \tag{7.3}$$

The suffix "in" represents ion fluxes percolating into the soil compartment in throughfall and the soil solution from the overlying horizon (e.g., throughfall for the O horizon), while the suffix "out" represents ion fluxes leaching out of the soil compartment in the soil solution.

Table 7.2 Representative processes of proton generation and calculation methods

H^+ budget	Representative reaction	Proton budget calculation
Proton-generating processes		
(1) H^+ input (e.g., acid rain)		H^+ input $= (H^+)_{in} - (H^+)_{out}$
(2) Cation excess uptake by plants	$Ca^{2+} + 2R\text{–}OH \rightarrow (R\text{–}O)_2Ca\ (org) + 2H^+$	$NPG_{Bio} = (Cat)_{bio} - (Ani)_{bio}$
(3) N transformation (e.g., nitrification)	$NH_4^+ + H_2O \rightarrow NO_3^- + 2H^+ + H_2O$	$NPG_{Ntr} = (NH_4^+)_{in} - (NH_4^+)_{out} + (NO_3^-)_{out} - (NO_3^-)_{in}$
(4) Dissociation of carbonic acid	$H_2CO_3 \rightarrow HCO_3^- + H^+$	$NPG_{Car} = (HCO_3^-)_{out} - (HCO_3^-)_{in}$
(5) Dissociation of organic acid	$2CH_2O + 3/2O_2 \rightarrow HC_2O_4^- + H^+ + H_2O$	$NPG_{Org} = (Org^{n-})_{out} - (Org^{n-})_{in}$
Proton-consuming processes		
(6) Weathering and cation exchange reaction	$n/2M_{2/n}O_{(s)} + nH^+ \rightarrow M^{n+} + n/2H_2O$	

The suffix "bio" represents ion fluxes ($kmol_c\ ha^{-1}\ year^{-1}$) caused by plant uptake, assuming that vegetation uptake was equal to the sum of wood increment and litterfall

The suffixes "in" and "out" represent ion fluxes entering the soil horizon (e.g., throughfall for the O horizon) and leaving the horizon, respectively

Cat and Ani represent cations (Na^+, K^+, Mg^{2+}, Ca^{2+}, Fe^{3+}, Al^{n+}) and anions (Cl^-, SO_4^{2-}, $H_2PO_4^-$), respectively

Net proton generation associated with the dissociation of organic acids (NPG_{Org}) is

$$NPG_{Org} = (Org^{n-})_{out} - (Org^{n-})_{in} \qquad (7.4)$$

where Org^{n-} represents the negative charge of organic anions.

Net proton generation associated with carbonic acid dissociation (NPG_{Car}) is

$$NPG_{Car} = (HCO_3^-)_{out} - (HCO_3^-)_{in} \qquad (7.5)$$

Net proton influx from the overlying horizon is

$$\{(H^+)_{in} - (H^+)_{out}\} \qquad (7.6)$$

7.2.8 Soil Acidification Rate

Soil acidification is defined as the decrease of acid-neutralizing capacity (ANC) of the solid phase of a soil (Van Breemen et al. 1984). $ANC_{s\ (ref\ pH=3)}$ ($cmol_c\ kg^{-1}$) is conceptually defined as the sum of basic cation equivalence minus the sum of strongly acidic anion equivalence at a reference soil pH of 3. In our study, soil ANC was calculated using the equation proposed by Brahy et al. (2000):

$$_{m}ANC_{s(refpH=3)} = 2(Na_2O) + 2(K_2O) + 2(CaO) + 2(MgO)$$
$$+ 6(Fe_2O_{3tot} - Fe_2O_{3d}) + 6(Al_2O_3) \qquad (7.7)$$

where parentheses denote total molar concentration. Iron-bearing silicates are considered to be ANC components, while ferric oxides are not taken into account, since at pH values above 3 they do not contribute to acid neutralization (Brahy et al. 2000). The soil acidification rate was calculated as the change in ANC components over a given period, i.e., ΔANC ($kmol_c\ ha^{-1}\ year^{-1}$). Using ion budget in each soil horizon, the soil acidification rate (ΔANC) is

$$\Delta ANC = \{(Anion)_{out} - (Anion)_{in}\} - \{(Cation)_{out} - (Cation)_{in}\}$$
$$- \{(Cation)_{bio} - (Anion)_{bio}\} \qquad (7.8)$$

The sum of acid load is compensated by the stoichiometric decrease of ANC. Theoretically, ΔANC, NPG_{Bio}, NPG_{Ntr}, NPG_{Org}, NPG_{Car}, and net proton influx from the overlying horizon have the following relationship:

$$-\Delta ANC = NPG_{Bio} + NPG_{Ntr} + NPG_{Org} + NPG_{Car}$$
$$+ \{(H^+)_{in} - (H^+)_{out}\}. \qquad (7.9)$$

7.3 Results

7.3.1 *Physicochemical Properties of Bedrocks and Soils*

The bedrocks and fresh serpentine collected in the KR1 site contained the high concentrations of Fe and Mg, while bedrocks of mudstone in the KR2, KR3, and RP sites contained the high concentrations of Al, Fe, and Si (Table 7.3). The C horizon sample of BB contains the higher concentrations of Si due to its origin of sandstone (Table 7.3).

In the BS and BB soils from sedimentary rocks (sandstone), all of the soil horizons are strongly acidic (pH 3.8–4.3), consistent with the low contents of basic cations and high Al saturation (Table 7.4). In the KR and RP soils from serpentine and sedimentary rocks (mudstone), the pH values (4.5–6.4) were higher than in the BS and BB soils (Table 7.4). Clay contents of the KR and RP soils were also higher than in the BS and BB soils (23–40 %). For Indonesian five sites, the DCB extractable Fe oxide (Fe_d) contents increased toward the serpentine belt and were highest in the KR1 soils from serpentine. The ANC increased with soil pH and clay content, varying from 309–643 $cmol_c\ kg^{-1}$ in the BS and BB soils to 1163–2012 $cmol_c\ kg^{-1}$ in the KR and RP soils (Table 7.4). The higher ANC values are mainly contributed by Mg and Fe in the KR1 soil and by Al in the KR2, KR3, and RP soils.

Table 7.3 Elemental composition of weathered rocks of the soil studied

Site	Na_2O[a]	K_2O[a]	MgO[a]	CaO[a]	Al_2O_3[a]	Fe_2O_3[a]	SiO_2[b]	P
	(%)							(mg kg^{-1})
Serpentine[c]	1.2	0.3	39.3	0.5	9.5	7.1	42.1	8
KR1	0.1	0.2	37.3	0.2	15.7	11.7	34.9	17
KR2	0.2	0.3	8.8	0.1	11.8	8.5	70.3	700
KR3	0.1	0.9	0.1	0.1	9.3	6.4	83.2	172
BB[d]	0.1	1.3	0.2	0.0	4.9	3.4	90.1	98
RP	0.3	4.4	0.0	0.0	21.8	8.8	64.6	281

[a]Oven-dried basis. Elemental analyses were made in hydrofluoric-sulfuric acid digestion
[b]$(SiO_2\%) = 100\% - \{(Na_2O\%) + (K_2O\%) + (MgO\%) + (CaO\%) + (Fe_2O_3\%) + (Al_2O_3\%)\}$
[c]Fresh serpentine rock sample collected from the bedrock outcrop close to KR1
[d]Rock sample could not be available. The data of the C horizon were presented for BB

The O horizons had only an Oi layer in the KR1, KR2, and RP soils, while they consisted of an Oea layer, as well as a Oi layer, in the acidic BS, BB, and KR3 soils (pH < 4.5) (Table 7.5). In the BB and BS soils, the O horizons are acidic (pH 4.5 to 5.0), consistent with lower concentrations of basic cations (17–63 cmol$_c$ kg^{-1}). The higher concentrations of basic cations in the O horizons of the KR and RP soils (87–129 cmol$_c$ kg^{-1}) are probably a consequence of parent materials rich in basic cations (Tables 7.3 and 7.5).

7.3.2 C Stock and Flow in Soils and Ecosystems

In five sites in Indonesia, the aboveground biomass (259.8 to 346.0 Mg C ha^{-1}) was greater than in RP, with the considerably lower exception of KR1 (134.2 Mg C ha^{-1}; Fig. 7.2). The fine root biomass was mainly distributed in the mineral soil horizons of all plots; however, it was also distributed in the O horizons of the KR3, BS, and BB soils (Fig. 7.3; 0.3–2.3 Mg C ha^{-1}). The C stock in the organic and mineral soil horizons is 2.9–4.1 and 26.6–74.7 Mg C ha^{-1}, respectively. The higher aboveground biomass and low C stock in the organic and mineral soil horizons in our study, as compared to temperate forests, are consistent with a previous report (Nakane 1980).

The rates of litterfall and wood increment in our study (3.6–4.8 and 5.1–11.1 Mg C ha^{-1}, respectively) are also higher than in temperate forests, owing to the higher primary production in tropical regions (Table 7.6).

7.3.3 Soil Solution Composition and Ion Fluxes in Throughfall and Soil Solution

Soil solution composition and ion fluxes are presented in Tables 7.7 and 7.8 (or Fig. 7.4), respectively. Soil solutions were strongly acidic in the BS and BB

Table 7.4 Physicochemical properties of the forest soils in Indonesia and Thailand

| Site | Horizon | Depth (cm) | pH | | Total | Total | CN ratio | | Exchangeable cation[a] | | | Particle size distribution[ac] | | | | Total content[a] | | | | | | | |
| | | | H_2O | KCl | C^a | N^a | | CEC^a | $Base^b$ | Al | H | Sand | Silt | Clay | $Fe_d{}^a$ | Na_2O | K_2O | MgO | CaO | Al_2O_3 | Fe_2O_3 | $_mANC_{(s)}{}^{ad}$ | $Base^{ae}$ |
					$(g\ kg^{-1})$				$(cmol_c\ kg^{-1})$			(%)			$(g\ kg^{-1})$	(%)						$(cmol_c\ kg^{-1})$	
KR1	A1	0–5	6.3	5.8	72.7	6.4	11	24.8	19.1	0.0	1.1	6	39	55	176	0.1	0.1	1.3	0.2	4.7	51.8	1358	80
	A2	5–20	6.2	5.7	23.8	2.6	9	11.4	5.0	0.0	1.1	14	45	41	196	0.2	0.1	1.2	0.0	5.8	60.0	1608	71
	B1	20–35	6.4	6.1	7.5	0.8	9	7.2	2.0	0.0	0.9	19	37	44	208	0.0	0.1	1.1	0.0	6.1	67.2	1817	57
	B2	35–50	6.4	5.8	5.7	0.7	8	7.0	1.8	0.0	0.7	16	37	47	217	0.1	0.1	1.2	0.0	6.3	68.7	1853	68
	B3	50–67+	6.4	6.2	4.8	0.6	8	7.7	2.7	0.0	0.8	38	33	28	216	0.1	0.1	0.9	0.0	6.2	65.7	1725	53
KR2	A	0–5	5.6	5.2	73.1	5.5	13	19.8	14.5	0.2	1.2	6	15	79	67	0.0	0.3	0.5	0.2	28.2	13.2	1837	39
	BA	5–20	4.6	3.9	24.9	2.9	9	13.7	1.4	2.9	2.4	5	18	77	71	0.2	0.3	0.3	0.0	31.3	14.0	2012	30
	Bt1	20–45	4.8	4.1	14.2	1.8	8	17.0	1.0	2.3	1.4	4	10	86	71	0.0	0.3	0.3	0.0	31.6	13.7	2011	23
	Bt2	45–70+	5.1	4.2	10.0	1.4	7	24.8	0.8	1.4	1.5	7	11	82	78	0.1	0.3	0.3	0.0	32.2	13.2	1996	25
KR3	A	0–4	4.5	3.7	38.7	3.5	11	20.8	1.8	4.9	2.8	28	24	48	30	0.1	0.2	0.4	0.2	17.6	6.8	1163	33
	BA	4–15	4.5	3.7	17.9	1.7	10	17.2	1.0	4.9	1.9	29	15	56	32	0.1	0.2	0.4	0.0	19.2	6.9	1245	30
	Bt1	15–30	4.5	3.7	11.1	1.2	9	19.7	1.2	6.3	1.5	28	15	57	34	0.0	0.2	0.4	0.0	20.0	7.2	1293	27
	Bt2	30–45	4.6	3.8	7.4	0.9	9	23.3	1.4	7.0	1.0	25	15	60	36	0.1	0.3	0.4	0.0	22.2	7.7	1430	30
	Bt3	45–50+	4.6	3.8	7.5	1.0	8	27.1	0.9	7.7	0.5	25	9	66	38	0.1	0.2	0.4	0.0	22.6	8.0	1459	30
BS	A	0–5	4.0	3.9	22.9	1.6	14	8.5	2.2	3.0	5.7	52	25	23	7	0.1	0.4	0.2	0.1	4.1	2.1	309	26
	BA	5–25	3.8	3.8	4.2	0.5	9	6.2	0.8	3.9	6.1	49	27	24	9	0.3	0.5	0.2	0.1	5.7	2.4	413	33
	B1	25–40	4.0	3.8	3.5	0.5	7	5.0	0.8	4.8	6.5	43	30	27	11	0.3	0.5	0.3	0.1	7.3	2.7	502	36
	Bt	40–50+	4.3	3.8	2.5	0.4	6	5.0	1.0	7.0	7.7	34	35	31	15	0.2	0.7	0.3	0.0	9.4	3.6	643	35

BB	A	0–5	4.2	3.4	36.1	2.3	16	13.0	1.2	6.7	8.5	42	31	27	9	0.2	0.6	0.2	0.1	4.9	3.4	400	32
	BA	5–20	4.1	3.7	10.7	0.9	13	8.3	0.8	5.5	7.2	45	31	24	11	0.2	0.4	0.2	0.1	5.8	1.5	369	30
	B1	20–37	4.1	3.8	7.1	0.6	11	8.5	1.1	5.7	8.2	43	33	24	11	0.2	0.5	0.3	0.1	6.7	2.1	440	32
	Bt	37–70	4.1	3.8	3.7	0.4	8	9.0	0.8	4.6	7.3	39	34	27	15	0.3	0.6	0.3	0.1	8.1	2.7	531	37
	BC	70+	4.2	3.8	3.3	0.4	8	13.6	0.9	6.7	7.5	34	25	40	18	0.1	0.6	0.3	0.0	9.4	3.9	630	28
RP	A	0–7	5.0	4.1	62.6	3.8	17	27.6	5.8	0.8	1.5	5	25	70	33	0.2	3.4	1.1	0.1	18.7	6.9	1324	139
	BA	7–20	4.9	3.9	19.8	1.5	13	19.9	2.0	1.5	1.6	4	23	73	36	0.3	3.8	1.3	0.0	23.1	8.1	1627	157
	Bt	20–45+	4.6	4.0	8.9	1.0	9	20.1	1.4	1.5	1.6	6	19	75	43	0.4	3.9	1.2	0.0	24.7	9.0	1716	158

[a]Oven-dried basis
[b]$Na^+ + K^+ + Mg^{2+} + Ca^{2+}$
[c]Clay ($<$0.002 mm), silt (0.002–0.05 mm), sand (0.05–2 mm)
[d]$2(Na_2O) + 2(K_2O) + 2(MgO) + 2(CaO) + 6(Fe_2O_{3tot} - Fe_2O_{3d}) + 6(Al_2O_3)$
[e]$2(Na_2O) + 2(K_2O) + 2(MgO) + 2(CaO)$

Table 7.5 Physicochemical properties of the O horizons in Indonesia and Thailand

Site	Horizon	Mass (Mg ha^{-1})	pH[a]	C (%)	N	C/N	Lignin (%)	Na[b] (g kg^{-1})	K[b]	Mg[b]	Ca[b]	Basic cation (cmol$_c$ kg^{-1})
KR1	Oi	8.2	6.5	43.5	1.1	38	35.0	0.5	5.6	2.3	7.2	87
KR2	Oi	7.8	7.4	47.1	1.5	32	28.0	0.4	4.3	4.7	11.3	129
KR3	Oi	4.0	5.9	44.0	1.1	41	42.0	0.5	5.2	3.7	9.0	108
	Oea	5.0	5.3	34.9	1.4	26	42.0	1.2	2.7	4.4	6.6	88
BS	Oi	6.0	5.0	49.3	1.2	41	44.0	0.4	5.5	2.3	4.4	63
	Oea	5.3	4.5	28.9	1.2	24	43.0	0.9	2.0	0.9	1.5	25
BB	Oi	4.5	5.0	53.2	1.4	39	46.0	0.7	3.2	2.4	1.5	35
	Oea	4.0	4.6	27.5	1.1	25	36.0	0.5	2.1	0.8	0.7	17
RP	Oi	6.0	5.9	43.1	0.7	61	24.0	0.7	5.7	4.6	10.0	123

[a]The pH was measured using the milled litter to solution (water) ratio of 1:20 for 1 h
[b]The cation concentrations were determined after nitric-sulfuric acid wet digestion

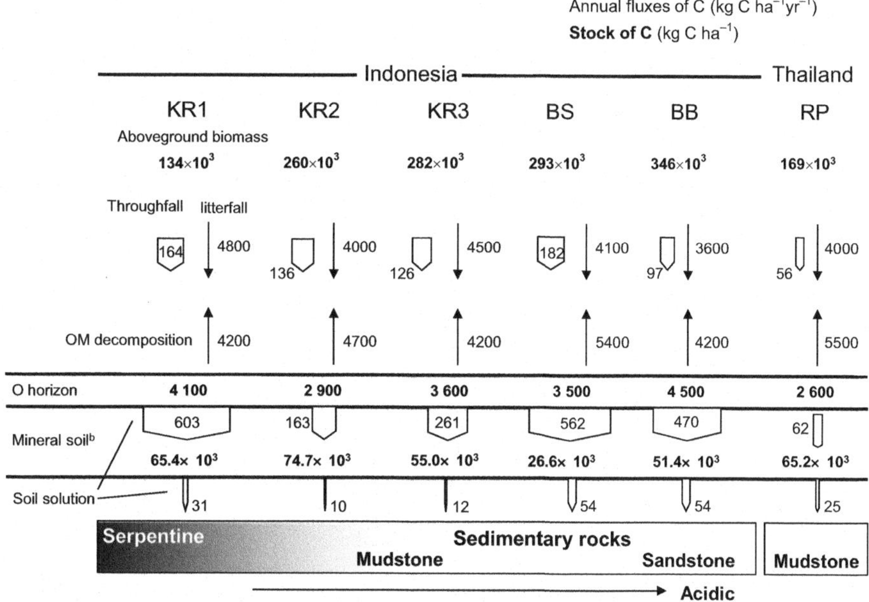

Fig. 7.2 Stocks and fluxes of C in tropical forest ecosystems. *Blank arrows* indicate DOC fluxes within ecosystems

Fig. 7.3 Depth distribution of fine root biomass in the tropical forest soils. *Bars* indicate standard errors ($n = 5$)

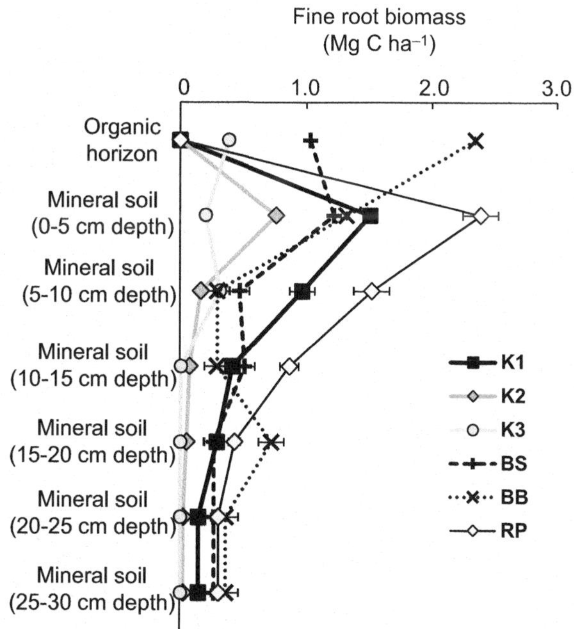

Table 7.6 Uptake of cations and anions by plants

Site		OM[a] production (Mg C ha^{-1} year^{-1})	Na (kg ha^{-1} year^{-1})	K	Ca	Mg	Fe	Al	Cl	S	P	Σcation (kmol$_c$ ha^{-1} year^{-1})	Σanion	NPG$_{Bio}$
Indonesia														
KR1	Wood increment	7.2	3	53	59	9	2	1	4	4	6	5.3	0.5	4.8
	Litterfall	5.7	5	64	82	26	0	2	1	8	4	8.4	0.6	7.7
KR2	Wood increment	9.8	3	55	133	10	0	1	2	5	5	9.1	0.6	8.6
	Litterfall	3.6	3	31	81	34	1	19	1	9	5	9.9	0.7	9.1
KR3	Wood increment	5.1	3	23	42	3	1	2	1	3	2	3.3	0.3	3.0
	Litterfall	4.5	5	27	57	33	2	2	1	8	4	6.8	0.7	6.1
BS	Wood increment	10.6	15	33	29	16	10	1	1	9	2	4.7	0.6	4.0
	Litterfall	4.1	4	45	36	19	2	3	1	5	3	5.0	0.4	4.6
BB	Wood increment	11.1	15	35	31	17	2	5	1	9	2	5.1	0.7	4.5
	Litterfall	3.6	5	23	15	17	2	4	1	5	3	3.5	0.5	3.1
Thailand														
RP	Wood increment	5.8	9	21	9	3	4	4	1	0	1	2.1	0.1	2.1
	Litterfall	3.4	3	23	41	19	1	1	1	2	2	3.9	0.4	3.5

[a]OM represents organic matter

soils (pH 3.9–4.4), while they were moderately acidic at the KR and RP soils (pH 5.0–6.2) (Table 7.7). The concentrations and fluxes of DOC in the O horizon were higher in the BS, BB, and KR1 soils compared to the KR2 and KR3 soils (Table 7.7; Fig. 7.2). The DOC concentrations were low throughout the soil profile of RP. Organic acids were the dominant anions in the O horizon solution in all plots except for the KR2 and RP soils (Table 7.7). The concentrations of organic acids and DOC in the soil solution decreased with depth (Table 7.7). From linear regression analysis between the concentrations of DOC and organic acids in soil solutions, the negative charge per 1 mole of DOC (0.09–0.17 mol_c) corresponds to one dissociated acidic functional group for 5.9–11.5 C atoms. The high ratios of DOC to charge in the soil solution suggest the presence of high molecular weight fulvic acids, which contain 7 C atoms for each acidic functional group (Thurman 1985).

Bicarbonate is present in moderately acidic soil solutions of the KR and RP sites, while it was negligible in acidic soil solutions of the BS and BB sites (Table 7.6). In the O horizon of the KR2 soil, bicarbonate was the dominant anion owing to the relatively high solution pH, while concentrations decreased with depth (Table 7.7).

Nitrogen in soil solution was dominated by nitrate in KR1, KR2, and RP soils, while DON and NH_4^+ also accounted for considerable fraction of TDN (Fig. 7.5). Nitrate concentrations were low (0.02–0.16 $mmol_c\ L^{-1}$) at all plots except for the A1 and A2 horizons of the KR1 soil (Table 7.7; 0.31–0.47 $mmol_c\ L^{-1}$), where the understory vegetation was the nitrogen-fixing *Bauhinia purpurea*. The major accompanying cations were K^+, Mg^{2+}, and Ca^{2+} in the KR1, KR2, and KR3 soils, while they were H^+, NH_4^+, and Al^{n+}, as well as basic cations, in the BS and BB soils (Table 7.7). The Si concentrations in the soil solution were highest in KR1 (0.14–0.63 $mmol\ L^{-1}$) compared to the other soils from sedimentary rocks (Table 7.7; 0.02–0.07 $mmol\ L^{-1}$).

7.3.4 Proton Budgets in Soils

Cation concentrations exceeded anion concentrations in litter and wood materials at all plots (Table 7.6). Excess cation charge was compensated for by the net proton load to the soil as NPG_{Bio}. NPG_{Bio} in each of the soil horizons was calculated by distributing it based on the distribution of the fine root biomass in the soil profiles (Fig. 7.3), according to Shibata et al. (1998). Based on the fluxes of solutes entering and leaving the soil horizon compartment (Table 7.8; Fig. 7.4) and plant uptake (Table 7.6; NPG_{Bio}) in each of the soil horizons, net proton generation and soil acidification rates were calculated based on proton budget theory. The example in the BS soils was shown in Fig. 7.6.

In the entire soil profiles, NPG_{Bio} was highest among the proton sources at all plots (Fig. 7.7). NPG_{Bio} was present mainly in the A and B horizons of soils at all plots (1.7–10.8 $kmol_c\ ha^{-1}\ year^{-1}$), while it was also present in the O horizons of the KR3, BS, and BB soils (Fig. 7.7; 1.5–3.1 $kmol_c\ ha^{-1}\ year^{-1}$). In the O horizons, proton

Table 7.7 Annual volume-weighed mean concentrations of ions in throughfall and soil solution

Site	Horizon	pH	DOC	IC	HCO_3^-	Cl^-	NO_3^-	SO_4^{2-}	Org^{n-b}	H^+	Na^+	NH_4^+	K^+	Mg^{2+}	Ca^{2+}	Fe^{2+}	Al^{n+c}	Si
			(mg C L^{-1})		(mmol$_c$ L^{-1})													(mmol L^{-1})
Indonesia																		
KR1	TF[a]	6.3	7.3	1.1	0.05	0.05	0.02	0.02	0.14	0.00	0.02	0.02	0.07	0.08	0.08	0.00	0.00	0.02
	O	6.1	31.6	2.0	0.08	0.17	0.10	0.06	0.27	0.00	0.04	0.03	0.20	0.21	0.20	0.00	0.00	0.14
	A1	5.7	13.4	0.8	0.03	0.09	0.47	0.08	0.16	0.00	0.03	0.04	0.15	0.38	0.21	0.00	0.00	0.27
	A2	6.2	5.4	4.3	0.16	0.10	0.24	0.07	0.15	0.00	0.04	0.04	0.07	0.38	0.23	0.00	0.00	0.63
KR2	TF[a]	6.2	6.2	1.3	0.05	0.02	0.03	0.05	0.07	0.00	0.02	0.01	0.05	0.06	0.08	0.00	0.00	0.02
	O	6.5	8.6	3.4	0.18	0.04	0.09	0.08	0.12	0.00	0.02	0.02	0.09	0.15	0.21	0.00	0.01	0.02
	A	5.9	5.7	0.7	0.03	0.02	0.07	0.02	0.05	0.00	0.01	0.01	0.02	0.06	0.07	0.00	0.01	0.05
	BA, Bt	5.1	1.7	0.2	0.00	0.01	0.06	0.03	0.03	0.01	0.02	0.01	0.01	0.03	0.03	0.00	0.01	0.07
KR3	TF[a]	6.1	5.7	1.1	0.04	0.03	0.01	0.03	0.07	0.00	0.02	0.01	0.03	0.05	0.07	0.00	0.00	0.02
	O	5.6	16.4	1.3	0.04	0.08	0.03	0.06	0.22	0.00	0.02	0.02	0.08	0.15	0.13	0.00	0.02	0.08
	A	5.0	10.1	0.4	0.01	0.06	0.06	0.04	0.18	0.01	0.02	0.03	0.06	0.12	0.08	0.00	0.02	0.05
	BA, Bt1	5.1	2.8	0.1	0.00	0.03	0.02	0.01	0.04	0.01	0.01	0.01	0.02	0.03	0.01	0.00	0.01	0.03

BS	TF[a]	5.2	9.0	0.4	0.01	0.04	0.04	0.06	0.10	0.01	0.03	0.04	0.08	0.05	0.04	0.00	0.01	0.02
	O	4.4	34.7	0.2	0.00	0.06	0.04	0.07	0.26	0.04	0.04	0.09	0.11	0.07	0.04	0.02	0.04	0.07
	A	4.2	17.2	0.1	0.00	0.05	0.04	0.05	0.20	0.06	0.04	0.04	0.07	0.05	0.03	0.01	0.04	0.07
	BA, B1	4.4	9.9	0.1	0.00	0.06	0.03	0.06	0.12	0.04	0.04	0.02	0.06	0.04	0.04	0.00	0.03	0.05
BB	TF[a]	5.5	4.7	0.2	0.00	0.07	0.02	0.03	0.09	0.00	0.04	0.03	0.07	0.03	0.03	0.00	0.00	0.01
	O	4.1	24.6	0.1	0.00	0.12	0.14	0.05	0.26	0.12	0.04	0.11	0.12	0.08	0.04	0.02	0.05	0.07
	A	4.0	19.1	0.1	0.00	0.10	0.16	0.06	0.21	0.13	0.04	0.06	0.11	0.08	0.04	0.01	0.07	0.07
	BA, B1	4.1	6.0	0.1	0.00	0.11	0.10	0.06	0.13	0.09	0.05	0.03	0.09	0.05	0.03	0.01	0.06	0.07
Thailand																		
RP	TF[a]	6.1	2.7	0.5	0.02	0.01	0.01	0.01	0.02	0.00	0.01	0.01	0.00	0.01	0.02	0.00	0.00	0.00
	A	6.2	3.9	0.7	0.04	0.04	0.14	0.05	0.13	0.00	0.03	0.02	0.05	0.08	0.14	0.01	0.05	0.06
	BA	6.1	3.1	0.7	0.05	0.04	0.14	0.05	0.09	0.00	0.04	0.02	0.04	0.06	0.15	0.01	0.03	0.05
	Bt	6.1	3.1	1.1	0.04	0.04	0.06	0.05	0.04	0.01	0.04	0.02	0.03	0.03	0.08	0.01	0.02	0.06

[a]TF represents throughfall

[b]Org^{n-} represents anion deficit, the negative charge of organic acids

[c]The total charge equivalent of Al ions was calculated as the equivalent sum of Al^{3+}, $AlOH^{2+}$, and $Al(OH)_2^{+}$

Table 7.8 Annual fluxes of solutes in throughfall and soil solution of Thailand and Indonesia

Site	Horizon	H_2O (mm year^{-1})	DOC (kg C ha^{-1} year^{-1})	IC	DON (kg N ha^{-1} year^{-1})	DIN	DON/TDN	DOC/DON	HCO_3^-	Cl^-	NO_3^-	SO_4^{2-}	Org^{n-b}	H^+ kmol$_c$ ha^{-1} year^{-1}	Na^+	NH_4^+	K^+	Mg^{2+}	Ca^{2+}	Fe^{2+}	Al^{n+c}	Σcation	Σanion	Si (kmol ha^{-1} year^{-1})
Indonesia																								
KR1	TF[a]	2240	164	24	15	12	56	11	1.05	1.12	0.45	0.38	3.19	0.01	0.51	0.38	1.56	1.83	1.86	0.01	0.04	6.19	6.19	0.34
	O	1907	603	38	21	34	39	29	1.56	3.32	1.86	1.07	5.19	0.01	0.82	0.53	3.74	3.93	3.77	0.08	0.06	14.22	13.00	2.67
	A1	1071	144	8	5	76	6	28	0.32	0.95	4.99	0.86	1.77	0.02	0.34	0.45	1.63	4.02	2.26	0.01	0.04	8.77	8.89	2.95
	A2	566	31	24	1	22	5	26	0.91	0.55	1.34	0.38	0.87	0.00	0.24	0.23	0.41	2.15	1.29	0.00	0.00	4.59	4.05	5.12
KR2	TF[a]	2211	136	28	4	12	27	31	1.06	0.52	0.59	1.00	1.55	0.01	0.42	0.27	1.05	1.25	1.66	0.01	0.06	4.73	4.72	0.40
	O	1922	166	66	4	29	13	39	3.52	0.71	1.74	1.46	2.31	0.01	0.37	0.32	1.75	2.94	3.98	0.01	0.11	9.49	9.74	0.41
	A	967	55	7	2	11	14	30	0.25	0.15	0.69	0.24	0.52	0.01	0.14	0.08	0.16	0.61	0.70	0.03	0.08	1.82	1.85	0.44
	BA, Bt	553	10	1	1	6	11	14	0.01	0.06	0.34	0.16	0.16	0.04	0.10	0.06	0.08	0.18	0.18	0.01	0.04	0.69	0.73	0.36
KR3	TF[a]	2205	126	24	6	8	41	23	0.80	0.67	0.28	0.66	1.65	0.02	0.37	0.28	0.68	1.03	1.62	0.00	0.04	4.05	4.05	0.34
	O	1594	261	20	7	11	39	36	0.62	1.27	0.55	0.88	3.45	0.04	0.32	0.27	1.34	2.36	2.11	0.07	0.25	6.77	6.78	1.20
	A	645	65	3	3	8	24	25	0.06	0.38	0.39	0.26	1.16	0.07	0.14	0.21	0.40	0.75	0.54	0.02	0.10	2.23	2.26	0.35
	BA, Bt1	416	12	0	1	2	32	16	0.00	0.11	0.08	0.06	0.15	0.03	0.04	0.03	0.07	0.11	0.06	0.01	0.04	0.39	0.39	0.11

BS	TF[a]	2031	182	9	15	21	42	12	0.18	0.81	0.72	1.30	2.06	0.12	0.67	0.77	1.57	0.94	0.81	0.06	0.14	5.07	5.08	0.34
	O	1619	562	3	19	29	40	30	0.02	1.02	0.61	1.18	4.16	0.59	0.66	1.43	1.72	1.14	0.62	0.30	0.64	7.11	6.99	1.03
	A	1196	206	2	9	13	40	23	0.00	0.54	0.52	0.65	2.34	0.73	0.47	0.42	0.87	0.66	0.35	0.11	0.46	4.07	4.06	0.77
	BA, B1	545	54	1	4	4	49	15	0.02	0.33	0.17	0.30	0.65	0.22	0.21	0.10	0.32	0.23	0.20	0.03	0.15	1.46	1.47	0.32
BB	TF[a]	2068	97	4	5	15	23	21	0.07	1.43	0.49	0.60	1.77	0.10	0.80	0.62	1.39	0.68	0.66	0.03	0.08	4.36	4.37	0.17
	O	1914	470	2	11	65	15	42	0.00	2.27	2.63	1.05	4.99	2.25	0.83	2.03	2.22	1.51	0.84	0.30	1.02	10.89	10.95	1.33
	A	1639	313	1	8	50	14	39	0.01	1.66	2.64	0.98	3.46	2.14	0.61	0.93	1.88	1.26	0.66	0.21	1.09	8.75	8.75	1.23
	BA, B1	893	54	1	2	17	10	28	0.00	1.00	0.92	0.52	1.18	0.78	0.41	0.27	0.77	0.48	0.30	0.08	0.55	3.62	3.62	0.64

Thailand

RP	TF[a]	2083	56	11	3	9	29	17	0.60	0.94	0.36	0.26	0.31	0.02	0.15	0.25	0.81	0.59	0.62	0.00	0.03	2.47	2.47	0.05
	A	1602	62	11	3	6	32	21	0.54	0.90	0.33	0.26	0.31	0.01	0.30	0.11	0.74	0.49	0.45	0.11	0.14	2.34	2.35	0.97
	BA	1162	36	8	2	5	29	17	0.34	0.55	0.29	0.14	0.14	0.01	0.19	0.08	0.43	0.34	0.24	0.07	0.09	1.45	1.47	0.62
	Bt	825	25	9	2	3	43	11	0.24	0.46	0.15	0.11	0.10	0.01	0.20	0.06	0.31	0.21	0.16	0.06	0.03	1.05	1.06	0.50

[a]TF represents throughfall. DIN represents sum of NH_4^+ and NO_3^-

[b]Org^{n-} represents anion deficit, the negative charge of organic acids

[c]The total charge equivalent of Al ions was calculated as the equivalent sum of Al^{3+}, $AlOH^{2+}$, and $Al(OH)_2^+$

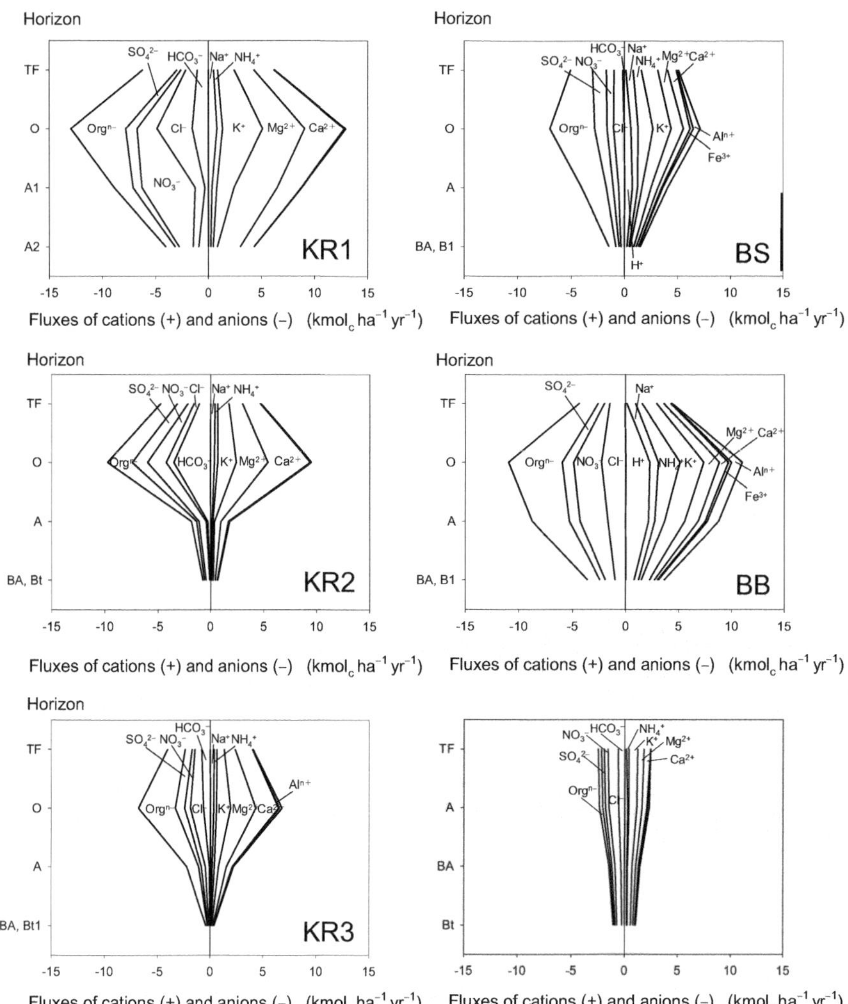

Fig. 7.4 Annual fluxes of cations and anions at each horizon. TF represents throughfall. O, A, A1, A2, BA, B1, Bt1, and Bt represent soil horizons

sources include NPG_{Org}, NPG_{Car}, and NPG_{Ntr}, as well as NPG_{Bio} (Fig. 7.7). NPG_{Org} is the largest proton source in the O horizons (NPG_{Org}, 1.8–3.2 $kmol_c$ ha^{-1} $year^{-1}$) except for the KR2 soil, where NPG_{Car} is higher than NPG_{Org} (0.8 $kmol_c$ ha^{-1} $year^{-1}$). In the moderately acidic O horizons of the KR1 and KR2 soils, protons were produced by the dissociation of carbonic acid (Fig. 7.7; NPG_{Car}, 0.5–2.5 $kmol_c$ ha^{-1} $year^{-1}$). In the A and B horizons, protons were consumed by mineralization and adsorption of organic acids and by protonation of carbonic acid (Fig. 7.7). In the RP soil, NPG_{Bio} was the dominant proton source, while NPG_{Org}, NPG_{Car}, and NPG_{Ntr} were small throughout the soil profile.

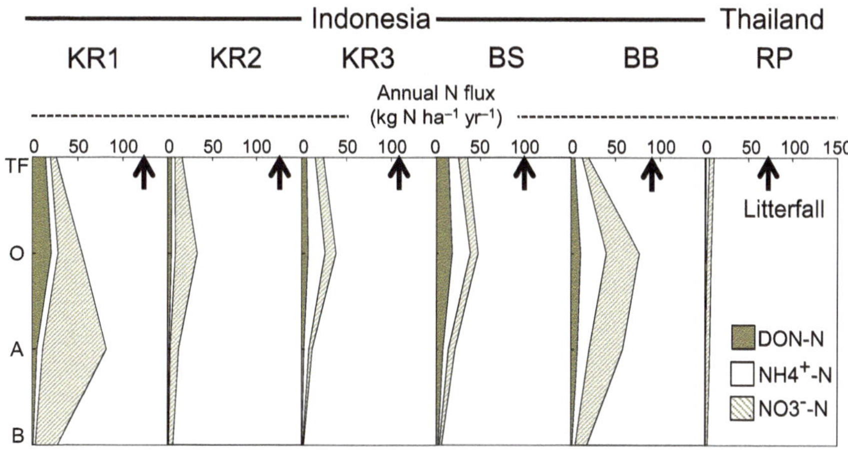

Fig. 7.5 Fluxes of N in litterfall, throughfall, and soil solution of tropical forest ecosystems. *Blank arrows* indicate litterfall N flux within ecosystems. TF, O, A, and B represent throughfall and O, A, and B horizons

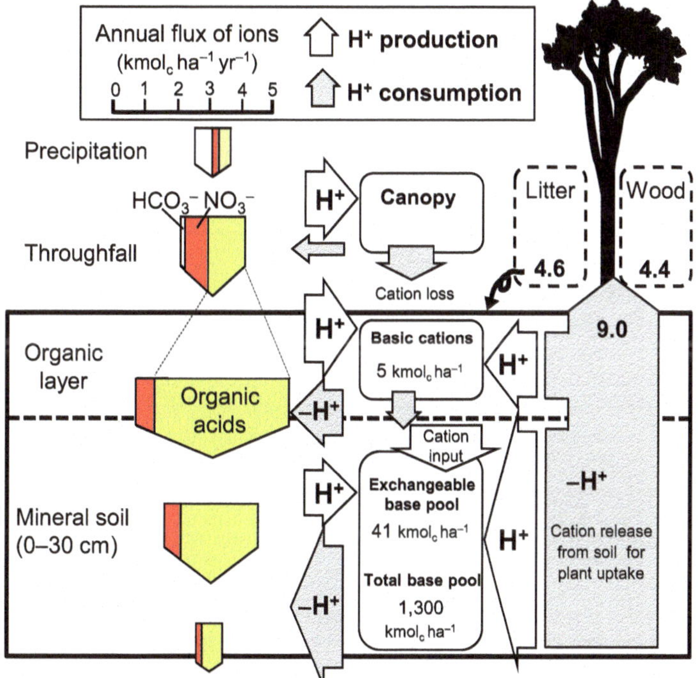

Fig. 7.6 Generation and consumption of protons in soil of the BS (Bukit Soeharto) in East Kalimantan, Indonesia. The *white arrows* indicate proton generation, whereas the *shaded arrows* indicate proton consumption (Data sources are Fujii et al. (2009a, 2010a))

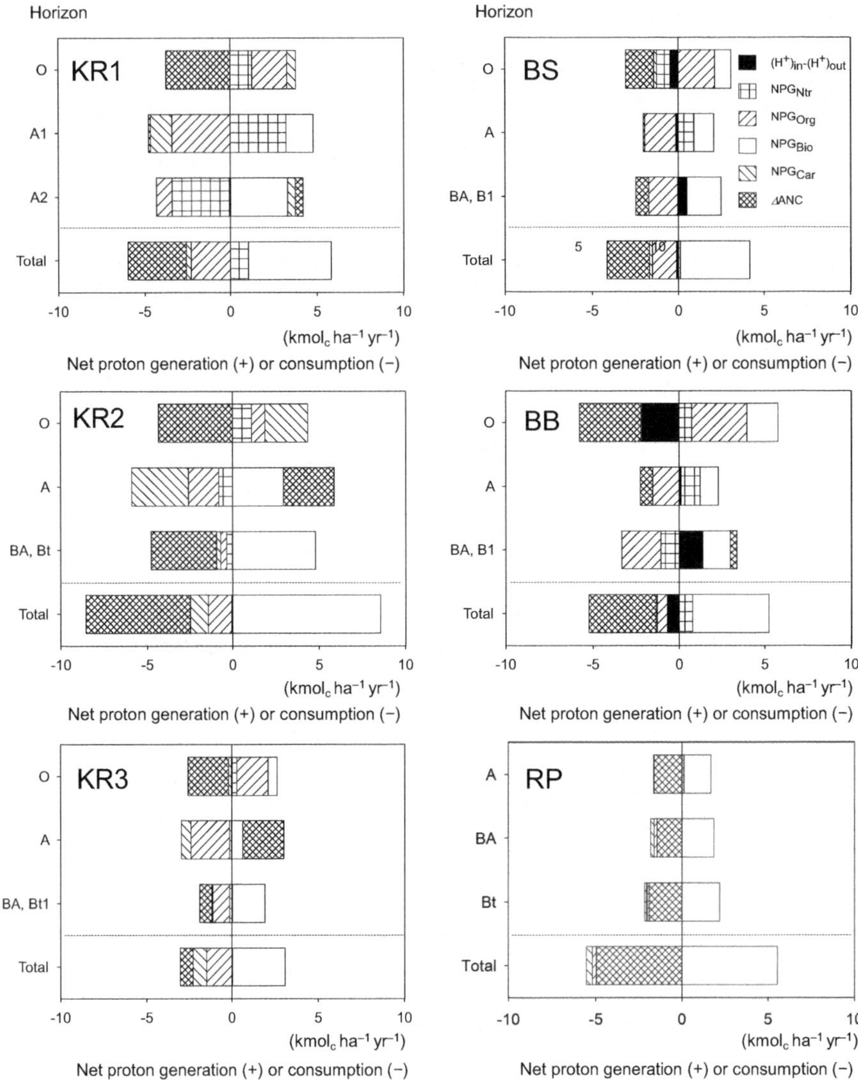

Fig. 7.7 Net proton generation and consumption in the soil profiles. TF represents throughfall. O, A, A1, A2, BA, B1, Bt1, and Bt represent soil horizons

Protons were produced by nitrification in the O horizons of the KR1, KR2, KR3, and BB soils (NPG$_{Ntr}$, 0.3–1.3 kmol$_c$ ha^{-1} year^{-1}), while they were consumed by nitrate uptake by vegetation or microorganisms in the A and B horizons. The low NPG$_{Ntr}$ in the RP soil indicated that proton generation due to nitrification appeared to be immediately compensated by nitrate uptake by plants in the same horizon. In the acidic O horizons of the BS and BB soils, an increase in the NH$_4^+$ flux (Fig. 7.2;

0.6 $kmol_c$ ha^{-1} $year^{-1}$) indicated that protons were mainly consumed by mineralization of organic N to NH_4^+ (Fig. 7.7; NPG_{Ntr}, -0.7 $kmol_c$ ha^{-1} $year^{-1}$). In the A horizons of the BS and BB soils, protons were released owing to the excess uptake of NH_4^+ over NO_3^- by biomass or adsorption of NH_4^+ on clays (Fig. 7.7; NPG_{Ntr}, 0.9–1.1 $kmol_c$ ha^{-1} $year^{-1}$). Exceptionally, in the A1 horizon of the KR1 soil, where understory vegetation was the nitrogen-fixing *Bauhinia purpurea*, protons were produced by nitrification (Table 7.4; NPG_{Ntr}, 3.2 $kmol_c$ ha^{-1} $year^{-1}$).

In the O horizons of the KR soils, acid load contributed mainly by NPG_{Ntr}, NPG_{Org}, and NPG_{Car} (3.5–4.3 $kmol_c$ ha^{-1} $year^{-1}$) was completely neutralized by basic cations (Fig. 7.8). On the other hand, in the O horizons of BS and BB, the intensive acid load contributed mainly by NPG_{Org} and NPG_{Bio} (3.2–7.0 $kmol_c$ ha^{-1} $year^{-1}$) was largely neutralized by basic cations, but a portion of protons were transported downward. The protons transported from the O horizon ($(H^+)_{in}$-$(H^+)_{out}$, 0.5–1.4 $kmol_c$ ha^{-1} $year^{-1}$) are neutralized in the B horizons of the BS and BB soils or are leached further downward (Figs. 7.7 and 7.8).

Using the bulk density and exchangeable acidity, the amounts of acidity accumulated in the soils in the form of exchangeable acidity were estimated to be 23–570 $kmol_c$ ha^{-1} $year^{-1}$. When we compare the sizes of exchangeable acidity in the soils and cumulative NPG needed for cation uptake for production of standing wood biomass (Fig. 7.9), the exchangeable acidity is greater than NPG attributable to wood production in the soils except for the KR1 and KR2 soils.

7.4 Discussion

7.4.1 Dominant Soil Acidification Processes in Tropical Forests

Proton budget theory allows us to quantify contribution of proton sources including plant uptake to soil acidification. In the tropical forests studied, NPG_{Bio} is a dominant proton source in the whole soil profiles (Fig. 7.7). The soil acidification rates of tropical forests are higher than those of temperate forests [0.1–4.6 $kmol_c$ ha^{-1} $year^{-1}$ from Bredemeier et al. (1990); Binkley (1992); Shibata et al. (2001), and Fujii et al. (2008)]. The higher acid load in tropical regions is considered to be caused by higher biomass production and resulting higher demand of basic cations (Table 7.6). Since NPG_{Bio} attributable to litter production would be neutralized by cations released from the fallen litter, soil acidification would be mainly caused by excess cation accumulation in wood (3.0–8.6 $kmol_c$ ha^{-1} $year^{-1}$) during forest growth.

Since protons react with silicate clay structure in the soils of Al buffering stages (soil pH < 5), Al in clay structures are solubilized. Assuming that most of acidity can be neutralized within the profiles, long-term acidification can be recorded as an increase in soil exchangeable Al^{3+}. When cumulative NPG_{Bio} attributable to

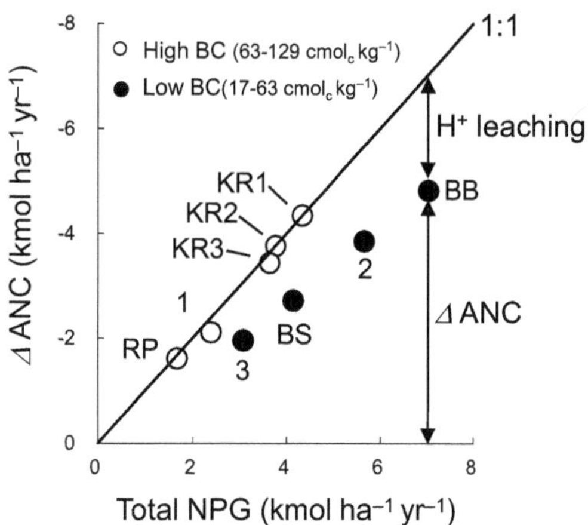

Fig. 7.8 Relationship between total NPG and ΔANC in the O horizons containing high or low concentrations of BC (basic cation). Data sources include five sites in Indonesia and one site in Thailand from the present study and three sites in Japan (1. Andisol, 2. Spodosol, 3. Inceptisol) from Fujii et al. (2008)

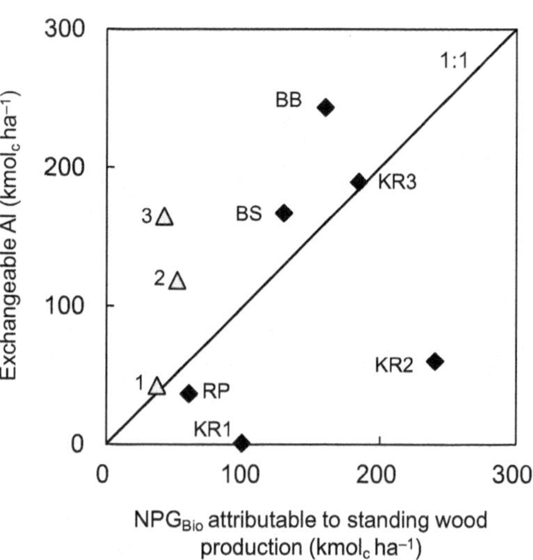

Fig. 7.9 Relationship between cumulative NPGBio attributable to wood production and exchangeable Al in the soils (0–30 cm). Data sources include five sites in Indonesia and one site in Thailand from the present study and three sites in Japan (1. Andisol, 2. Spodosol, 3. Inceptisol) from Fujii et al. (2008)

standing wood production and exchangeable Al^{3+} are compared, the amounts of soil exchangeable Al^{3+} are comparable to NPG$_{Bio}$ attributable to wood production (Fig. 7.9). This also supports high contribution of plants to soil acidification. The exchangeable Al^{3+} was smaller than NPG$_{Bio}$ in the KR1 and KR2 soils derived from serpentine and mudstone. This indicates buffering capacities of these soils are derived from the greater amounts of primary minerals as well as weathering of silicate minerals (Table 7.3). Our data support the idea that higher rates of cation cycling in plant-soil system result in consistently high acid loads to soils under tropical forest (Figs. 7.6 and 7.7).

7.4.2 Factors Regulating Proton Generation and Consumption in Soil Profiles

The dynamics of protons within the soil profiles can be quantified by application of proton budgets to each of soil horizons (Fig. 7.6). In all the soil profiles, the contribution of NPG_{Ntr}, NPG_{Car}, and NPG_{Org} to acidification is minor (Fig. 7.7). This is consistent with the fact that complete cycles of C and N are balanced with no net proton fluxes in forest ecosystems (Binkley 1987). However, translocation of the temporary acids (carbonic and organic acids and nitric acids), as well as distribution of fine root biomass (Fig. 7.3), contributes to heterogeneity of proton generation and consumption throughout the soil profiles, which varied from soil to soil, depending on geology and climate (Fig. 7.7).

Organic acid dissociation is a common proton-generating process in the O horizons of the forest soils studied. DOM-associated proton generation accounts for 18–77 % of total proton generation in the O horizons (Fig. 7.7). The large contribution of organic acids to acidification in the KR1, BS, and BB soils arises from the substantial fluxes of DOM production in the O horizon. This is primarily caused by the greater fluxes of precipitation and C input and quality of the foliar litter (Fujii et al. 2009b), as discussed in the following section.

On the other hand, carbonic acid dissociation is also a proton-generating process in the less acidic O horizons of the KR1 and KR2 soils (Table 7.5; $5 < pH$) because of their weakly acidic nature. Although NPG_{Car} is negligible in RP, bicarbonate can be a dominant anion in soil solution. This is consistent with substantial proton generation by carbonic acid dissociation in soils at neutral pH reported by Johnson et al. (1983), Van Breemen et al. (1984), and Gower et al. (1995). Although NPG_{Car} associated with active root and microbial respiration has generally been recognized as a dominant soil-acidifying process in tropical regions, the process is dominant only in moderately acidic and neutral soils.

Proton generation by nitrification is generally the dominant process involved in NPG_{Ntr} in the O horizons, while proton consumption by mineralization of organic N to NH_4^+ is also involved in NPG_{Ntr} in the highly acidic O horizons of the BS and BB soils (Figs. 7.5 and 7.7). These differences are dependent on the balance between mineralization, nitrification, and NH_4^+ and NO_3^- uptake by plants and microorganisms (Fig. 7.3). The rates of N mineralization by microorganisms are generally dependent on C/N ratio of substrates or soil environments and pH (Booth et al. 2005). Judging from most of the tropical soils and litters studied exhibiting relatively narrow C/N ratios, it was considered that ammonification is not a limiting step. Nitrification can be retarded by acidic conditions, as shown in NH_4^+ leaching from the highly acidic O horizons of the KR3, BS, and BB soils (Fig. 7.5).

NPG_{Bio} is the dominant proton-generating process in the mineral soil horizons of the KR1 and KR2 soils, while NPG_{Bio} is present in the O horizons of the KR3, BS, and BB soils (Figs. 7.3 and 7.7). Soil acidity and vegetation types could be the factors controlling the distribution of fine roots and thus NPG_{Bio}. In acidic soils, a fine root and ectomycorrhizal system is developed in the O horizon (Fujimaki et al.

2004). Dipterocarpaceae, which is the dominant vegetation on the BS and BB soils, has fine roots and ectomycorrhizal systems developed in the O horizons of acidic Ultisols (Ashton 1988). The high NPG_{Bio} in the O horizons of the KR3, BS, and BB soils arises from the presence of a fine root mat (0.3–2.3 Mg C ha^{-1} $year^{-1}$; Table 7.3), which is related to soil acidity (pH < 4.5) and ectomycorrhizal associations of Dipterocarpaceae in the BS and BB soils. Soil acidity and vegetation have a strong influence on the intensity and distribution of acids.

Release of cationic components is the principal mechanism of acid neutralization in organic and mineral soil horizons (Van Breemen et al. 1984). In the O horizons, the extent of acid neutralization varies with basic cation contents. The higher concentrations of basic cations in the O horizons of the KR soils (87–129 $cmol_c$ kg^{-1}) are considered to result in complete acid neutralization (Fig. 7.8). In the O horizons of the BS and BB soils, lower basic cation concentrations (17–63 $cmol_c$ kg^{-1}), as well as the intensive acid load, could result in incomplete acid neutralization and net eluviation of protons, Al, and Fe (Fig. 7.8).

In mineral soil horizons, the extents of acid neutralization depend on ANC of soils and their parent materials. Based on both published data and those from our study, soil ANC is variable, depending on parent materials and the extent of soil acidification or weathering and clay migration (Fig. 7.10). Our data show that parent materials have a strong influence on soil ANC with the BS and BB soils from sandstone having less ANC and a lower pH than the KR soils from serpentine and mudstone (Fig. 7.10). In the KR soils from serpentine or mudstone, their high ANC suggests that their acidity is completely neutralized by basic cation release (Figs. 7.8 and 7.10). In the BS and BB soils from sandstone, acidity is not completely neutralized due to their low ANC (Figs. 7.8 and 7.10). Thus, parent materials have a strong influence on acid neutralization processes through their effects on basic cation contents in the O horizons and on soil ANC.

7.4.3 Significance of Dissolved Organic Matter Flux in Acidification and Nutrient Cycling in Tropical Forests of Southeast Asia

Dissolved organic matter plays roles in cation mobilization and soil acidification (Guggenberger and Zech 1994). In forest soils, DOM is an intermediate by-product of litter decomposition by microorganisms (Guggenberger and Zech 1994). The formation of thick O horizons has typically been considered to lead to a sizable production of DOM in cool and humid climates (Michalzik et al. 2001). In tropical forests, the rapid mineralization of litter to CO_2 has been hypothesized to result in low concentrations of DOM in soil solutions (Johnson 1977). However, in the KR1, BS, and BB sites, a large flux of DOM is produced from the thin O horizon (Fig. 7.2; Fujii et al. 2009b, 2011c).

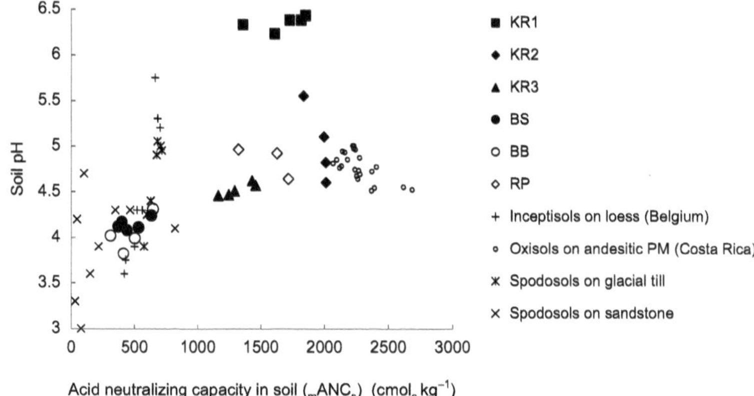

Acid neutralizing capacity in soil ($_m$ANC$_s$) (cmol$_c$ kg^{-1})

Fig. 7.10 Relationship between soil pH and acid-neutralizing capacity of soil ($_m$ANC$_s$). The data sources include Andriesse (1969) and Kleber et al. (2007). PM represents parent materials

Data synthesis indicated that large DOC fluxes from the O horizons in tropical forests can be caused by high precipitation and C input (sum of throughfall and litterfall) (Michalzik et al. 2001; Fujii et al. 2009b). The proportion of DOC flux relative to C input increased with decreasing pH (Fig. 7.11a), suggesting that the sizable production of DOC in the O horizons is common to acidic soils (pH < 4.3) in both temperate and tropical forests (Fujii et al. 2009b). The dominant DOM fractions leached from the O horizons are recalcitrant high-molecular-weight humic substances (Guggenberger and Zech 1994; Qualls and Bridgham 2005). The larger DOC flux at lower pH was considered to result from the release of aromatic compounds via lignin solubilization (Guggenberger and Zech 1993; Fujii et al. 2011b). Lignin solubilization is enhanced by the high activity of fungal enzymes [e.g., lignin peroxidase, Mn peroxidase; Fujii et al. (2012)], which contrasts with low microbial activity of cellulose degradation in the acidic soils (Hayakawa et al. 2014).

The DOC fluxes in the O horizon of KR1 were high despite high soil pH (Fig. 7.2), and this is an exceptional case of DOC flux-pH relationship (Fig. 7.11a). Within the five tropical forests in East Kalimantan, the magnitude of DOC leaching from the O horizons increased with decreasing P concentrations in the foliar litters (Fig. 7.11b; Fujii et al. 2011c). Low P concentrations in the foliar litter, as well as a high lignin concentration, could reduce DOC biodegradability and increase DOC leaching from the O horizons (Wieder et al. 2008). P concentrations in the foliar litter can account for DOC flux at local scale.

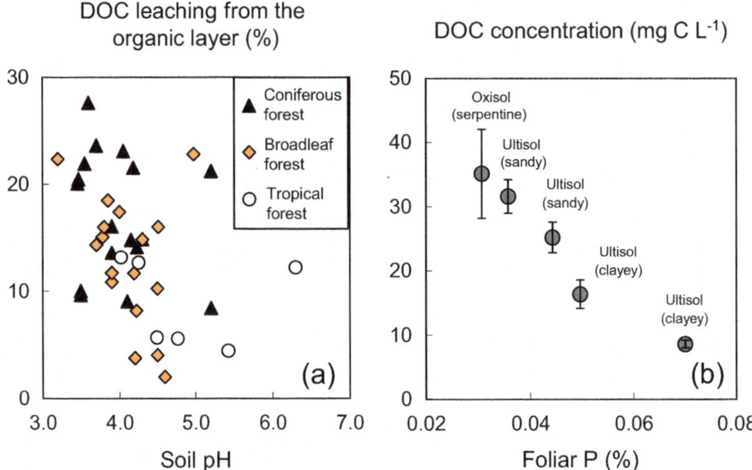

Fig. 7.11 Relationships between soil pH and DOC leaching from the O horizons (**a**) and foliar P concentrations and DOC concentrations in the O horizon leachate in five Indonesian forests (**b**). The DOC leaching was calculated as the proportion of DOC flux from the O horizons relative to C input (throughfall DOC and litterfall C). Data sources are Fujii et al. (2009b, 2011c)

7.4.4 *Implication of Proton Budgets for Pedogenetic Acidification*

The acid load is consistently higher in tropical regions than in temperate regions (Fig. 7.3). This supports the presence of strongly weathered soils in the humid tropics from the viewpoint of acidification (Eyre 1963). The kind, intensity, and distribution of acid load could vary with parent materials and vegetation, and therefore, weathering reactions and pedogenesis could also differ among tropical soils. The effects of parent materials on the dominant acidifying processes and pedogenesis can be characterized using the fluxes of Si, Al, and Fe and proton budgets within the soil profiles (Table 7.8; Fig. 7.7).

Judging from the low concentrations of Al and Fe in the moderately acidic and neutral soil solutions of the KR1 and KR2 soils (Table 7.7), the accumulated Al and Fe oxides appear to arise from in situ weathering rather than eluviation/illuviation processes. In the KR1 soil, the substantial fluxes of Si (2.67–3.55 kmol ha^{-1} year^{-1} from Table 7.8) and the high contents of Fe oxides throughout the soil profiles (Table 7.4) support the concept of ferralitization, which implies an absolute loss of Si (c) and a relative accumulation of Al and Fe oxides (Cornu et al. 1998). Dissolution of olivine by acids [$Mg_{1.6}Fe_{0.4}SiO_4$ (olivine) + 4H$^+$ = 1.6 Mg^{2+} + 0.4Fe^{2+} + H$_4$SiO$_4$] and Si leaching are considered to result in desilication, shown by a decrease of Si content from 45 % for serpentine to 9 % in the KR1 soil (Effendi et al. 2000). The fluxes of Si leaching from the KR1 soil (3.55 kmol Si ha^{-1} year^{-1}) are higher than those of Oxisols from sedimentary rocks under Amazonian forests (1.1 kmol Si ha^{-1} year^{-1}) because of the higher dissolution rates of serpentine (olivine) than quartz and kaolinite (Cornu et al. 1998). This is consistent with the rapid formation of Oxisols (ferralitization) from easily weatherable serpentine, as compared to sedimentary rocks (Pfisterer et al. 1996).

In the KR2 and KR3 soils from sedimentary rocks, no net loss of Si occurs. In the KR2 soil, the high rates of NPG_{Car} and minor contribution of NPG_{Org} to soil acidification have contributed to incongruent dissolution of Fe-rich parent materials, which results in accumulation of Al and Fe oxides throughout the profile. This process in the KR2 soil is similar to brunification, which implies accumulation of Al and Fe oxides owing to incongruent dissolution by weak acids (e.g., carbonic acid) in brown forest soils formed under temperate forest (Ugolini et al. 1990; Fujii et al. 2008).

In the highly acidic BS and BB soils, the intensive acid load contributed by NPG_{Org} and NPG_{Bio} in the O horizons results in net eluviation of protons, Al, and Fe (Fig. 7.7). These acidification processes are similar to podzolization (Cronan and Aiken 1985; Guggenberger and Kaiser 1998; Fujii et al. 2008), which involves the complexing of Al and Fe with organic acids and their translocation downward. However, translocation of Al and Fe in the BS and BB soils is different from podzolization because of the absence of spodic B horizons. The degree of podzolization is considered to be controlled by the ANC and Fe contents in the parent materials (Duchaufour and Souchier 1978; Do Nascimento et al. 2008). The higher ANC and Fe content in the BS and BB soils (Table 7.4; 309–643 $cmol_c$ kg^{-1} and 3.6–3.9 %Fe_2O_3, respectively), as compared to the typical values for the tropical Spodosols (av. 291 $cmol_c$ kg^{-1} and < 2 %Fe_2O_3, respectively; Fig. 7.11), is considered to reduce the mobility of organic acids and the degree of podzolization.

7.4.5 Ecological Significance of Soil Acidification in Tropical Forests

Proton budgets showed that there are similarities and dissimilarities in soil acidification patterns between tropical forests. The similarity of soil acidification processes between tropical forests is high cation demand by plants. Plants can take up the large amounts of basic cations mobilized through acidification (Fig. 7.6). This leads to a decrease in ANC in most of the soils, but buffering mechanisms are different between parent materials (Fig. 7.10). Acidification leads to accumulation of exchangeable Al in the highly acidic Ultisol soils on sandstone, while it leads to loss of basic cations in the soils on mafic parent materials (Fig. 7.9).

One of dissimilarities in soil acidification processes between tropical forest soils is flux of DOM (or organic acids) from the O horizon. Organic acids are well known to play roles in Al detoxification and P solubilization (Jones 1998). The high concentrations of DOM might be responses of plants and microorganisms to high Al toxicity in the highly acidic Ultisols containing the high concentrations of exchangeable Al and soil solution Al ions (Tables 7.4 and 7.7).

In addition, DOM can transport N and P in organic form (Qualls et al. 1991). The proportions of DON to TDN in the O horizon solutions (13–40 %) are smaller than temperate coniferous forests (Table 7.8; Fig. 7.4). These values are within the range of tropical forests and temperate broad-leaved forests exhibiting narrow CN ratio of the foliar litters (Fuji et al. 2013).

Because DOM leached from the O horizons is stabilized by sorption onto clays in the mineral soil horizons, leaching loss from the soil is minimal (Fig. 7.6). Once DOM is stabilized in the mineral horizons, soil organic matter functions as a reservoir and slow-release source of N, P, and bases (Kalbitz et al. 2000). The production of tannin-rich litter is hypothesized to be an adaptive strategy of coniferous trees for minimizing the leaching loss of N from nutrient-limited forests (Northup et al. 1995a, b). In the P-limited tropical forests of Southeast Asia, DOM can increase P solubility in the surface soil through the competition for sorption sites by organic anions and can minimize loss of dissolved organic P through sorption in the subsoil (Fig. 7.11 b).

There might exist two different mechanisms of acidification that drive tight nutrient cycling in tropical forests. Tropical forests in Central America develop efficient nutrient cycles through carbonic acid leaching in the moderately acidic soils (Johnson 1977). The soil acidification process in the moderately acidic soils derived from mafic parent materials (high ANC) in our study are close to this pattern. On the other hand, tropical forests on sedimentary rocks (low ANC) appear to develop DOM (or organic acid)-driven nutrient cycling on the highly acidic soils to acquire bases and P and minimize their losses.

7.5 Conclusions

The proton budgets in the soils showed quantitatively that soil acidification is an ongoing process in tropical forests due to high primary productivity. The dominant soil-acidifying process is excessive accumulation of cations over anions in woody biomass. This finding is common in a variety of the soils under tropical forests. Although soil acidification has typically been recognized as one of land degradation, acidification by plants and microorganisms can be one of nutrient acquisition strategies that promote cation mobilization through mineral weathering and cation exchange reaction. When proton generation and consumption were analyzed for each of soil horizons, proton-generating processes are variable depending ANC of parent materials even within tropical forests. In the O horizons of the highly acidic soils, dissociation of organic acids and plant uptake contributed to intensive acidification of surface horizons. In the less acidic soils, production of carbonic acid (bicarbonate) and nitric acid can contribute to leaching of cations. In the mineral soil horizons, protons generated by plant uptake are consumed by mineralization and sorption of organic acids, nitrate uptake by plants, and basic cation or Al/Fe release from the soil. The spatiotemporal variation in roots and acids can cause different pathways of pedogenesis, incipient podzolization (Al eluviation/

illuviation) and ferralitization (in situ weathering). The production of DOM, the sources of organic acids, can be enhanced by the lignin-rich and P-poor foliar and woody litters and the high activities of fungal enzymes (peroxidases) in the highly acidic soils. The differences in acid-neutralizing capacities of parent materials and climatic patterns can generate the variability in soil acidity, and plant and microbial feedbacks can further reinforce the patterns of soil acidification and nutrient cycling. Understanding of soil acidification mechanisms in different geological and climatic conditions is helpful to minimize the potential impacts of land use changes (conversion of forests to agricultural lands) on soil fertility or to accelerate restoration of the ecosystems damaged by human disturbances (e.g., fires).

References

Andriesse JP (1969) A study of the environment and characteristics of tropical podzols in Sarawak (East-Malaysia). Geoderma 2:201–227

Ashton PS (1988) Dipterocarp biology as a window to the understanding of tropical forest structure. Annu Rev Ecol Syst 19:347–370

Binkley D (1992) Proton budgets. In: Johnson DW, Lindberg SE (eds) Atmospheric deposition and forest nutrient cycling. Ecol Stud 91:450–466, Springer, New York

Binkley D, Richter D (1987) Nutrient cycles and H+ budgets of forest ecosystems. Adv Ecol Res 16:2–51

Booth MS, John MS, Edward R (2005) Controls on nitrogen cycling in terrestrial ecosystems: a synthetic analysis of literature data. Ecol Monogr 75:139–157

Brahy V, Deckers J, Delvaux B (2000) Estimation of soil weathering stage and acid neutralizing capacity in a toposequence Luvisol-Cambisol on loess under deciduous forest in Belgium. Eur J Soil Sci 51:1–13

Bredemeier M, Matzner E, Ulrich B (1990) Internal and external proton load to forest soils in Northern Germany. J Environ Qual 19:469–477

Busman LM, Dick BP, Tabatabai MA (1983) Determination of total sulfur and chlorine in plant materials by ion chromatography. Soil Sci Soc Am J 47:1167–1170

Connell J, Lowman MD (1989) Low-diversity tropical rain forests: some possible mechanisms for their existence. Am Nat 134:88–119

Cornu S, Lucas Y, Ambrosi JP, Desjardins T (1998) Transfer of dissolved Al, Fe and Si in two Amaznian forest environments in Brazil. Eur J Soil Sci 49:377–384

Cronan CS, Aiken GR (1985) Chemistry and transport of soluble humic substances in forested watersheds of the Adirondack park, New York. Geochim Cosmochim Acta 49:1697–1705

Do Nascimento NR, Fritsch E, Bueno GT, Bardy M, Grimaldi C, Melfi AJ (2008) Podzolization as a deferralitization process: dynamics and chemistry of ground and surface waters in an Acrisol-Podzol sequence of the upper Amazon Basin. Eur J Soil Sci 59:911–924

Duchaufour P, Souchier B (1978) Roles of iron and clay in genesis of acid soils under a humid, temperate climate. Geoderma 20:15–26

Effendi S, Miura S, Tanaka N, Ohta S (2000) Serpentine soils on catena in the southern part of East Kalimantan, Indonesia. In: Guhardja E, Fatawi M, Sutisna M, Mori T, Ohta S (eds) Rainforest ecosystems of East Kalimantan, Ecological studies, vol 140. Springer, Tokyo

Eyre SR (1963) Vegetation and Soils: a world picture. Edward Arnold, London

Fujii K (2014) Soil acidification and adaptations of plants and microorganisms in Bornean tropical forests. Ecol Res 29:371–381

Fujii K, Funakawa S, Hayakawa C, Kosaki T (2008) Contribution of different proton sources to pedogenetic soil acidification in forested ecosystems in Japan. Geoderma 144:478–490

Fujii K, Funakawa S, Hayakawa C, Sukartiningsih FS, Kosaki T (2009a) Quantification of proton budgets in soils of cropland and adjacent forest in Thailand and Indonesia. Plant Soil 316:241–255

Fujii K, Uemura M, Funakawa S, Hayakawa C, Sukartiningsih FS, Kosaki T, Ohta S (2009b) Fluxes of dissolved organic carbon in two tropical forest ecosystems of East Kalimantan, Indonesia. Geoderma 152:127–136

Fujii K, Hartono A, Funakawa S, Uemura M, Sukartiningsih FS, Kosaki T (2011a) Distribution of Ultisols and Oxisols in the serpentine areas of East Kalimantan, Indonesia. Pedologist 55:63–76

Fujii K, Funakawa S, Shinjo H, Hayakawa C, Mori K, Kosaki T (2011b) Fluxes of dissolved organic carbon and nitrogen throughout andisol, spodosol and inceptisol profiles under forest in Japan. Soil Sci Plant Nutr 57:855–866

Fujii K, Hartono A, Funakawa S, Uemura M, Kosaki T (2011c) Fluxes of dissolved organic carbon in three tropical secondary forests developed on serpentine and mudstone. Geoderma 163:119–126

Fujii K, Uemura M, Hayakawa C, Funakawa S, Kosaki T (2012) Environmental control of lignin peroxidase, manganese peroxidase, and laccase activities in forest floor layers in humid Asia. Soil Biol Biochem 57:109–115

Fujii K, Hayakawa C, Funakawa S, Sukartiningsih FS, Kosaki T (2013) Fluxes of dissolved organic carbon and nitrogen in cropland and adjacent forest in a clay-rich Ultisol of Thailand and a sandy Ultisol of Indonesia. Soil Tillage Res 126:267–275

Fujimaki R, Tateno R, Hirobe M, Tokuchi N, Takeda H (2004) Fine root mass in relation to soil N supply in a cool temperate forest. Ecol Res 19:559–562

Gower C, Rowell DL, Nortcliff S, Wild A (1995) Soil acidification: comparison of acid deposition from the atmosphere with inputs from the litter/soil organic layer. Geoderma 66:85–98

Guggenberger G, Kaiser K (1998) Significance of DOM in the translocation of cations and acidity in acid forest soils. Zeitschrift für Pflanzenernährung, Düngung und Bodenkunde 161:95–99

Guggenberger G, Zech W (1993) Dissolved organic carbon control in acid forest soils of the Fichtelgebirge (Germany) as revealed by distribution patterns and structural composition analyses. Geoderma 59:109–129

Guggenberger G, Zech W (1994) Dissolved organic carbon control in forest floor leachate: simple degradation products or humic substances? Sci Total Environ 152:37–47

Hallbäcken L, Tamm CO (1986) Changes in soil acidity from 1927 to 1982–1984 in a forest area of south-west Sweden. Scand J For Res 1:219–232

Hayakawa C, Funakawa S, Fujii K, Kadono A, Kosaki T (2014) Effects of climatic and soil properties on cellulose decomposition rates in temperate and tropical forests. Biol Fertil Soils 50:633–643

Jackson ML (1958) Soil chemical analysis. Prentice-Hall, Inc., Englewood Cliffs

Johnson DW (1977) Carbonic acid leaching in a tropical temperate, subalpine, and northern forest soil. Arct Alp Res 9(4):329–343

Johnson FL, Risser PG (1974) Biomass, annual net primary production, and dynamics of six mineral elements in a post oak-blackjack oak forest. Ecology 55:1246–1258

Johnson DW, Richter DD, Van Miegroet H, Cole DW (1983) Contributions of acid deposition and natural processes to cation leaching from forest soils: a review. J Air Pollut Control Assoc 33:1036–1041

Jones DL (1998) Organic acids in the rhizosphere—a critical review. Plant Soil 205:25–44

Kalbitz K, Solinger S, Park JH, Michalzik B, Matzner E (2000) Controls on the dynamics of dissolved organic matter in soils: a review. Soil Sci 165:277–304

Kleber M, Schwendenmann L, Veldkamp E, Rößner J, Jahn R (2007) Halloysite versus gibbsite: silicon cycling as a pedogenetic process in two lowland neotropical rain forests soils of La Selva, Costa Rica. Geoderma 138:1–11

Klinge R (2001) Simulation of water drainage of a rain forest and forest conversion plots using a soil water model. J Hydrol 246:82–95

Krug EC, Frink CR (1983) Acid rain on acid soil: a new perspective. Science 221:520–525

McDowell WH (1998) Internal nutrient fluxes in a Puerto Rican rain forest. J Trop Ecol 14:521–536

Mehra OP, Jackson ML (1960) Iron oxide removal from soils and clays by a dithionite-citrate system buffered with sodium bicarbonate. Clay Clay Miner 7:317–327

Michalzik B, Kalbitz K, Park JH, Solinger S, Matzner E (2001) Fluxes and concentrations of dissolved organic carbon and nitrogen—a synthesis for temperate forests. Biogeochemistry 52:173–205

Mualem Y, Dagan G (1978) Hydraulic conductivity of soils: unified approach to the statistical models. Soil Sci Soc Am J 42:392–395

Nakane K (1980) Comparative studies of cycling of soil organic carbon in three primeval moist forests. Jpn J Ecol 3:155–172

Northup RR, Dahlgren RA, Yu Z (1995a) Intraspecific variation of conifer phenolic concentration on a marine terrace soil acidity gradient: a new interpretation. Plant Soil 171:255–262

Northup RR, Yu Z, Dahlgren RA, Vogt KA (1995b) Polyphenol control of nitrogen release from pine litter. Nature 377:227–229

Pfisterer U, Blume HP, Kanig M (1996) Genesis and dynamics of an oxic dystrochrept and a typic haploperox from ultrabasic rock in the tropical rain forest climate of south-east Brazil. Z Pflanzenernährung Düng Bodenkd 159:41–50

Qualls RG, Bridgham SD (2005) Mineralization rate of ^{14}C-labeled dissolved organic matter from leaf litter in soils of a weathering chronosequence. Soil Biol Biochem 37:905–916

Qualls RG, Haines BL, Swank WT (1991) Fluxes of dissolved organic nutrients and humic substances in a deciduous forest. Ecology 72:254–266

Sanchez, P.A., Logan, T.J. 1992. Myths and science about the chemistry and fertility of soils in the tropics. In: Lal, R., Sanchez, P.A. (eds) Myths and science of soils in the tropics, SSSA Special Publication 29. ASA and SSSA, Madison, pp 35–46.

Shibata H, Kirikae M, Tanaka Y, Sakuma T, Hatano R (1998) Proton budgets of forest ecosystem on canogenous regosols in Hokkaido. Water Air Soil Pollut 105:63–72

Shibata H, Satoh F, Sasa K, Ozawa M, Usui M, Nagata O, Hayakawa Y, Hatano R (2001) Importance of internal proton production for the proton budget in Japanese forested ecosystems. Water Air Soil Pollut 130:685–690

Soil Survey Staff (2006) Keys to soil taxonomy, Tenth edn. United States Department of Agriculture Natural Resources Conservation Service, Washington, DC

Thurman, E.M., 1985. Humic substances in groundwater. In: Aiken, G.R., McKnight, D.M., Wershaw, R.L., MacCarthy, P. (Eds.), Humic substances in soil, sediment, and water. Wiley, New York, pp 87–103.

Ugolini FC, Sletten RS (1991) The role of proton donors in pedogenesis as revealed by soil solution studies. Soil Sci 151:59–75

Ugolini FC, Sletten RS, Marrett DJ (1990) Contemporary pedogenic processes in the artic: brunification. Sci du sol 28:333–348

Van Breemen N, Mulder J, Driscoll CT (1983) Acidification and alkalization of soils. Plant Soil 75:283–308

Van Breemen N, Driscoll CT, Mulder J (1984) Acidic deposition and internal proton in acidification of soils and waters. Nature 307:599–604

Van Genuchten MT (1980) A closed-form equation for predicting the hydraulic conductivity of unsaturated soils. Soil Sci Soc Am J 44:892–898

Vitousek PM, Howarth RW (1991) Nitrogen limitation on land and in the sea: how can it occur? Biogeochemistry 13:87–115

Wieder WR, Cleveland CC, Townsend AR (2008) Tropical tree species composition affects the oxidation of dissolved organic matter from litter. Biogeochemistry 88:127–138

Yamakura T, Hagihara A, Sukardjo S, Ogawa H (1986) Aboveground biomass of tropical rain forest stands in Indonesia Borneo. Vegetation 68:71–82

Chapter 8
Savannazation of African Tropical Forest Critically Changed the Soil Nutrient Dynamics in East Cameroon

Soh Sugihara

Abstract Central Africa contains the second largest area of contiguous moist tropical forest in the world, and its conservation is one of the most critical global environmental problems. Deforestation is still ongoing, however, due to illegal and/or legal logging and slash-burn agriculture, and as a result, savannazation is widely occurring. Because deforestation affects soil conditions, it is important to understand the effects of vegetation on soil nutrient dynamics in order to implement sustainable land management practices and support future afforestation. In this chapter, I explain and discuss the different nutrient stocks and flows in forest and savanna oxisols in Central Africa, based on field surveys and experiments in East Cameroon. First, I describe the difference in soil nutrient stocks between the forest and savanna: soil C stock is mostly the same, but the soil carbon to nitrogen (C:N) ratio is substantially smaller in forest than in savanna, indicating N-rich conditions in forest. There is also lower soil K stock in forest, indicating that K deficiency could be one of the limiting factors for afforestation. Second, I describe the difference in soil nutrient flow between forest and savanna in relation to soil microbes. There is a significant positive correlation between soil moisture and microbial biomass phosphorus (MBP) in forest, indicating the importance of organic P mineralization for MBP, whereas in savanna, there is a significant positive correlation between soil N availability and MBP, indicating N limitation for MBP. These results suggest that forest is an N-rich and P-limited ecosystem, whereas savanna is an N-limited ecosystem.

Keywords Land use • Oxisols • Soil microbial dynamics • Soil physico-chemical properties • Vegetation

S. Sugihara (✉)
Tokyo University of Agriculture and Technology, Saiwai-cho 3-5-8, 183-8509 Fuchu, Tokyo, Japan
e-mail: sohs@cc.tuat.ac.jp

© Springer Japan KK 2017
S. Funakawa (ed.), *Soils, Ecosystem Processes, and Agricultural Development*,
DOI 10.1007/978-4-431-56484-3_8

8.1 What Happened to the African Tropical Forest: Past Climate Change and Human Activity

Central Africa contains the second largest area of contiguous moist tropical forest of the world, covering about two million km^2(Duveiller et al. 2008; Mayaux et al. 2005). Forest conservation in this area is a global issue because tropical forest (and indeed all forest) ecosystems can act as carbon sinks (Laporte et al. 2007; Lewis et al. 2009), and they also support high and distinctive biodiversity levels (Brooks et al. 2001; Mertens and Lambin 2000). In the Guineo-Congolese region, which includes the northern extreme of the African tropical forest, the forests are surrounded by periforest savannas that constitute a forest-savanna transition zone (Duveiller et al. 2008; Sankaran et al. 2005). According to recent studies dealing with past climate change, substantial tropical forest loss occurred between 3000 and 2200 years ago, and savanna widely replaced the forest, possibly owing to extended periods of drought during this interval, i.e., the cause was a change in seasonality rather than a decrease in rainfall (Bayon et al. 2012; Maley and Brenac 1998; Schefub et al. 2005). After this period, with further change in climate, the forest again expanded, but not to the same extent as in the early and middle Holocene (Guillet et al. 2001; Schwartz et al. 1996). At present, the savanna vegetation in the forest-savanna transition zone cannot be maintained naturally because there is sufficient rainfall (ca. 1500 mm $year^{-1}$)to support forest growth. Many studies have indicated that the savanna vegetation is largely maintained by anthropogenic disturbance such as fire and/or clearance by livestock farmers, i.e., pastoralists and agriculturalists (Bucini and Hanan 2007; Sankaran et al. 2008; Staver et al. 2011). Moreover, these human activities have caused critical deforestation in the forest-savanna transition zone (Mertens and Lambin 2000; Struhsake et al. 2005), resulting in severe land degradation because nutrient-poor soils (oxisols) are widely distributed in this region (Nguetnkam and Dultz 2011; Yemefack et al. 2005). To implement sustainable land management practices for forest and savanna vegetation that support both agricultural people and livestock farmers, it is necessary to understand soil physicochemical properties and nutrient status in areas covered by both vegetation types. Although several studies have evaluated the mechanisms of forest expansion in this region based on ^{13}C and ^{14}C data (Guillet et al. 2001; Schwartz et al. 1996), information on soil physicochemical properties and nutrient status is quite limited, especially for N, P, and other minerals (K, Ca, and Mg) that control the net primary production and vegetational succession in this region (Bond 2010; Sankaran et al. 2005; Wright et al. 2011).

In this chapter, I will first discuss the effect of vegetation on soil physicochemical characteristics, and then I will examine the effect of vegetation on the soil stock of C, N, P, and other minerals and on soil C, N, and P flow in relation to soil microbial biomass. Finally, I will propose methods for sustainable land management for each vegetation type based on these results.

8.2 Description of Study Sites

The data, which are presented in Chap. 9, were collected in an eastern province of Cameroon in the forest-savanna transition zone (Guillet et al. 2001; Mertens and Lambin 2000) (Fig. 8.1). This area is located in the northern part of the Southern Cameroon Plateau, which is the dominant geographical feature of Cameroon. Metamorphic rocks are widely distributed as parent material, and oxisols are generally distributed in this region (Fig. 8.2; Soil Survey Staff 2006). The elevation is ca. 650–700 m. Based on our field measurements from 2008 to 2012, the mean annual temperature was ca. 23.5 °C and the annual rainfall was ca. 1350–1550 mm. Rainfall distribution is usually bimodal, with a short rainy season from September to November and a long rainy season from March to July. A clear dry season restricts the invasion of tropical forest and may also support the economic livelihood of livestock farmers, who depend on savanna vegetation for grazing (Klop and van Goethem 2008). Based on our vegetation survey, the dominant vegetation in the forest includes *Albizia zygia*, which predominates in the tree layer, and *Myrianthus arboreus*, which is also an abundant species. The dominant species in the savanna are the perennial grasses *Pennisetum setaceum*, *Imperata cylindrica*, and *Chromolaena odorata* (Guillet et al. 2001; Mertens and Lambin 2000). Aboveground biomass values in the forest and savanna were ca. 230 and 13 Mg ha^{-1}, respectively (Shibata et al. 2012).

8.3 Effect of Vegetation on Soil Physicochemical Characteristics

To evaluate the physicochemical characteristics of the soils, I collected 52 soil samples from five pits in each of the two different vegetation types in this area (forest and savanna), and evaluated the effect of vegetation type on soil physicochemical properties (pH, soil texture, cation-exchange capacity, bulk density, and crystalline and non-crystalline Al and Fe) (Sugihara et al. 2014).

Table 8.1 presents the physicochemical properties of the soil samples taken from the forest and savanna in the East Cameroon region ($N = 5$ for each vegetation type). Both the forest and savanna soils were acidic, and soil pH (H_2O) was lower in the forest than in the savanna, although the soil pH (KCl) was similar. Clay content was quite high (ca. 60–75 % at 40–80 cm depth) in soils from both vegetation types. Surprisingly, clay content in surface soil (0–5, 5–10 cm) was significantly lower in savanna than in forest, as assessed by Student's t-test, and a repeated measures analysis of variance (RM-ANOVA) also indicated clay content in savanna was significantly lower than that in forest. Bulk density of surface soil from forest (0.98 g cm^{-3}) was significantly less than that from savanna (1.25 g cm^{-3}), which is consistent with the report of Maquere et al. (2008). Exchangeable cations (Na$^+$, K$^+$, Ca^{2+}, Mg^{2+}) were very low, and there was no difference between the vegetation types for each cation and

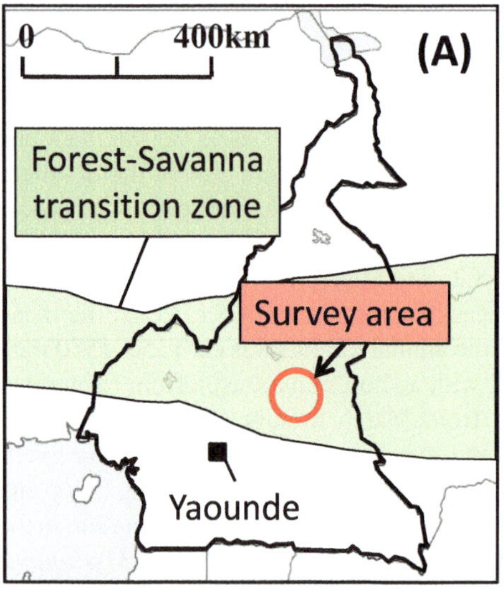

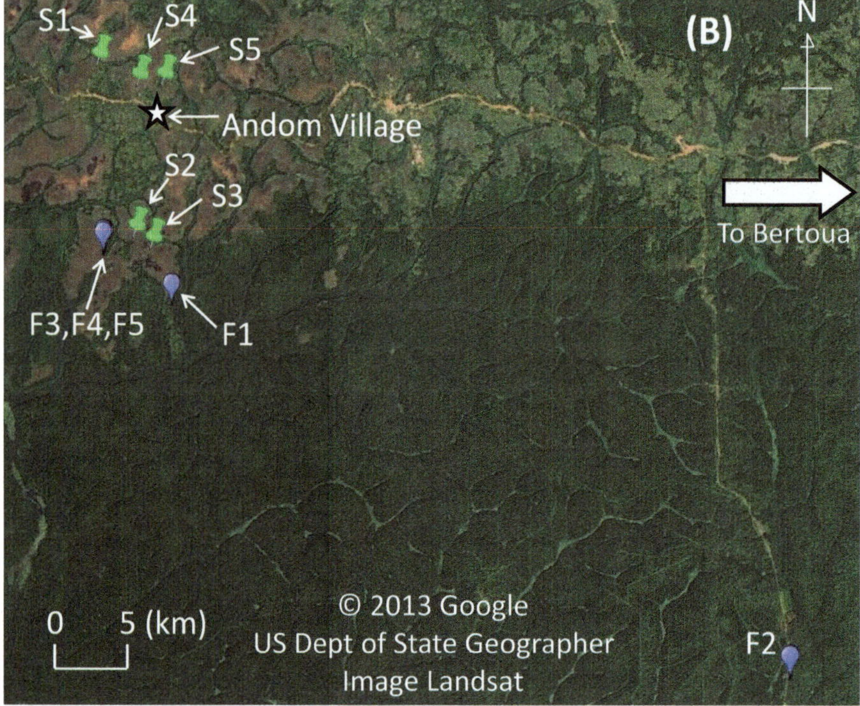

Fig. 8.1 Map of survey area in Cameroon (**a**) and detailed place of soil profile at forest (F1–F5; *blue circles*) and savanna (S1–S5; *green* pins). (**b**) F3, F4, and F5 were within 500 m. (**b**) was made based on the Google Earth map

Site: F3 in Fig.3.3.1   Location: N 04°31′ 08.10″ E 13°16′ 36.32″   Altitude: 676 m   Vegetation: Forest

Horizon	Depth (cm)	Description
O	3-0	
A	0-5	Dark reddish brown (5YR3/4); moist; weak medium sub angular blocky; friable; many medium and many fine roots; no coarse fragment; abrupt wavy boundary to;
AB	5-15	Dark reddish brown (5YR4/4); moist; moderate medium sub angular blocky; friable; many medium and many fine roots; no coarse fragment; clear wavy boundary to
BA	15-30	Dark reddish brown (2.5YR4/4); moist; moderate medium sub angular blocky; friable; common medium and many fine roots; no coarse fragment; clear wavy boundary to
Bt1	30-60	Dark reddish brown (2.5YR3/6); moist; moderate medium sub angular blocky; firm; very few faint clay cutan; common fine roots; no coarse fragment; gradual wavy boundary to
Bt2	60-90+	Reddish brown (2.5YR4/6); moist; moderate medium sub angular blocky; firm; very few faint clay cutan; common fine roots; no coarse fragment

Site: S5 in Fig.3.3.1   Location: N 04°32′ 22.45″ E 13°15′ 52.71″   Altitude: 685 m   Vegetation: Savanna

Horizon	Depth (cm)	Description
A1	0-10	Dark reddish brown (5YR3/2); moist; moderate medium sub angular blocky; friable; few coarse and many medium and fine roots; no coarse fragment; clear wavy boundary to
A2	10-20	Dull reddish brown (5YR3/3); moist; moderate medium sub angular blocky; very friable; common medium and common fine roots; no coarse fragment; clear wavy boundary to
BA	20-40	Dull reddish brown (2.5YR3/6); moist; moderate medium sub angular blocky; friable; common fine roots; no coarse fragment; gradual wavy boundary to
Bt1	40-60	Dark reddish brown (2.5YR3/6); moist; moderate medium sub angular blocky; friable; very few faint clay cutan; common fine roots; no coarse fragment; gradual wavy boundary to
Bt2	60-85+	Reddish brown (2.5YR4/8); moist; moderate medium sub angular blocky; friable; very few faint clay cutan; common fine roots; no coarse fragment

Fig. 8.2 Forest and savanna vegetation and each soil profile with description (Sugihara et al. 2014)

Table 8.1 Soil physicochemical characters at forest (For) and savanna (Sav) in Andom, Cameroon, and F values according to repeated measures analysis of variance (RM-ANOVA) ($N = 5$, Sugihara et al. 2014)

	Soil pH (H_2O)		Soil pH (KCl)		Clay (%)		Bulk density (g cm^{-3})		TEB$^※$ (cmol$_c$ kg^{-1} soil)		CEC (cmol$_c$ kg^{-1} soil)		CEC/Clay (cmol$_c$ kg^{-1} clay)	
	For	Sav	For	Sav	For	Sav	For	Sav	For	Sav	For	Sav	For	Sav
0–5 cm	4.7	5.1	4.3	4.3	58.0a	45.8b	0.98b	1.25a	4.9	2.6	10.9	9.0	18.7	19.4
5–10 cm	4.6b	5.0a	4.2	4.3	58.7a	46.1b	1.17	1.26	1.8	2.4	10.5	8.8	18.3	18.9
10–20 cm	4.5	4.7	4.2	4.2	59.0	50.4	1.24	1.31	1.3	1.4	9.3	7.8	15.9	15.9
20–40 cm	4.6	4.8	4.3	4.3	69.7	61.0	1.28	1.33	1.1	1.0	8.5	7.1	12.3	11.8
40–60 cm	4.6b	4.9a	4.4	4.5	73.8	68.1	1.28	1.33	0.9	1.2	7.1	7.2	9.4	10.6
60–80 cm	4.8	4.9	4.5	4.7	74.9	69.0	–	–	0.9	1.4	7.8	7.5	10.3	10.9
RM-ANOVA	F value		F value		F value		F value		F value		F value		F value	
Vegetation (V)	5.1	$P = 0.056$	0.4	$P = 0.56$	5.4	*	3.4	$P = 0.10$	0.1	$P = 0.73$	1.0	$P = 0.36$	0.06	$P = 0.82$
Soil depth (SD)	2.5	*	9.5	***	55.1	***	10.8	***	10.7	***	2.7	*	12.4	***
V*SD	0.9	$P = 0.52$	0.8	$P = 0.56$	1.4	$P = 0.25$	2.9	*	2.9	*	0.3	$P = 0.88$	0.07	$P = 0.99$

	TC (g kg^{-1} soil)		TN (g kg^{-1} soil)		TP (g kg^{-1} soil)		TP$_o$* (mg kg^{-1} soil)		TC:TN		TC:TP$_0$		TN:TP$_0$	
	For	Sav	For	Sav	For	Sav	For	Sav	For	Sav	For	Sav	For	Sav
0–5 cm	24.5	20.2	2.2a	1.3b	0.44	0.34	100.5	86.4	11.0b	15.7a	247	253	22.7a	16.1b
5–10 cm	14.9	19.3	1.3	1.2	0.38	0.33	79.9	82.7	11.5b	15.6a	187	252	16.4	16.1
10–20 cm	12.8	14.5	1.1	0.9	0.39	0.30	75.3	67.4	12.0b	15.3a	176	231	14.9	15.0
20–40 cm	8.4	8.6	0.7	0.7	0.38a	0.26b	46.9	34.9	11.5b	13.0a	200	271	17.4	20.8
40–60 cm	6.6	6.6	0.6	0.5	0.37a	0.27b	32.0a	18.0b	11.3	12.1	273	438	20.8	35.7
60–80 cm	5.6	5.2	0.5	0.4	0.36a	0.27b	28.8a	14.6b	11.0	11.7	217	497	19.4	41.8
RM-ANOVA	F value		F value		F value		F value		F value		F value		F value	
Vegetation (V)	0.03	$P = 0.87$	3.3	$P = 0.11$	5.0	$P = 0.054$	0.8	$P = 0.39$	47.2	***	3.5	$P = 0.10$	1.5	$P = 0.26$
Soil depth (SD)	50.2	***	54	***	7.2	***	42.2	***	18.7	***	2.3	$P = 0.65$	3.5	**
V*SD	2.3	$P = 0.06$	7.2	***	1.1	$P = 0.40$	0.5	$P = 0.77$	14.1	***	1.5	$P = 0.21$	2.5	*

※ TEB total exchangeable base ($Na^+ + K^+ + Ca^{2+} + Mg^{2+}$)

*TP$_o$ = labile-P$_o$ (NaHC0$_3$-P$_0$) + slowly-P$_o$ (NaOH-P$_o$)

Different letters indicate significant difference between forest and savanna at each horizon (Student's t-test; $P < 0.05$)

* $P < 0.05$, ** $P < 0.01$, *** = $P < 0.001$

for total exchangeable bases. Cation-exchange capacity and (cation-exchange capacity)/clay values for both vegetation types were also small, consistent with the established characteristics of oxisols (Soil Survey Staff 2006; Yemefack et al. 2005), and there was no clear difference between the vegetation types.

Our results show that most soil physicochemical properties are not affected by vegetation type. Bayon et al. (2012) indicated that soil erosion occurred when Central African forest was replaced by savanna approximately 3000–2200 years ago, resulting in a decrease in clay content, and the consequent soil degradation limited forest recovery after that. I also found lower surface-layer clay content in savanna soil compared with forest soil, possibly indicating a lower water-holding capacity in savanna than in forest, although the clay content in savanna (45.8–46.1 %) was sufficiently high to mitigate the effects of severe droughts, which prevent the reforestation of savanna. In addition, most soil physicochemical properties, as well as the rainfall distribution, were the same, and soil fertility and climate may not prevent the reforestation of savanna; wild/controlled fire could therefore be a factor that maintains the savanna vegetation in this region (Sakurai et al. 1998; Staver et al. 2011). Future study is needed to evaluate this possibility. Because I collected samples from only five sites for each vegetation type in this study, it will be necessary to analyze more samples to fully determine whether the relatively low surface-layer clay content, on the one hand, or wild/controlled fire, on the other, can inhibit forest recovery in this region.

8.4 Effect of Vegetation on the Soil Stock of C, N, P, and Other Minerals

To evaluate the stock of C, N, P, and other minerals in the soil profile, I used the same soil samples as in the previous section and analyzed the total (T) amounts of C, N, P, Na, K, Ca, and Mg in the soil. In case of P, I also analyzed the inorganic P (P_i) and organic P (P_o) separately.

Table 8.1 presents the data on soil C, N, and P through the soil profile. TC in each soil layer was similar for both vegetation types, but TN was relatively large through the forest soil profile compared with savanna; this was especially the case for surface soil (0–5 cm; $P < 0.05$). Thus, there was clearly a lower soil carbon to nitrogen (C:N) ratio in forest (11.0–12.0) than in savanna (13.0–15.7) in the upper soil (<40 cm depth; $P < 0.05$, RM-ANOVA). TP and TP_o were higher in forest than in savanna, especially in the deeper soil (>40 cm depth). To determine the availability of soil P for plants, I calculated the ratios $TC:TP_o$ and $TN:TP_o$, which are widely used to estimate the mineralization potential of organic P. Both $TC:TP_o$ and $TN:TP_o$ were apparently larger in the savanna soil, but the differences were not statistically significant. The C, N, and P values in the forest soil profile (80 cm soil depth) were 87.9 ± 5.5 (standard error, $N = 5$), 7.7 ± 0.3, and 3.7 ± 0.4 Mg ha^{-1}, respectively, and those in savanna were 98.6 ± 7.1, 7.1 ± 0.5, and 3.1 ± 0.2 Mg ha^{-1}, respectively. Thus, based on the soil profile (0–80 cm depth), vegetation type did not significantly affect soil C, N, and P.

A lower C:N ratio in forest soil than that in savanna soil has been reported in tropical humid and subhumid regions of Southeast Asia and South America (Lilienfein et al. 2001; Yonekura et al. 2010). Guillet et al. (2001) also observed that the C:N ratio at 0–20 cm soil depth was clearly lower in forest compared with savanna at a study site in eastern Cameroon that is ca. 40 km away from our study site. In our survey region, the main forest species is *Albizia zygia*, which can fix N effectively (Baggie et al. 2000), while the main savanna species cannot fix N. The fixed N therefore contributes to the low C:N ratio in forest soil. In support of this, I found substantial nitrate leaching below 30 cm soil depth in the forest, although little nitrate leaching was observed at the same depth in savanna soil (Shibata et al. 2012). Lilienfein et al. (2001) also observed much nitrate leaching and high N concentrations in soil under planted forest, but they found little N leaching in Cerrado in Brazil. As a result, they recorded a lower C:N ratio in soil in young planted forest (20 years) than in Cerrado in deeper soil (1.2–2.0 m). In our present study, N that leached from the surface to deeper soils in the old forest (>50 years) contributed to the observed low soil C:N ratio throughout the soil profile (Table 8.1). Based on the vertical distribution of the soil C:N ratio in savanna, the grass roots of the savanna seem to be mainly distributed from 0 to 40 cm, resulting in a higher soil C:N ratio. Therefore, in terms of vegetational succession or forest recovery, the high C:N ratio of savanna soil may indicate an N-limited condition, which is the first step in succession from savanna to young forest, and the low C:N ratio of forest soil may indicate an N-saturated (and possibly P-limited) condition, which is the next step in succession to mature forest (Kitayama et al. 2000, 2004; Vitousek 1984). The C:N ratio, C:P ratio, and N:P ratios of leaf litter in the forest were 16.2 ± 0.2 (standard error, $N = 22$), 475.3 ± 18.3, and 29.3 ± 1.2, respectively, whereas, in the savanna, the values were 21.7 ± 0.1 (standard error, $N = 10$), 284.4 ± 22.8, and 13.1 ± 1.1, respectively (unpublished data). These data are also consistent with our idea of an N-limited condition in savanna and an N-saturated (and possibly P-limited) condition in forest (McGroddy et al. 2004; Ratnam et al. 2008).

A low C:N ratio in surface soil indicates an immediate supply of mineralized N for crop growth after land clearing, e.g., by slash and burn, whereas a high C:N ratio in surface soil would supply mineralized N gradually to crops during cultivation (Bengtsson et al. 2003; Janssen 1996). Therefore, in terms of land management for farmers, N-mineralizable forest vegetation would be suitable for cropland compared with N-unmineralizable savanna vegetation. On the other hand, savanna in this region seems to be useful for raising livestock, although controlled fire is necessary for efficient grazing, and this possibly prevents the forest from invading savanna areas.

With respect to C sequestration in tropical ecosystems, many studies have assessed the effect of deforestation and/or afforestation on soil C content. Generally, afforestation in the relatively humid tropics decreases soil C, although afforestation in the relatively dry tropics increases soil C (Berthrong et al. 2012; Jackson et al. 2002). In the present study, accumulated soil C values were 9.3 Mgha^{-1} (80 cm soil depth) greater in savanna than in forest, although this difference wasnot statistically significant. This is because the greater bulk density in savanna contributes to greater accumulation of soil C (Maquere et al. 2008). However, when the aboveground biomass C of forest in this region is included (e.g., ca. 100–130 Mg ha^{-1} of C in the present forest; Shibata et al. 2012), reforestation and forest conservation must contribute more to C sequestration than savanna in this ecosystem.

Table 8.2 presents the values for Na_t, K_t, Ca_t, and Mg_t content in soils from each layer of both vegetation types. Na_t, K_t, Ca_t, and Mg_t were 2.1–3.3, 1.7–1.8, 1.7–3.8, and 4.2–5.2 $cmol_c$ kg^{-1}, respectively, in forest soil, and 1.8–2.8, 2.5–2.6, 1.5–2.4, and 4.5–5.4 $cmol_c$ kg^{-1} in savanna. TRB was quite low: 9.7–13.3 $cmol_c$ kg^{-1} in forest soil and 10.7–12.5 $cmol_c$ kg^{-1} in savanna soil. There was no clear difference between the soils of the two vegetation types, with the exception of K_t. In surface soil (0–20 cm depth), K_t serves as an important nutrient pool (Bond 2010; Johnson et al. 2001); K_t in forest soil was ca. 1590 kg ha^{-1} (20 cm depth), which was ca. 930 kg ha^{-1} (20 cm depth) lower than in savanna soil (ca. 2520 kg ha^{-1} at 20 cm). Johnson et al. (2001) found that the aboveground biomass of Brazilian tropical forest accumulates ca. 75–150 kg ha^{-1} of K in secondary forest and ca. 350 kg ha^{-1} of K in mature forest. Therefore, plant uptake of soil K must contribute to a smaller K_t in forest than in savanna, although the amounts seem to be relatively small to explain our observed difference in total K_t between forest and savanna (930 kg ha^{-1} of K at 20 cm soil depth). Because K^+ is particularly prone to leaching in forest ecosystems (Markewitz et al. 2004; Tripler et al. 2006), and I previously observed considerable nitrate leaching in the Cameroon forest (Shibata et al. 2012), leaching of K^+ with nitrate would also contribute to a smaller K_t in forest. The levels of other minerals, i.e., Na, Ca, and Mg, were 1290, 980, and 1260 kg ha^{-1}, respectively, in forest soil (20 cm depth) and 1440, 1020, and 1450 kg ha^{-1} in savanna soil (20 cm depth); the differences in values between vegetation types were not significant. Bond (2010) indicated that the nutrients in relatively short supply were Ca and K but not P, based on the nutrient stock of a tropical forest with nutrient-poor soil. Based on my soil mineral analysis, the K supply in soils throughout the Cameroon region may limit future forest maturity and forest recovery after deforestation.

8.5 Effect of Vegetation on Soil C, N, and P Flow in Relation to Soil Microbes

Soil microbes play an important role in the biogeochemical cycle as decomposers and "nutrient sink-sources" (Singh et al. 1989; Wardle 1992; Wardle et al. 2004), and they are very sensitive to environmental conditions, such as quality and

Table 8.2 Amount of total minerals (Na, K, Ca, and Mg) of soil and total residual base (TRB) of soil at each layer at forest (For) and savanna (Sav) in Andom, Cameroon, and F values according to repeated measures analysis of variance (RM-ANOVA) ($N = 5$, Sugihara et al. 2014)

	Na_t		K_t		Ca_t		Mg_t		$TRB*$	
($cmol_c\ kg^{-1}$ soil)	For	Sav	For	Sav	For	Sav	For	Sav	For	Sav
0–5 cm	2.4 (0.7)	2.5 (0.4)	1.8 (0.2) b	2.5 (0.2) a	3.8 (0.9)	2.4 (0.3)	5.2 (0.1)	4.9 (0.5)	13.3 (0.7)	12.3 (1.0)
5–10 cm	2.1 (0.7)	2.6 (0.3)	1.8 (0.2) b	2.5 (0.2) a	1.7 (0.4)	2.3 (0.3)	4.2 (0.3)	4.9 (0.4)	9.7 (1.5)	12.3 (1.0)
10–20 cm	2.6 (1.1)	2.4 (0.7)	1.7 (0.2) b	2.5 (0.3) a	1.7 (0.5)	1.6 (0.5)	4.4 (0.5)	4.5 (0.7)	10.4 (2.0)	11.1 (2.0)
20–40 cm	2.4 (0.3)	2.6 (0.4)	1.7 (0.2) b	2.6 (0.3) a	2.1 (0.5)	1.5 (0.3)	4.5 (0.2)	5.0 (0.5)	10.7 (0.9)	11.8 (1.3)
40–60 cm	2.5 (0.5)	2.8 (0.5)	1.7 (0.3) b	2.6 (0.2) a	2.2 (0.6)	1.7 (0.4)	4.6 (0.3)	5.4 (0.5)	11.0 (1.4)	12.5 (1.4)
60–80 cm	3.3 (1.0)	1.8 (0.5)	1.7 (0.3)	2.5 (0.2)	2.6 (0.9)	1.4 (0.5)	5.1 (0.6)	5.0 (0.3)	12.7 (2.6)	10.7 (1.2)
RM-ANOVA	F value		F value		F value		F value		F value	
Vegetation (V)	0.03	$P = 0.87$	6.42	*	0.82	$P = 0.39$	0.34	$P = 0.58$	0.08	$P = 0.79$
Soil depth (SD)	0.12	$P = 0.99$	0.24	$P = 0.94$	2.66	*	1.98	$P = 0.10$	1.00	$P = 0.43$
V*SD	1.46	$P = 0.23$	0.65	$P = 0.66$	1.52	$P = 0.21$	1.46	$P = 0.22$	1.42	$P = 0.24$

*TRB $= Na_t + K_t + Ca_t + Mg_t$
Different letters indicate significant difference between forest and savanna at each horizon (Student's t-test; $P < 0.05$)
Values in parentheses indicate standard errors of the mean ($N = 5$)
*$P<0.05$

quantity of the soil substrate (Liu et al. 2006; Wang et al. 2003), temperature and moisture (Fang and Moncrief 2001; Hamel et al. 2006), and soil chemistry (Baath and Anderson 2003; Morenom et al. 1999). As a result, many researchers have used soil microbial characteristics as indicators of soil fertility in various ecosystems or under various land use regimes (Joergensen and Emmerling 2006; Bastida et al. 2008), or as indicators of limiting nutrients and other environmental factors (Cleveland and Liptzin 2007; Gnankambary et al. 2008). Because the size of microbial biomass (MB) is mainly controlled by the amount of substrate, MB carbon (MBC) has been used as a good indicator of soil fertility (Joergensen and Emmerling 2006). Using the MB C:N, C:P, and N:P ratios, it is possible to predict the nutrient that limits soil microbial dynamics and so limits the ecosystem as a whole (Aponte et al. 2010; Galicia and Garcia-Oliva 2004). Chen et al. (2003) compared the seasonal fluctuation of MBC and MB phosphorus (MBP) in forest and grassland in New Zealand and found a faster MBP turnover rate in forest than in savanna, indicating the importance of MBP as a P pool in P-limited forest ecosystems. Therefore, I compared the nutrient flow in forest (FOR) and savanna (SAV) in this area by time-course measurement of the MBC, MBN, and MBP values in both vegetation types.

In FOR, MBC varied from 342.7 to 536.8 mg C kg^{-1} soil (CV 13.9 %), MBN varied from 39.3 to 63.5 mg N kg^{-1} soil (CV 16.4 %), and MBP varied from 24.8 to 45.6 mg P kg^{-1} soil (CV 17.5 %) during the experimental period (Fig. 8.3). In SAV, MBC varied from 335.2 to 414.4 mg C kg^{-1} soil (CV 7.0 %), MBN varied from 43.6 to 76.1 mg N kg^{-1} soil (CV 17.0 %), and MBP varied from 26.9 to 41.6 mg P kg^{-1} soil (CV 12.9 %) (Fig. 8.3). MBC and MBN fluctuated during the experimental period in both FOR and SAV, but MBP did not fluctuate in either FOR or SAV (Table 8.3). According to a repeated measures ANOVA, the MBC in FOR was greater than in SAV, whereas MBN in FOR was less than in SAV. There was no significant difference in MBP between FOR and SAV.

The MB C:N, C:P, and N:P ratios in FOR fluctuated from 6.7 to 11.0 (CV 17.3 %), from 8.4 to 17.5 (CV 23.4 %), and from 1.1 to 1.8 (CV 16.0 %), respectively (Fig. 8.4). In SAV, the MB C:N, C:P, and N:P ratios fluctuated from 5.1 to 8.6 (CV 18.0 %), from 9.3 to 14.4 (CV 14.8 %), and from 1.2 to 2.6 (CV 24.1 %), respectively (Fig. 8.4). All ratios in FOR and SAV fluctuated substantially during the experimental period. This was especially true for the MB C:N and C:P ratios, which clearly increased in the dry period in FOR, as sampled on 28 Jan 2012 and 20 Mar 2012, but not in SAV. The MB C:N ratio was significantly higher in FOR than in SAV, whereas the MB N:P ratio was significantly lower in FOR than in SAV (Table 8.3). The MB C:P ratios in FOR and SAV were not significantly different ($P < 0.05$).

Table 8.4 shows the results of a Pearson correlation analysis of the soil microbial values and soil environmental conditions such as soil moisture, air and soil temperature, substrate, and soil pH. In FOR, soil moisture was positively correlated with MBP and negatively correlated with the MB C:P ratio ($P < 0.05$). Ext-N was also negatively correlated with MBN ($P < 0.05$). In addition, I observed a weak negative correlation between soil moisture and the MB C:N ratio ($r = -0.64$,

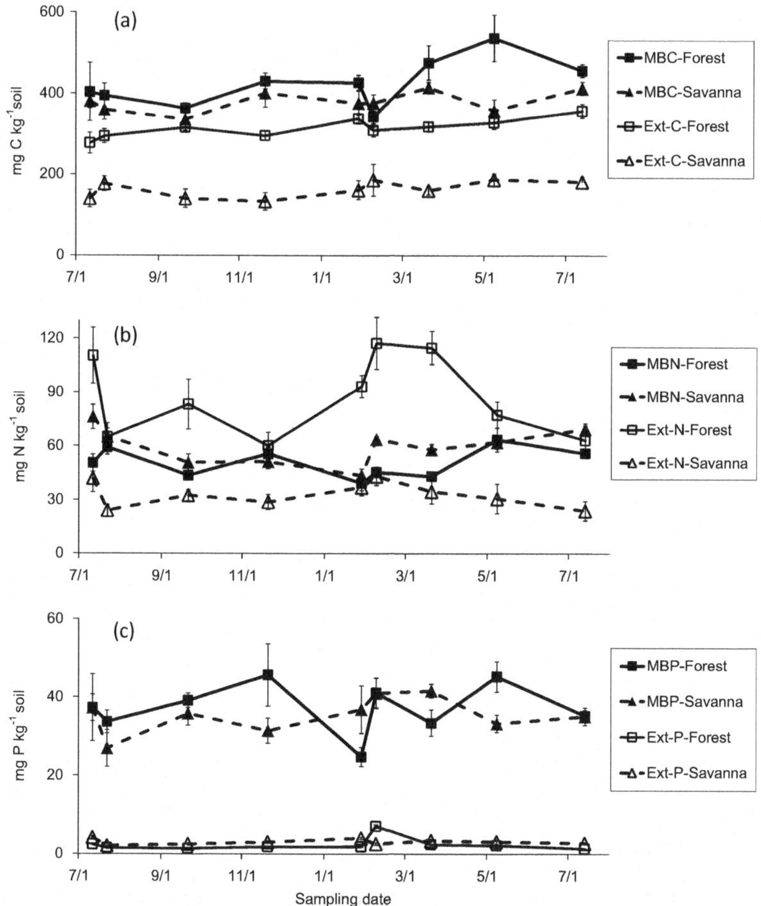

Fig. 8.3 Seasonal fluctuation of soil microbial biomass (MB) carbon (C) (**a**), nitrogen (N) (**b**), and phosphorus (P) (**c**), as well as extractable (Ext) C (**a**), N (**b**), and P (**c**) in the forest and savanna during the experimental period (July 2011–2012) in eastern Cameroon (Sugihara et al. 2015). *Bars* indicate the standard error ($N = 3$)

$P = 0.06$) and weak positive correlations between air temperature and MBC ($r = 0.56$, $P = 0.12$), between soil pH and MBN ($r = 0.64$, $P = 0.06$), and between soil temperature and MBP ($r = 0.58$, $P = 0.10$). On the other hand, in SAV, only Ext-N was significantly correlated with soil MB values ($P < 0.05$), including a positive correlation with MBP and a negative correlation with the MB C:P ratio. Moreover, I observed a weak negative correlation between soil moisture and the MB C:N ratio ($r = -0.57$, $P = 0.11$) and a weak positive correlation between soil temperature and MBP ($r = 0.55$, $P = 0.13$).

In the present study, I found that the maintained MBC was larger in FOR than in SAV. Because the Ext-C, which is a substrate for soil microbes, was consistently greater in FOR than in SAV (Fig. 8.3), it contributed to the larger MBC in FOR

Table 8.3 Summary of average values and results of repeated measures ANOVA for the effects of vegetation and seasonal fluctuation on each value ($N = 9$ for averages and $N = 54$ for repeated measures ANOVA, Sugihara et al. 2015)

	MBC	MBN (mg kg^{-1} soil)	MBP	MB C:N ratio	MB C:P ratio	MB N:P ratio	Soil pH (1:5 water)
Forest	425.5	50.7	37.3	8.6	11.9	1.4	4.3
Savanna	379.2	60.0	35.4	6.5	11.0	1.7	5.6
Vegetation (V)	$F = 12.7$	$F = 27.3$	NS	$F = 557.2$	NS	$F = 29.3$	$F = 801.4$
	$P = 0.02$	$P = 0.006$		$P < 0.001$		$P = 0.006$	$P < 0.001$
Seasonal variation (S)	$F = 3.0$	$F = 8.0$	NS	$F = 19.6$	$F = 3.8$	$F = 5.9$	$F = 5.9$
	$P = 0.01$	$P < 0.001$		$P < 0.001$	$P = 0.003$	$P < 0.001$	$P < 0.001$
S*V	NS[a]	$F = 2.9$	NS	$F = 4.1$	$F = 3.1$	NS	NS
		$P = 0.02$		$P = 0.002$	$P = 0.01$		
	Ext-C	Ext-N (mg kg^{-1} soil)	Ext-P	Ext-C:N ratio	Ext-C:P ratio	Ext-N:P ratio	GMC[b] (%)
Forest	314.6	87.1	2.4	3.9	166.5	42.7	21.5
Savanna	163.1	32.9	3.0	5.3	58.5	11.4	18.9
Vegetation (V)	$F = 187.2$	$F = 264.5$	$F = 36.4$	$F = 15.2$	$F = 252.7$	$F = 336.4$	$F = 30.9$
	$P < 0.001$	$P < 0.001$	$P = 0.004$	$P = 0.02$	$P < 0.001$	$P < 0.001$	$P < 0.01$
Seasonal variation (S)	$F = 5.7$	$F = 12.7$	$F = 29.3$	$F = 18.6$	$F = 32.0$	$F = 12.4$	$F = 48.3$
	$P < 0.001$	$P < 0.001$	$P < 0.001$	$P < 0.001$	$P < 0.001$	$P < 0.001$	$P < 0.001$
S*V	NS	$F = 4.2$	$F = 33.3$	$F = 3.2$	$F = 29.8$	$F = 19.6$	$F = 5.0$
		$P = 0.002$	$P < 0.001$	$P = 0.01$	$P < 0.001$	$P < 0.001$	$P < 0.001$

[a]NS means not significant
[b]GMC means gravimetric moisture contents of soil

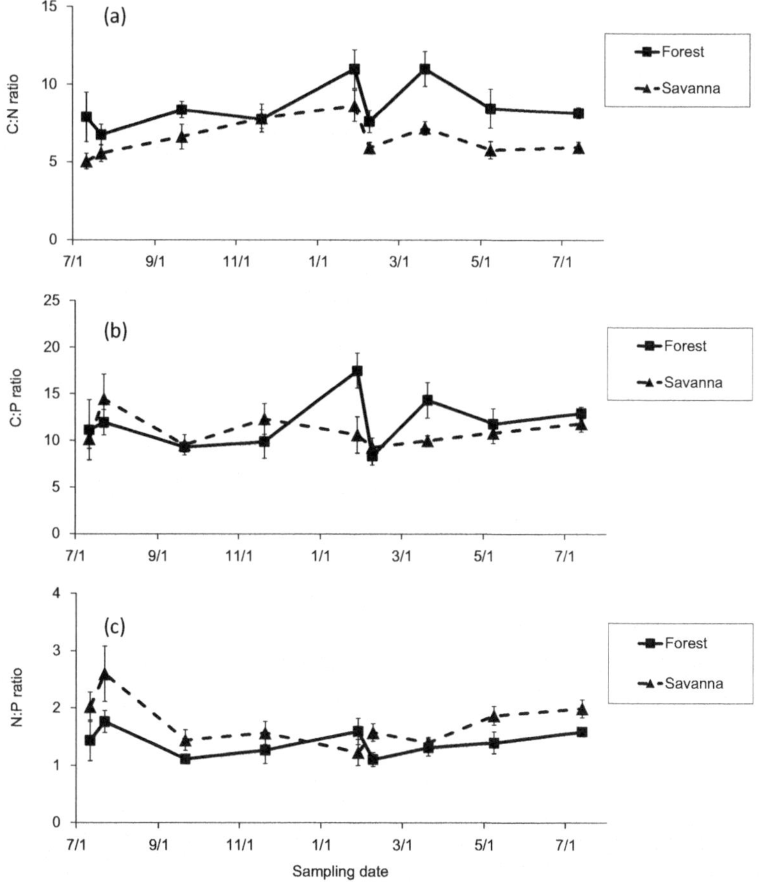

Fig. 8.4 Seasonal fluctuation of soil microbial biomass (MB) carbon to nitrogen (C:N) ratio (**a**), C to phosphorus (P) ratio (**b**), and N:P ratio (**c**) in the forest and savanna during the experimental period (July 2011–2012) in eastern Cameroon (Sugihara et al. 2015). *Bars* indicate the standard error ($N = 3$)

(Zaman et al. 1999). However, MBN was smaller in FOR than in SAV, even though the Ext-N was substantially larger in FOR than in SAV. As a result, the MB C:N ratio was significantly higher in FOR than in SAV, indicating a lower microbial N content in FOR than in SAV. This result is quite interesting because of its apparent contradiction: a low MBN level and high MB C:N ratio in FOR, which nevertheless has rich soil N. This could be explained by the existence of different microbial communities in soils of different chemical compositions in FOR and SAV. Many studies have found that fungi dominate in forest ecosystems and bacteria dominate in savanna ecosystems because of the different soil pH levels (Aciego and Brookes 2008; Rousk et al. 2010). Generally, fungi have higher MB C:N ratios than bacteria (Anderson and Domsch 1980); fungi are therefore the dominant species in FOR,

Table 8.4 Correlation coefficient between soil microbial factors and the environmental and soil nutrient factors for each vegetation during the experimental period (July 2011–2012) in eastern Cameroon ($N = 9$ for each vegetation, Sugihara et al. 2015)

Forest	MBC	MBN	MBP	MB C:N ratio	MB C:P ratio	MB N:P ratio
Soil moisture	NS[a]	NS	0.68*	−0.64[#]	−0.74*	NS
Air temperature	0.56[#]	NS	NS	NS	NS	NS
Soil temperature	NS	NS	0.58[#]	NS	NS	NS
Ext-C	NS	NS	NS	NS	NS	NS
Ext-N	NS	−0.67*	NS	NS	NS	NS
Ext-P	NS	NS	NS	NS	NS	NS
Soil pH	NS	0.64[#]	NS	NS	NS	NS
Savanna	MBC	MBN	MBP	MB C:N ratio	MB C:P ratio	MB N:P ratio
Soil moisture	NS	NS	NS	−0.57[#]	NS	NS
Air temperature	NS	NS	NS	NS	NS	NS
Soil temperature	NS	NS	0.55[#]	NS	NS	NS
Ext-C	NS	NS	NS	NS	NS	NS
Ext-N	NS	NS	0.75*	NS	−0.78*	NS
Ext-P	NS	NS	NS	NS	NS	NS
Soil pH	NS	NS	NS	NS	NS	NS

*Indicates $P < 0.05$
#Indicates $P < 0.15$
[a]NS means not significant

where the MB C:N ratio is high, whereas bacteria are the dominant species in SAV, where the MB C:N ratio is low. Baath and Anderson (2003) found that soil pH affects soil microbial communities and concluded that the fungi-to-bacteria ratio increases as the soil pH decreases. In the current study, I found a significantly lower soil pH level in FOR than in SAV (Table 8.1), and a positive correlation between the soil pH level and MBN occurred only in FOR (Table 8.4), indicating that there is a soil pH-limited MBN value in FOR. However, the substrate quality should also be considered, although I did not evaluate it in the current study (Bossuyt et al. 2001; Six et al. 2006). Lauber et al. (2008) found that in the southeastern United States, the fungi-to-bacteria ratio was higher in forest than in grassland soil because of their different substrate qualities, including high C:N content in forest and low C: N content in grassland. Our results indicated that soil pH (and possibly litter quality) could be the primary factor affecting both soil microbial composition (including the fungi-to-bacteria ratio) and the MBN dynamics between the different vegetation types, as opposed to the amount of available N substrate, such as Ext-N (Rousk et al. 2010). Because I did not evaluate the soil microbial community or the litter quality in either vegetation type, further studies are necessary to assess the effects of soil pH and litter quality on the soil microbial community.

The MBP dynamics in FOR and SAV were similar and, unlike MBC and MBN, did not fluctuate significantly throughout the experimental period (Table 8.3). This indicates that soil microbes play an important role as a stable P nutrient pool in oxisols under both vegetation types (Oberson et al. 2001). Kaspari et al. (2008) revealed that the tropical rain forest P uptake is 6.4 kg ha^{-1} annually in Panama, and P uptake was estimated at 6.6 kg ha^{-1} annually in the present FOR (unpublished data). Tondoh et al. (2013) also reported that *Chromolaena odorata*, which is the same plant species present in our SAV, takes up 3.6 kg P ha^{-1} annually in Ivory Coast, and P uptake was estimated at 2.1 kg ha^{-1} annually in SAV in the present study (unpublished data). Ext-P (2.3 and 4.0 kg P ha^{-1} at 0–10 cm in FOR and SAV, respectively) was similar to, or rather less than, the annual P uptake in each vegetation type, whereas MBP (35.4 and 47.4 kg P ha^{-1} at 0–10 cm in FOR and SAV, respectively) was greater than the annual P uptake in each vegetation type. This indicates the significance of soil microbes as a stable nutrient pool for vegetational P uptake in this region. Interestingly, there are different limiting factors for MBP in FOR and SAV (Table 8.4). Because I observed a positive correlation between Ext-N and MBP only in SAV, the amount of N substrate must limit MBP in SAV, indicating that N limits soil P dynamics. However, only soil moisture positively correlated with MBP in FOR, possibly indicating (1) no substrate limitations for MBP and (2) the importance of soil organic matter (SOM) decomposition for MBP in FOR. Because Ext-P in FOR did not correlate with MBP (Table 8.4), it appears that the amount of P substrate did not limit the MBP variations in FOR. However, the amount of Ext-P (1.3–7.0 mg P kg^{-1} soil) seems to be too small to explain the MBP variations (24.7–45.6 mg P kg^{-1} soil). This suggests that the current analytical method of determining Ext-P (the resin method) is not sufficient to evaluate MBP dynamics, at least in this region and/or soil. Khan and Joergensen (2012) found that $NaHCO_3$-extractable organic P was closely related to MBP, based on laboratory analyses using various soil samples. Because of soil moisture, which controls the SOM decomposition rate and which was positively correlated with MBP in FOR, the decomposed organic P should be an important P source for soil microbes in FOR, indicating a P substrate limitation in FOR. Our present results suggest that, in terms of soil microbial dynamics, forests are N-saturated, P-limited ecosystems, and savannas are N-limited ecosystems in this region.

I found greater soil N availability and lower soil pH in FOR than in SAV, and I found positive relationships between soil moisture and MBP in FOR and between Ext-N and MBP in SAV, indicating a P-limited FOR ecosystem and a N-limited SAV ecosystem. In addition, contrary to the rich soil N substrate conditions in FOR, I found a lower MBN value and higher MB C:N ratio in FOR than in SAV. This indicates that substrate conditions should not be the primary controlling factor for MBN dynamics in this area. Other soil environmental values indicate that soil pH is the primary controlling factor affecting microbial communities (the fungi:bacteria ratio) in each vegetation type. Further studies are necessary to assess the effects of soil pH and other factors, such as litter quality, on soil microbial communities in forest and savanna in this region.

8.6 Conclusions and Remarks

In this chapter, I explained the effect of vegetation on soil nutrient stock and flow in oxisols in Central African forest and savanna. Soil C stock was mostly the same in the two vegetation types, but soil N availability was substantially greater in forest than in savanna due to the N-fixing trees in forest. Low soil pH in forest led to smaller soil K stocks and a higher microbial C:N ratio in forest than in savanna, indicating the significant role of soil pH on the flow and stocks of soil nutrients. Based on a series of studies, I conclude that the forest ecosystem in this region is N rich and P limited, while the savanna ecosystem is N limited. This basic information should serve as an incentive for local farmers to select forest vegetation as a fallow vegetation after slash-burn agriculture and for the afforestation of savanna areas.

References

Aciego PJC, Brookes PC (2008) Relationship between soil pH and microbial properties in a UK arable soils. Soil Biol Biochem 40:1856–1861

Anderson JPE, Domsch KH (1980) Quantities of plant nutrients in the microbial biomass of selected soils. Soil Sci 130:211–215

Aponte C, Maranon T, Garcia LV (2010) Microbial C, N and P in soils of Mediterranean oak forest: influence of season, canopy cover and soil depth. Biogeochemistry 101:77–92

Baath E, Anderson TH (2003) Comparison of soil fungal/bacterial ratios in a pH gradient using physiological and PLFA-based techniques. Soil Biol Biochem 35:955–963

Baggie I, Zapata F, Sanginga N, Danso SKA (2000) Ameliorating acid infertile rice soil with organic residue from nitrogen fixing trees. Nutr Cycl Agroecosyst 57:183–190

Bastida F, Zsolnay A, Hernandez T, Garcia C (2008) Past, present and future of soil quality indices: a biological perspective. Geoderma 147:159–171

Bayon G, Dennielou B, Etoubleau J, Ponzevera E, Toucanne S, Bermell S (2012) Intensifying weathering and land use in iron age Central Africa. Science 335:1219–1222

Bengtsson G, Bengtsson P, Mansson KF (2003) Gross nitrogen mineralization-, immobilization-, and nitrification rates as a function of soil C/N ratio and microbial activity. Soil Biol Biochem 35:143–154

Berthrong ST, Pineiro G, Jobbagy EG, Jackson RB (2012) Soil C and N changes with afforestation of grasslands across gradients of precipitation and plantation age. Ecol Appl 22(1):76–86

Bond WJ (2010) Do nutrient-poor soils inhibit development of forests? a nutrient stock analysis. Plant Soil 334:47–60

Bossuyt H, Denef K, Six J, Frey SD, Merckx R, Paustian K (2001) Influence of microbial populations and residue quality on aggregate stability. Appl Soil Ecol 16:195–208

Brooks T, Balmford A, Burgess N, Fjeldsa J, Hansen LA, Moore J, Rahbek C, Williams P (2001) Toward a blueprint for conservation in Africa. Bioscience 51(8):613–624

Bucini G, Hanan NP (2007) A continental-scale analysis of tree cover in African savannas. Glob Ecol Biogeogr 16:593–605

Chen CR, Condron LM, Davis MR, Sherlock RR (2003) Seasonal changes in soil phosphorus and associated microbial properties under adjacent grassland and forest in New Zealand. For Ecol Manag 177:539–557

Cleveland CC, Liptzin D (2007) C:N:P: stoichiometry in soil: is there a "Redfield ratio" for the microbial biomass. Biogeochemistry 85:235–252

Duveiller G, Defourny P, Desclee B, Mayaux P (2008) Deforestation in Central Africa: estimates at regional, national and landscape levels by advanced processing of systematically-distributed landsat extracts. Remote Sens Environ 112:1969–1981

Fang C, Moncrief JB (2001) The dependence of soil CO_2 efflux on temperature. Soil Biol Biochem 33:155–165

Galicia L, Garcia-Oliva F (2004) The effects of C, N and P additions on soil microbial activity under two remnant tree species in a tropical seasonal pasture. Appl Soil Ecol 26:31–39

Gnankambary U, Ilstedt U, Nyberg G, Hien V, Malmer A (2008) Nitrogen and phosphorus limitation of soil microbial respiration in two tropical agroforestry parklands in the south-Sudanese zone of Burkina Faso: the effects of tree canopy and fertilization. Soil Biol Biochem 40:350–359

Guillet B, Achoundong G, Happi JY, Beyala VKK, Bonvallot J, Riera B, Mariotti A, Schwartz D (2001) Agreement between floristic and soil organic carbon isotope (13C/12C, 14C) indicators of forest invasion of savannas during the last century in Cameroon. J Trop Ecol 17:809–832

Hamel C, Hanson K, Selles F, Cruz AF, Lemks R, McConkey B, Zentner R (2006) Seasonal and long-term resource-related variations in soil microbial communities in wheat-based rotations of the Canadian prairie. Soil Biol Biochem 38:2104–2116

Jackson RB, Banner JL, Jobbagy EG, Pockman WT, Wall DH (2002) Ecosystem carbon loss with woody plant invasion of grassland. Nature 418:623–626

Janssen BH (1996) Nitrogen mineralization in relation to C:N ratio and decomposability of organic materials. Plant Soil 181:39–45

Joergensen RG, Emmerling C (2006) Methods for evaluating human impact on soil microorganisms based on their activity, biomass, and diversity in agricultural soils. J Plant Nutr Soil Sci 169:295–309

Johnson CM, Vieira ICC, Zarin DJ, Frizano J, Johnson AH (2001) Carbon and nutrient storage in primary and secondary forests in eastern Amazonia. For Ecol Manag 147:245–252

Kaspari M, Garcia MN, Harms KE, Santana M, Wright J, Yavitt JB (2008) Multiple nutrients limit litterfall and decomposition in a tropical forest. Ecol Lett 11:35–43

Khan KS, Joergensenm RG (2012) Relationships between P fractions and the microbial biomass in soils under different land use management. Geoderma 173-174:274–281

Kitayama K, Majalap N, Aiba S (2000) Soil phosphorus fractionation and phosphorus-use efficiencies of tropical rainforests along altitudinal gradients of Mount Kinabalu, Borneo. Oecologia 123:342–349

Kitayama K, Aiba S, Takyu M, Majalap N, Wagai R (2004) Soil phosphorus fractionation and phosphorus-use efficiency of a Bornean tropical montane rain forest during soil aging with podzolization. Ecosystems 7:259–274

Klop E, van Goethem J (2008) Savanna fires govern community structure of ungulates in Bénoué National Park, Cameroon. J Trop Ecol 24:39–47

Laporte NT, Stabach JA, Grosch R, Lin TS, Goetz SJ (2007) Expansion of industrial logging in Central Africa. Science 316:1451

Lauber CL, Strickland MS, Bradford MA, Fierer N (2008) The influence of soil properties on the structure of bacterial and fungal communities across land-use types. Soil Biol Biochem 40:2407–2415

Lewis SL, Gonzalez GL, Sonke B, Baffoe KA, Baker TR et al (2009) Increasing carbon storage in intact African tropical forests. Nature 457:1003–1006

Lilienfein J, Wilcke W, Thomas R, Vilela L, Lima SC, Zech W (2001) Effects of *Pinus caribaea* forests on the C, N, P, and S status of Brazilian savanna Oxisols. For Ecol Manag 147:171–182

Liu HS, Li LH, Han XG, Huang JH, Sun JX, Wang HY (2006) Respiratory substrate availability plays a crucial role in the response of soil respiration to environmental factors. Appl Soil Ecol 32:284–292

Maley J, Brenac P (1998) Vegetation dynamics, palaeoenvironments and climatic changes in the forest of western Cameroon during the last 28,000 years B.P. Rev Palaeobot Palynol 99:157–187

Maquere V, Laclau JP, Bernoux M, Saint-Andre L, Goncalves JLM, Cerri CC, Piccolo MC, Ranger J (2008) Influence of land use (savanna, pasture, Eucalyptus plantations) on soil carbon and nitrogen stocks in Brazil. Eur J Soil Sci 59:863–877

Markewitz D, Davidson E, Moutinho P, Nepstad D, Moutinho P, Nepstad D (2004) Nutrient loss and redistribution after forest clearing on a highly weathered soil in Amazonia. Ecol Appl 14 (4):451–466

Mayaux P, Holmgren P, Achard F, Eva H, Stibig HJ, Branthomme A (2005) Tropical forest cover change in the 1990s and options for future monitoring. Philo. Trans Royal Soc B 360:373–384

Mertens B, Lambin EF (2000) Land cover change trajectories in Southern Cameroon. Ann Assoc Am Geogr 90(3):467–494

McGroddy ME, Daufresne T, Hedin LO (2004) Scaling of C:N:P stoichiometry in forests worldwide: implications of terrestrial redfield-type ratios. Ecology 85(9):2390–2401

Morenom JL, Hernandez T, Garcia C (1999) Effect of a cadmium-contaminated sewage sludge compost on dynamics of organic matter and microbial activity in an arid soil. Biol Fertil Soils 28:230–237

Nguetnkam JP, Dultz S (2011) Soil degradation in central north Cameroon: water-dispersible clay in relation to surface charge in Oxisol A and B horizons. Soil Tillage Res 113:38–47

Oberson A, Friesen DK, Rao IM, Buhler S, Frossard E (2001) Phosphorus transformations in an Oxisol under contrasting land-use systems: the role of the soil microbial biomass. Plant Soil 237:197–210

Ratnam J, Sankaran M, Hanan NP, Grant RC, Zambatis N (2008) Nutrient resorption patterns of plant functional groups in a tropical savanna: variation and functional significance. Oecologia 157:141–151

Rousk J, Brookes PC, Baath E (2010) Investigating the mechanisms for the opposing pH relationships of fungal and bacterial growth in soil. Soil Biol Biochem 42:926–934

Sakurai K, Tanaka S, Ishizuka S, Kanzaki M (1998) Difference in soil properties of dry evergreen and dry deciduous forests in the Sakaerat environmental research station. Tropics 8(1/2):61–80

Sankaran M, Hanan NP, Scholes RJ, Tatnam J, Augustin DJ et al (2005) Determinants of woody cover in African savannas. Nature 438:846–849

Sankaran M, Ratnam J, Hanan N (2008) Woody cover in Africa savannas: the role of resources, fire and herbivory. Glob Ecol Biogeogr 17:236–245

Schefub E, Schouten S, Schneider RR (2005) Climate controls on central African hydrology during the past 20,000 years. Nature 437:1003–1006

Schwartz D, de Foresta H, Mariotti A, Balesdent J, Massimba JP, Girardin C (1996) Present dynamics of the savanna-forest boundary in the Congolose Mayombe: a pedological, botanical and isotopic (13C and 14C) study. Oecologia 106:516–524

Shibata M, Sugihara S, Funakawa S, Araki S (2012) A comparison of soil solution composition from forest and savanna vegetation in eastern Cameroon. In: Proceeding of the 8th international symposium on plant-soil interactions at low pH. Bangaroru, India, 2012. pp 52–53

Singh JS, Raghubanshi AS, Singh RS, Srivastava SC (1989) Microbial biomass act as a source of plant nutrients in dry tropical forest and savanna. Nature 338:499–500

Six J, Frey SD, Thiet RK, Batten KM (2006) Bacterial and fungal contributions to carbon sequestration in agroecosystems. Soil Sci Soc Am J 70:555–569

Soil Survey Staff (2006) Keys to soil taxonomy, 10th edn. United States Department of Agriculture Natural Resources Conservation Service, Washington, DC

Staver AC, Archibald S, Levin S (2011) Tree cover in sub-Saharan Africa: rainfall and fire constrain forest and savanna as alternative stable states. Ecology 92(5):1063–1072

Struhsake TT, Struhsaker PJ, Siex KS (2005) Conserving Africa's rain forest: problems in protected areas and possible solutions. Biol Conserv 123:45–54

Sugihara S, Shibata M, Mvondo Ze A, Araki S, Funakawa S (2014) Effect of vegetation on soil C, N, P and other minerals in Oxisols at the forest-savanna transition zone of central Africa. Soil Sci Plant Nutr 60:45–59

Sugihara S, Shibata M, Mvondo Ze A, Araki S, Funakawa S (2015) Effect of vegetation on soil microbial C, N and P dynamics in a tropical forest and savanna of Central Africa. Appl Soil Ecol 87:91–98

Tondoh JE, Kone AW, N'Dri JK, Tamene L, Brunet D (2013) Changes in soil quality after subsequent establishment of *Chromolaena odorata* fallows in humid savannahs, Ivory Coast. Catena 101:99–107

Tripler CE, Kaushal SS, Likens GE, Walter MT (2006) Patterns in potassium dynamics in forest ecosystems. Ecol Lett 9:451–466

Vitousek PM (1984) Litterfall, nutrient cycling, and nutrient limitation in tropical forests. Ecology 65(1):285–298

Wang WJ, Dalal RC, Moody PW, Smith CJ (2003) Relationship of soil respiration to microbial biomass, substrate availability and clay content. Soil Biol Biochem 35:273–284

Wardle DA (1992) A comparative assessment of factors which influence microbial biomass carbon and nitrogen levels in soils. Biol Rev 67:321–358

Wardle DA, Bardgett RD, Klironomos JN, Setala H, van der Putten W, Wall DH (2004) Ecological linkages between aboveground and belowground biota. Science 304:1629–1633

Wright SJ, Yavitt JB, Wurzburger N, Turner BL, Tanner EVJ et al (2011) Potassium, phosphorus, or nitrogen limit root allocation, tree growth, or litter production in a lowland tropical forest. Ecology 92(8):1616–1625

Yemefack M, Rossiter DG, Njomgang R (2005) Multi-scale characterization of soil valiability within an agricultural landscape mosaic system in southern Cameroon. Geoderma 125:117–143

Yonekura Y, Ohta S, Kiyono Y, Aksa D, Morisada K, Tanaka N, Kanzaki M (2010) Changes in soil carbon stock after deforestation and subsequent establishment of "Imperata" grassland in the Asian humid tropics. Plant Soil 329:495–507

Zaman M, Di HJ, Cameron C, Frampton CM (1999) Gross nitrogen mineralization and nitrification rates and their relationships to enzyme activities and the soil microbial biomass in soil treated with dairy shed effluent and ammonium fertilizer at different water potentials. Biol Fertil Soils 29:178–186

Chapter 9
Ecosystem Processes of Ferralsols and Acrisols in Forest-Soil Systems of Cameroon

Makoto Shibata

Abstract Tropical African forests are dominated by Ferralsols and leguminous species, while Southeast Asian forests are dominated by Acrisols/Alisols and Dipterocarpaceae. Hence, their ecological processes can differ, depending on soil acidity and nitrogen (N) availability. To provide an overview of the carbon (C) and N dynamics, as well as soil acidification processes on Ferralsols and Acrisols, in tropical African forests, we quantified soil respiration and element fluxes through different flow paths (as precipitation, throughfall, litterfall, litter leachate, and soil solutions) and analyzed proton budgets in two secondary forested sites in Cameroon. Our results demonstrate that at Mvam Village (MV; Acrisols), N was mostly taken up within the O horizon, which has a dense root mat, while half of the input N leached down to the mineral horizon at the Andom Village (AD; Ferralsols) site. Nitrification was the main proton-generating process in the canopy and the O horizon of AD, and it caused a large amount of cation leaching, which resulted in the accumulation of basic cations because of the high proton consumption rates in the A horizon. In contrast, because of the dense root mat at MV, the excess cation uptake by plants in the O horizon made the largest contribution to proton generation, which resulted in intensive acidification of the surface soil. Our results suggest that ecosystem processes differ depending on soil type (i.e., soil acidity). Thus, legumes growing on Ferralsols in tropical African forests have unique plant-soil interactions via active nitrification in the O horizon.

M. Shibata (✉)
Graduate School of Global Environmental Studies, Kyoto University, Kitashirakawaoiwake-cho, Sakyo-ku, Kyoto 606-8502, Japan
e-mail: makotosbt@gmail.com

© Springer Japan KK 2017
S. Funakawa (ed.), *Soils, Ecosystem Processes, and Agricultural Development*,
DOI 10.1007/978-4-431-56484-3_9

9.1 Introduction

The environmental conditions in the tropical forests of Central Africa differ from those of other tropical regions with respect to soil type and vegetation. The most widely distributed soils in the Congo basin are clayey Ferralsols (Jones et al. 2013), which have low nutrient availability (van Wambeke 1992). This contrasts with the tropical forests of Asia, where Acrisols and Alisols dominate. Acrisols and Alisols, which are strongly acidic, are accompanied by undulating topography and granitic materials (Jones et al. 2013). In terms of vegetation, the leguminous family (Fabaceae) dominates the lowland tropical rainforests of Africa and the Neotropics (but not those of Asia, which are dominated by the Dipterocarpaceae) (Crews 1999; Doyle and Luckow 2003). In particular, in terms of biomass, the proportion of woody legumes is much higher in Africa than in the Neotropics, although species richness is slightly higher in the Neotropics (Yahara et al. 2013). Because of their N_2-fixing symbioses with rhizobia, many native leguminous species could play an important role in ecosystem processes in relation to carbon (C) and nitrogen (N) dynamics (Diabate et al. 2005; Knops et al. 2002), as well as soil acidification processes via N transformation. The presence of legumes, combined with symbiotic N_2 fixation, often has a positive effect on ecosystem N pools, which can significantly increase aboveground biomass (Pons et al. 2007; Spehn et al. 2002). Considering that previous studies showed that the N concentration pattern in foliage is mainly controlled by phylogeny, not geologic or edaphic source variation, in Amazonian, Bornean, and Australian tropical forests (Asner et al. 2009, 2012, 2014), legumes, with their N-demanding lifestyle and high foliar N content (Fyllas et al. 2009), can be expected to be widely distributed in African tropical forests, which may result in high N availability in African soils. Therefore, in the tropical forests of Africa that are dominated by the Fabaceae, ecological processes, which are closely tied to the water flow that links vegetation and soil (McDowell 1998), could differ from those in other tropical forests that have fewer legumes.

One of the well-known mechanisms by which plants recycle N and other nutrients is the development of a superficial root mat. The importance of a root mat, which efficiently (>99 %) captures dissolved nutrients before they leach down to the mineral soil, was first proposed for Ferralsols and Podzols in the Amazon by Stark and Jordan (1978). This mechanism retains N in these ecosystems, which limits N losses through plant uptake. In contrast, root growth is retarded in acidic soil that has high Al^{3+} and H^+ concentrations (De Graaf et al. 1997; Murach and Ulrich 1988). In addition, nitrification, which is a key process that leads to ecosystem N losses via NO_3^- leaching underground or denitrification into the atmosphere, is often inhibited at low pH (Jordan et al. 1979). Thus, soil acidity (pH), in combination with root mat formation, may affect N transformation processes.

In this chapter, we provide an overview of N fluxes through different flow paths (as precipitation, throughfall, litterfall, litter leachate, and soil solutions), as well as the soil C cycle and soil acidification processes, of two secondary forested sites that are situated on clay-rich Ferralsols and Acrisols in Cameroon, Central Africa.

9.2 Description of the Study Sites

This study was conducted from March 2010 to May 2012 in secondary forested sites in eastern (Andom Village (AD)) and southern (Mvam Village (MV)) parts of the Republic of Cameroon, which is located to the northwest of Congolian lowland forests (Fig. 9.1), where a semi-deciduous forest forms the climax vegetation. Site descriptions are shown in Table 9.1. Both sites lie between tropical monsoon (Am) and tropical savanna (Aw) climates in the Köppen system, and they support tropical moist forests in the Holdridge life zone. Approximately 400 km apart, the sites are located on the South Cameroon Plateau, which is the dominant geographical feature of Cameroon, with an average elevation of approximately 650–700 m asl. The soils and geology differ at the two sites. The soil at AD is reddish brown (2.5YR 4/6) in the subsurface horizon, and it is classified as Typic Kandiudox (Soil Survey Staff 2014) or Acric Ferralsols (Vetic) (IUSS 2014) that have developed on Neoproterozoic (Pan-African) granitoids. The soil at MV is yellowish brown (10YR 5/6–8) in the subsurface horizon, and it is classified as Typic Paleudults (Soil Survey Staff 2014) or Haplic Acrisols (Vetic) (IUSS 2014) that developed on Archean Congo Craton (Oliveira et al. 2006; Toteu et al. 2006). Although the distribution of rainfall is similarly bimodal for both sites, their floristic composition differs slightly: at AD, the canopy is dominated by *Albizia zygia* (Fabaceae) and species belonging to the Urticaceae and Moraceae, whereas at MV, the canopy is dominated by *Piptadeniastrum africanum* (Fabaceae) and species belonging to the Malvaceae, Burseraceae, and Euphorbiaceae, and it has

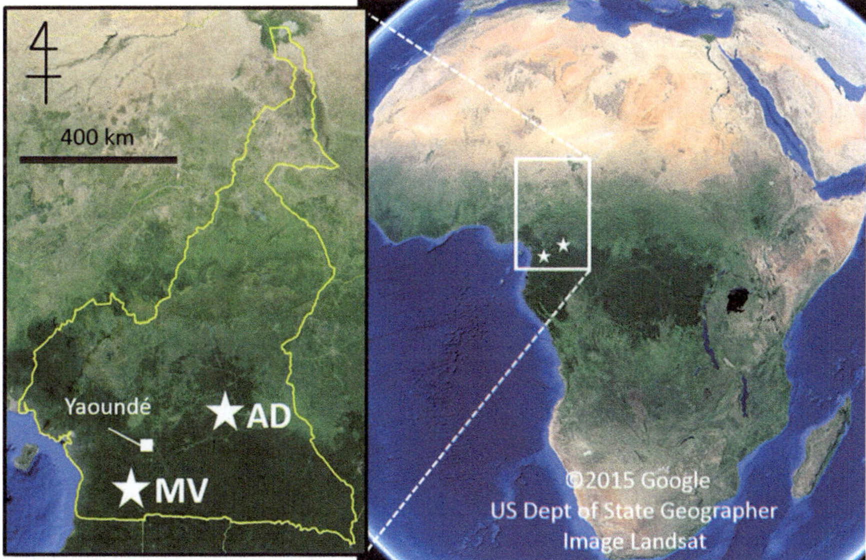

Fig. 9.1 Google Earth image of the study sites in Cameroon (AD and MV) located on the northwestern edge of Congolian lowland forests

Table. 9.1 Site description

Site	AD	MV
Coordinates	N 4° 32′ E 13° 15′	N 2° 52′ E 11° 09′
Annual air temp, (°C)	22.9	23.8
Annual soil temp, (°C)	22.7	23.2
Annual rainfall (mm)	1531	1650
Elevation (m)	677	607
Soil type	Typic Kandiudox[a]	Typic Paleudults[a]
	Lixic Ferralsols (Vetic)[b]	Haplic Acrisols (Vetic)[b]
Geological substrate	Neoproterozoic (Pan-African) granitoids	Archean Congo Craton
Vegetation	*Albizia zygia*	*Piptadeniastrum africanum*
	Myrianthus arboreus	*Ceiba pentandra*

[a]Classified according to soil taxonomy (Soil Survey Staff 2014)
[b]Classified according to WRB soil classification system (IUSS 2014)

more evergreen taxa. The dominant vegetation at both sites consists of N_2-fixing species.

9.3 General Physicochemical Properties of the Soils and Environmental Factors

The mineral soils at MV were strongly acidic, with pH (H_2O) values of 3.5–4.0, whereas those at AD were moderately acidic with pH (H_2O) values of 4.2–4.9, which is consistent with the lower base saturation and higher Al saturation at MV (Table 9.2). Exchangeable Al^{3+} in the mineral soil (0–30-cm depth) was 126 $kmol_c$ ha^{-1} at MV and 59.8 $kmol_c$ ha^{-1} at AD. The pH values of the O horizon were a few units higher than those of the mineral horizons at each site. A dense fine-root mat, which could be peeled back from the soil like a carpet, was observed only at MV (Fig. 9.2). Fine-root biomass in the O horizon was 1.5 Mg ha^{-1} (54 % of the mineral horizon, 0–30-cm depth) at MV and 0.3 Mg ha^{-1} (12 % of the mineral horizon, 0–30-cm depth) at AD (Fig. 9.3); the large fine-root biomass may account for the higher organic matter content in the surface horizon at MV. Clay contents in the soils were consistently high at MV (47–67 %) and AD (54–73 %) and increased with increasing soil depth. Silt contents in the soils were consistently low, ranging from 3 to 8 % through the horizons at both sites, which were the result of a long period of weathering. C/N ratios in the mineral horizon were equally and consistently low throughout the soil profiles at both sites, whereas the C/N ratio in the O horizon was lower at AD than at MV (N was more abundant in the O horizon at AD) (Table 9.2). Kaolin minerals were dominant at both sites, whereas 1.4-nm minerals, which might contribute to a slightly higher CEC, were recognized only at MV (data not shown).

Table. 9.2 Physicochemical properties of soils at the study sites

Site	Horizon	Depth (cm)	pH H$_2$O	KCl	Particle size distribution[a] Sand (%)	Silt	Clay	Total C (g kg^{-1})	Total N	C/N (g g^{-1})	CEC/ clay[a]	Exchangeable cations Bases[b] (cmol$_c$ kg^{-1})	Al	H	Base saturation[c] (%)	Bulk density (g cm^{-3})
AD	O	+2–0	7.1[d]	–	–	–	–	438	26	17	–	–	–	–	–	–
Oxisols/	A1	0–10	4.2	3.9	40	5	55	19	1.6	11	15	2.3	2.3	0.7	43	1.0
Ferralsols	A2	10–20	4.2	4.0	39	6	54	14	1.1	12	11	1.6	2.5	0.7	33	1.1
	Bt1	20–35	4.8	4.3	29	4	68	9.5	0.9	11	10	1.8	0.9	0.7	52	1.2
	Bt2	35–50	4.3	4.3	25	4	70	6.9	0.6	11	8.7	1.1	1.5	0.3	38	–
	Bt3	50–80 +	4.9	4.7	23	3	73	5.7	0.6	8.9	8.8	1.7	0.4	0.4	69	–
MV	O	+3–0	5.4[d]	–	–	–	–	440	20	22	–	–	–	–	–	–
Ultisols/	A	0–10	3.5	3.3	45	8	47	30	2.5	12	26	1.3	4.5	2.6	16	0.9
Acrisols	BA	10–25	3.7	3.6	41	7	51	12	1.1	10	14	0.7	3.8	1.3	12	1.2
	Bt1	25–50	4.0	3.8	29	7	64	6.8	0.7	9.1	17	0.6	3.0	0.6	14	1.3
	Bt2	50–75 +	4.0	3.7	28	6	67	5.7	0.7	8.5	32	0.6	2.4	0.7	16	–

[a]Clay($<$0.002 mm); silt(0.002–0.05 mm); sand(0.05–2 mm)

[b]$Na^+ + K^+ + Ca^{2+} + Mg^{2+}$

[c]$(Na^+ + K^+ + Ca^{2+} + Mg^{2+})/(Na^+ + K^+ + Ca^{2+} + Mg^{2+} + Al^{3+} + H)$

[d]The pH was measured using the milled litter to water ratio of 1:20 for 1 h

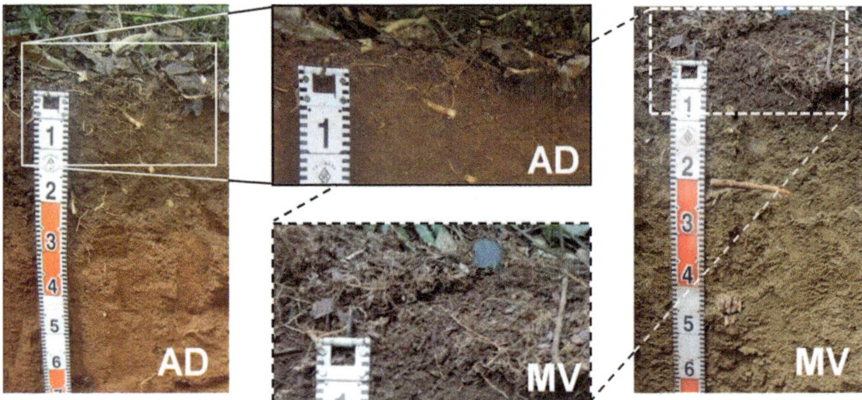

Fig. 9.2 Soil profile and O horizon. Ferralsols in AD and Acrisols in MV

Fig. 9.3 Depth distribution of fine-root biomass in the tropical forest soils. Bars indicate standard errors ($n = 5$)

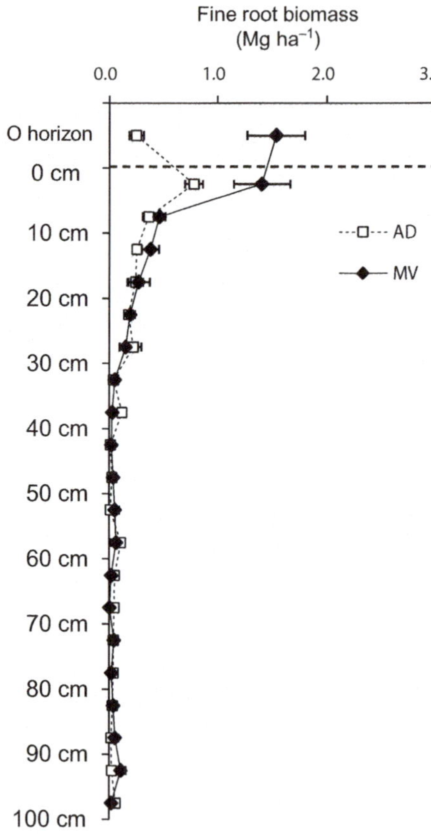

Figures 9.4 and 9.5 show the daily precipitation and half-hourly soil temperatures at 5-cm depth, seasonal fluctuations of volumetric water content (VWC), and daily water fluxes at 0-, 15-, and 30-cm soil depths for each site. The total rainfall amount during the experimental period (April 2010 to May 2012) was 3029 mm at AD and 3329 mm at MV. There are four distinct seasons at both sites: a minor wet season from mid-March to mid-July, a minor dry season until the end of August, a major wet season that starts in September and lasts until the middle of November, and a major dry season that lasts until the next short rainy season. Soil VWC seasonally fluctuated and were lowest in the dry season, although the degree was moderate at MV. Mean soil temperature (5-cm depth) during the experimental period was 22.6 °C at AD and 23.2 °C at MV. Little seasonal fluctuation of soil temperature was observed, although it sometimes dropped to approximately 17 °C during the dry season at AD possibly because of the adjacent forest-savanna ecotone to the north (Fig. 9.1).

9.4 Carbon Cycles

As shown in Table 9.3, C was stored as aboveground biomass (AD, 122 Mg C ha^{-1}; MV, 152 Mg C ha^{-1}), which is comparable to that of other tropical forests (55–218 Mg C ha^{-1}) (Alves et al. 2010; Gaston et al. 1998). The fine-root biomass C at AD (0.13 Mg C ha^{-1}) was significantly (*t*-test, $p < 0.0001$) lower (17 %) than that at MV (0.72 Mg C ha^{-1}) because of the existence of the dense root mat at the latter site. The C stocks in the O horizon were low at both sites (2.0 Mg C ha^{-1} at each site), which is in agreement with those of tropical forests in Cameroon (Peh et al. 2012; 3.5 Mg dry mass ha^{-1}). The higher aboveground biomasses (AD, 122 Mg C ha^{-1}; MV, 152 Mg C ha^{-1}) and lower organic matter stocks in the O horizon and mineral soil (AD, 47 Mg C ha^{-1}; MV, 53 Mg C ha^{-1}), compared with temperate forests, are consistent with previous reports (Fujii et al. 2008, 2011a).

Major soil C dynamics were similar at both sites. Most of the C inputs to the O horizon of each site, as litterfall, root litter, and throughfall, were mostly mineralized to CO_2 because the input was balanced by the organic matter decomposition rate, and only a portion of the input C leached into the mineral horizon as dissolved organic C (DOC) (AD, 7 % of the input; MV, 5 % of the input; Tables 9.3 and 9.4), as reported by Fujii et al. (2009a). DOC in the forest floor is considered to be released mainly by microorganisms as they decompose litter via solubilization and mineralization (Guggenberger and Zech 1994). Considering the lower DOC fluxes from throughfall, the O horizon, not the canopy, was the main source of DOC, NO_3-N, and dissolved organic N (DON) at both sites, as has generally been reported, except for a wet Costa Rican tropical forest (Schwendenmann and Veldkamp 2005) where the annual precipitation is much higher (4200 mm). The DOC fluxes from the O horizon and root litter are important C sources for mineral soils (Neff and Asner 2001). This was also the case in this study because of the small fluxes from the 30-cm depth (AD, 18 kg C ha^{-1} year^{-1}; MV, 29 kg C ha^{-1} year^{-1}; Table 9.4),

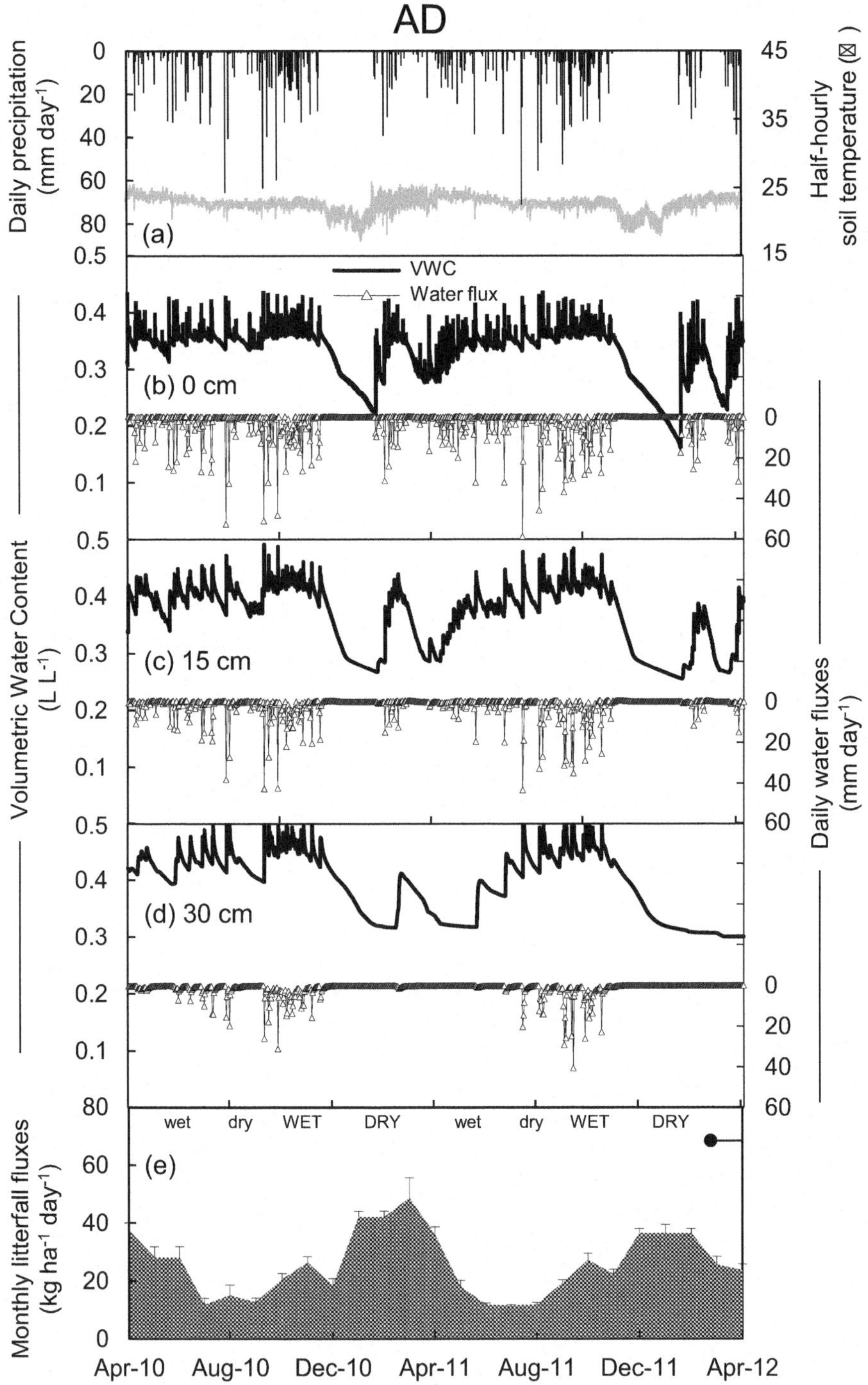

Fig. 9.4 Daily precipitation and half-hourly soil temperature at a 5-cm soil depth (**a**); seasonal fluctuations of VWC and daily water fluxes calculated by HYDRUS at 0- (**b**), 15- (**c**), and 30-cm

although the proportion of DOC fluxes from the O horizon to a depth of 30 cm was higher in MV (AD, 5.3 % of that at the 0-cm depth; MV, 13 % of that at the 0-cm depth; Table 9.4). Although, in mineral soil horizons, the concentrations and fluxes of DOC in the soil solution decrease generally via mineralization and adsorption onto mineral surfaces (Kaiser and Zech 2000; Kalbitz et al. 2000), adsorption is the dominant factor (Qualls and Haines 1992), especially in clay-rich Acrisols and Ferralsols where the amounts of clay minerals or Al and Fe (hydr)oxide adsorbents are high, compared with sandy materials (Fujii et al. 2011b). Therefore, the low DOC fluxes from the 30-cm depth in our study are attributed to highly weathered, clayey Ferralsols and Acrisols, where adsorption is the dominant process that controls the DOC fluxes in the soil profile. The slightly higher flux from, as well as the higher proportion of DOC at, the 30-cm depth in MV might contribute to the development of the deep BA horizon at this site. Our results show that the mineral soil appeared to be the major sink for DOC and DON at each site.

9.5 Nitrogen Cycles

The N stock in the mineral soil was 3.8 Mg N ha^{-1} at AD and 4.7 Mg N ha^{-1} at MV; the N stocks in the organic horizon were low (AD, 120 kg N ha^{-1}; MV, 91 kg N ha^{-1}), which are comparable to those of other tropical forests (Markewitz et al. 2004). The fine-root N biomasses at AD and MV were 7.8 kg and 38 kg N ha^{-1}, respectively.

As expected, the large amounts of N input from leaf litterfall at both sites (AD, 237 kg N ha^{-1} year^{-1}; MV, 209 kg N ha^{-1} year^{-1}) indicate that both forest ecosystems are N-rich, although the leaf C/N ratio at AD (17) was slightly, but significantly, lower than that at MV (20) (Table 9.5). Total dissolved N (TDN) fluxes beneath the forest floor differed substantially between the two sites (Fig. 9.6). At AD, the TDN flux increased by nearly 90 kg N ha^{-1} year^{-1} from throughfall to the 0-cm soil depth, and approximately 80 % of it was NO_3-N, which is susceptible to leaching. The TDN fluxes beneath the forest floor accounted for almost half of the annual N input. At the 30-cm soil depth, the TDN fluxes decreased to 13 % of that at the 0-cm depth (5 % of the input), and NO_3-N still dominated the TDN. Thus, the results indicate that almost half of the input N leaches down to the mineral soil and is taken up as NO_3-N by plants in the mineral horizon, as well as in the O horizon at AD, which has moderately acidic, clayey Ferralsols. Such a high N availability and a leaky, open N cycle at AD are not caused by anthropogenic N deposition, but can be explained by the dominance of leguminous species that are capable of symbiotic N_2 fixation, which suggests that NO_3-N is not only dominant

---

Fig. 9.4 (continued) soil depths (**d**); and monthly fluctuation of litterfall fluxes (**e**) at AD (wet, minor wet season; dry, minor dry season; WET, major wet season; and DRY, major dry season). Bars indicate standard errors

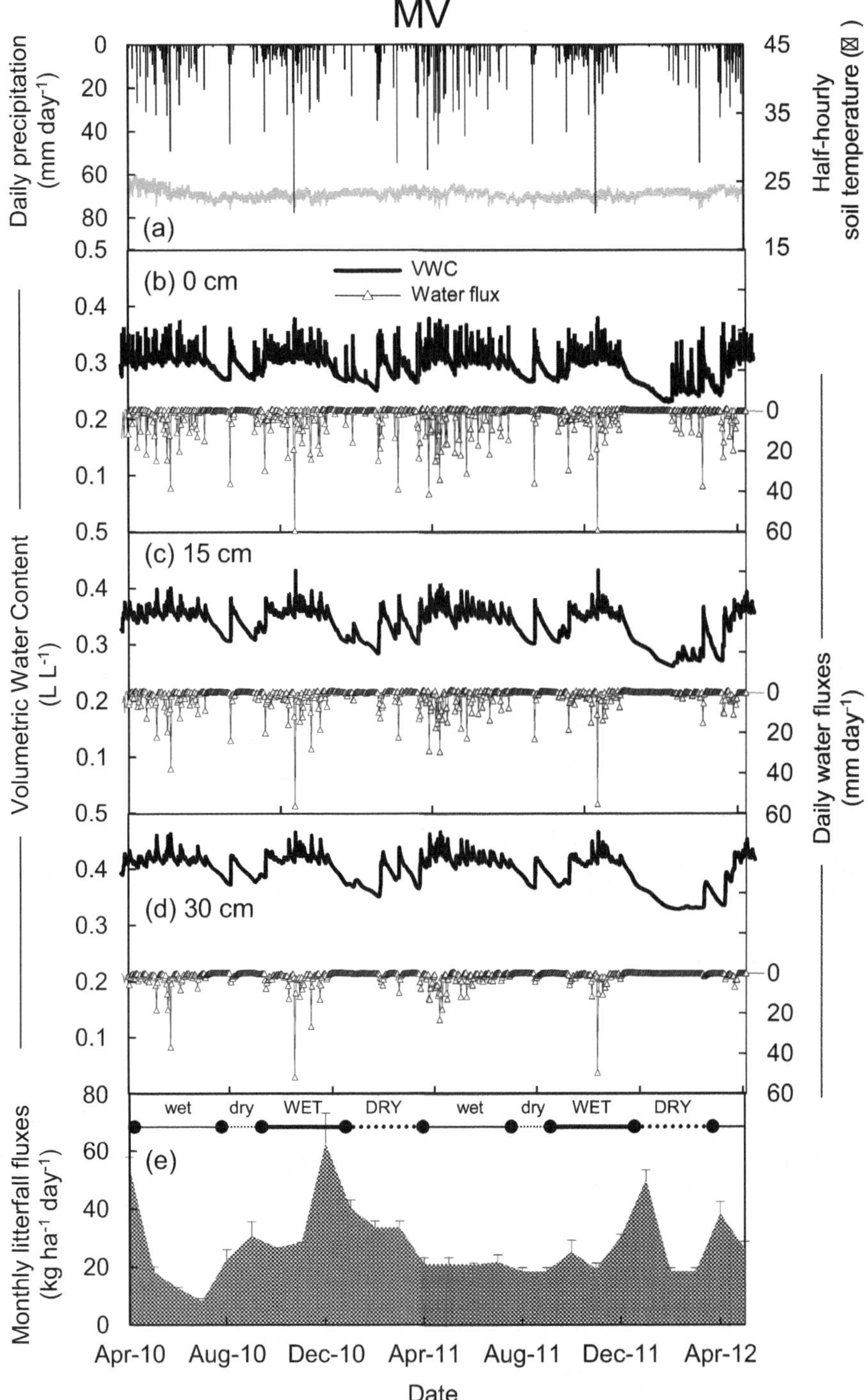

Fig. 9.5 Daily precipitation and half-hourly soil temperature at a 5-cm soil depth (**a**); seasonal fluctuations of VWC and daily water fluxes calculated by HYDRUS at 0- (**b**), 15- (**c**), and 30-cm soil depths (**d**); and monthly fluctuation of litterfall fluxes (**e**) at MV (wet, minor wet season; dry, minor dry season; WET, major wet season; and DRY, major dry season). Bars indicate standard errors

Table. 9.3 Annual flow and stock of carbon

		AD	MV
C stock (Mg C ha^{-1})			
Aboveground biomass		122 (−)	152 (−)
Fine-root biomass			
O horizon		0.1 (0.0)	0.7 (0.1)
Mineral horizon (0–30 cm)		1.0 (0.1)	1.3 (0.1)
Total		1.1 (0.1)	2.0 (0.2)
Soil organic matter			
O horizon		2.0 (0.4)	2.0 (0.6)
Mineral soil (0–30 cm)		45 (−)	51 (−)
C flow (Mg C ha^{-1} year^{-1})			
Litterfall	(a)	4.5 (0.5)	4.5 (0.6)
Root litter[a]	(b)	0.2 (0.0)	0.4 (0.0)
C input[b]	(c)	4.7 (0.5)	4.9 (0.6)
Whole soil respiration		6.7 (0.9)	7.1 (0.8)
Organic matter decomposition	(d)	4.1 (0.6)	4.8 (0.6)
C budget in soil[c]	(e)	0.6 (0.8)	0.1 (0.9)

[a]The annual rates of root litter were assumed to be 20 % of the fine-root biomass (Nakane 1980)
[b]C input was calculated as the sum of litterfall and root litter (c = a + b)
[c]C budget in the soil was calculated as the difference between organic matter decomposition and C input (e = c − d)

in forest ecosystems that receive large anthropogenic N inputs. Such a high dissolved inorganic N (DIN) flux beneath the forest floor is exceptional among other tropical forests, except for wet tropical forests in Costa Rica (Schwendenmann and Veldkamp 2005) and the Amazon (Markewitz et al. 2004) that consist of Ferralsols and are dominated by leguminous species. In contrast, at MV, the TDN flux of throughfall was almost the same as that of the litter leachate, and it only accounted for 11 % of the input, indicating that 89 % of the mineralized N within the O horizon was absorbed by plants and/or soil microbes. The TDN fluxes at the 30-cm depth decreased to 2 % of the input, and N species were almost evenly distributed because of the low amount of NO_3-N, compared with that at AD. NH_4^+ leaching from the highly acidic O horizon was minor at MV, which contrasts with the findings of KR3, BS, and BB from Southeast Asia in Chap. 7 that have strongly acidic O horizon, where nitrification could be retarded by acidic condition. Thus, at MV, which has strongly acidic, clayey Acrisol soils, the results indicate that N is mostly taken up within the O horizon that contains a dense root mat. Consequently, N is better conserved at MV. Taking account of the slightly lower N concentration of leaf litter, which controls (i.e., slows down) net N release rates (Parton et al. 2007), as well as the small flux of leachable NO_3-N beneath the forest floor, compared with AD, plants at MV might take up more N as NH_4-N, although this will need to be clarified by future studies.

The different N-acquiring strategies of each ecosystem could be attributed to the dense root mat within the O horizon at MV. Dense root mats with associated

Table. 9.4 Annual fluxes of water and solutes and volume-weighted mean pH and DOC/DON ratios leached from each stratum (PP, precipitation; TF, throughfall). The figures in parentheses represent the standard error

		Water		DOC	NO_3^-	NH_4^+	DON	DOC/DON
		mm year^{-1}	pH	kg C ha^{-1} year^{-1}	kg N ha^{-1} year^{-1}			g g^{-1}
AD	PP	1460	6.4 (0.1)	83 (7.9)	1.1 (0.2)	4.3 (0.6)	5.0 (0.3)	18 (2.0)
	TF	1224	5.9 (0.1)	114 (9.4)	24 (1.7)	4.3 (0.6)	8.1 (0.5)	14 (1.4)
	0 cm	1047	6.5 (0.2)	335 (14)	92 (6.9)	3.3 (0.5)	25 (1.1)	12 (0.7)
	15 cm	700	5.1 (0.1)	26 (1.0)	14 (1.2)	1.4 (0.2)	2.1 (0.2)	12 (1.1)
	30 cm	514	5.4 (0.2)	18 (2.4)	12 (1.3)	1.7 (0.3)	1.5 (0.3)	12 (2.9)
MV	PP	1573	–	–	–	–	–	–
	TF	1226	6.0 (0.1)	102 (3.8)	20 (1.5)	5.6 (0.6)	8.0 (0.6)	13 (1.0)
	0 cm	887	4.8 (0.1)	228 (18)	12 (1.5)	5.0 (0.6)	10 (0.8)	23 (2.5)
	15 cm	575	5.3 (0.1)	41 (1.5)	3.7 (0.8)	1.5 (0.3)	2.9 (0.2)	14 (1.2)
	30 cm	471	5.5 (0.1)	29 (0.9)	2.6 (0.4)	1.3 (0.3)	2.0 (0.2)	14 (1.6)

Table. 9.5 Mean annual nutrient flux of litterfall in from April 2010 to May 2012. The figures in parentheses represent standard errors

		Dry mass	C	N	C/N
		Mg ha^{-1} year^{-1}		kg ha^{-1} year^{-1}	g g^{-1}
AD	Leaf	7.6 (0.8)	3.8 (0.4)	217 (20)	17
	Total	9.0 (1.0)	4.5 (0.5)	237 (20)	19
MV	Leaf	7.7 (0.9)	3.8 (0.5)	190 (27)	20
	Total	9.1 (1.2)	4.5 (0.6)	209 (30)	21

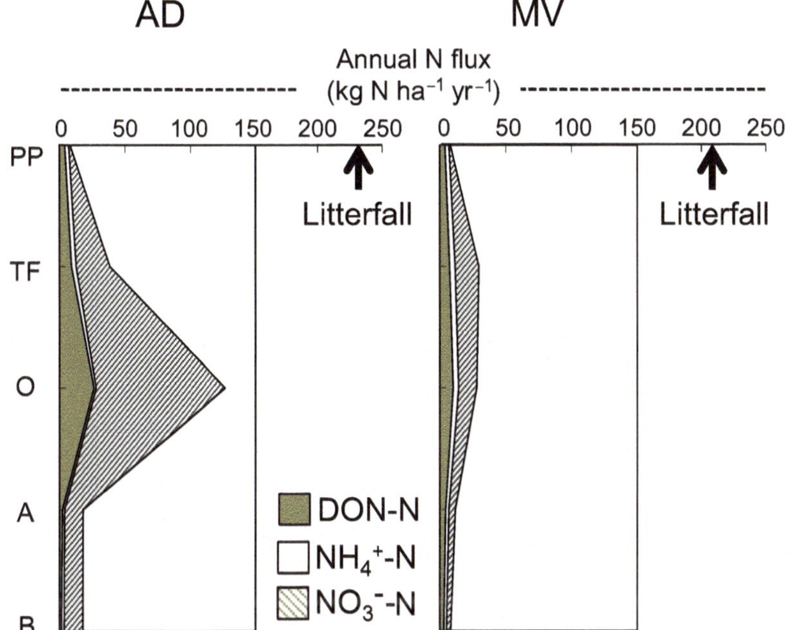

Fig. 9.6 Fluxes of N in litterfall, throughfall, and soil solution of tropical forest ecosystems. Blank arrows indicate litterfall N flux within ecosystems. PP, TF, O, A, and B represent precipitation, throughfall, and the O, A, and B horizons

ectomycorrhizae are believed to remove NH_4-N from the soil (Aber et al. 1985; Chandler 1985), prevent an increase in pH and subsequent nitrification (Jordan et al. 1979), and take up nutrients directly from litter (Sayer et al. 2006). These highly effective mechanisms for recycling nutrients may occur at MV because of a high exchangeable Al concentration in the mineral soil (Dezzeo and Chacón 2006; Kingsbury and Kellman 1997) or high water availability in the surface soil (Sayer et al. 2006). Fine roots, which often produce a fibrous and well-drained litter layer that may reduce the moisture content (Kingsbury and Kellman 1997), are reported to be very sensitive to changes in moisture levels (Persson 1980). Additionally, a strongly superficial root distribution may increase fine-root mortality during dry spells (Joslin and Henderson 1987). As a result, the majority of fine roots in the litter likely die during the dry season, as the production of new fine roots is costly. This also might be the reason that a root mat was not observed at AD, which is subject to a more intensive dry period. Overall, our data reveal a different picture of plant-soil N dynamics in two Central African tropical forests that have Ferralsol and Acrisol soils.

9.6 Dissolved Organic Matter Dynamics

The DOC and DON concentrations were positively and significantly associated for each stratum at both sites ($r = 0.63–0.87$, $p < 0.001$, $n = 79–109$), which is indicative of tight C and N cycling (Solinger et al. 2001). Since the volume-weighted mean DOC/DON ratios of the soil solution beneath the forest floor significantly differed and well reflected the litterfall C/N ratio for each site (Tables 9.4 and 9.5), litter decomposition was the primary source of the DOC and DON in the mineral soils (Homann and Grigal 1992; Kalbitz et al. 2000). This likely affects the subsequent rate at which organic N is mineralized into bioavailable inorganic N (NH_4-N and NO_3-N), which is regarded as a key process for soil-based N release (Davidson et al. 1992). Campbell et al. (2000) found a significantly negative relationship between the annual DIN flux and the stream water DOC/DON ratio, and they suggested that the DOC/DON ratio could be an effective predictor of N leaching and N status. This fact supports the view that AD is a N-leaky ecosystem with an extremely low DOC/DON ratio (12 ± 0.7) in the forest floor leachate, while at MV, the net N mineralization rate in the forest floor is supposed to be slower because of its higher DOC/DON ratio (23 ± 2.5), implying that it has a conservative nutrient cycle.

Unlike the forest floor leachate, the DOC/DON ratios in the mineral soil were similarly close to the C/N ratios of the soil organic matter at both sites. At AD, the DOC/DON ratio of the forest floor leachate was constant in the mineral soil because of the originally low ratios. This could be due to a weaker adsorption of dissolved organic matter (DOM) with low DOC/DON ratios (Kalbitz et al. 2000). In contrast, at MV, the DOC/DON ratio of the forest floor leachate decreased in the mineral soil. This could be due to the presence of older and more decomposed organic matter in the deeper soil horizon, or because DON was less reactive than DOC, resulting in a decreasing DOC/DON ratio with increasing soil depth (Sleutel et al. 2009). This might also suggest that the organic matter being translocated downward is enriched in N, as was reported by Schoenau and Bettany (1987) and Qualls et al. (1991). The incorporation of extra N in DOM occurs at MV and is characteristic of decomposition and humification (Qualls and Haines 1992). Thus, it was quantitatively shown that the role of DOM in C and N cycling varies depending on soil type or the litter C/N ratio (Fujii et al. 2011a).

9.7 Proton Generation and Consumption Processes

Based on the ion fluxes in each horizon (Fig. 9.7; the canopy and O, A, and B horizons) and the excess cation uptake by vegetation, we calculated carbonic acid dissociation and protonation (NPG_{Car}); organic anion dissociation, mineralization, and adsorption to soil (NPG_{Org}); N mineralization, nitrification, and N uptake by plants (NPG_{Ntr}); plant excess cation uptake (NPG_{Bio}); and cation release and accumulation (ΔANC) (Fig. 9.8), as described in Chap. 7. At both MV and AD, in the entire ecosystems, including the forest canopy, NPG_{Bio} and ΔANC accounted for almost all of the proton generation and consumption, respectively. This is consistent with the fact that complete C and N cycles are balanced by net proton fluxes of 0 in forest ecosystems (Binkley 1987). However, horizontal heterogeneity of proton generation and consumption was observed in both ecosystems, and AD was more dynamic in this regard.

In the canopy and the O horizon in AD, NPG_{Ntr} and NPG_{Org} were the largest proton-generating processes. In terms of proton cycling, the canopy and O horizon had qualitatively similar functions, although the canopy contribution was approximately one-third of that of the O horizon. Therefore, the canopy at AD can be regarded as an incipient O horizon in the sky. Then, in the A horizon, protons generated in the canopy and O horizon were almost completely consumed by plant nitrate uptake and the mineralization and/or adsorption of organic anions, as well by NPG_{Car}, which resulted in a positive ΔANC. This indicates that a net accumulation of cations leached from the upper horizon and that C may have accumulated because of the adsorption of organic anions in the clay-rich soils of AD. Finally, in the B horizon in AD, 98 % of protons were generated by plant cation uptake, and 95 % of proton consumption resulted from the release of basic cations from the mineral soil. This indicates that the proton cycles related to C and N cycles were almost complete from the canopy to the A horizon in AD, which resulted in minor

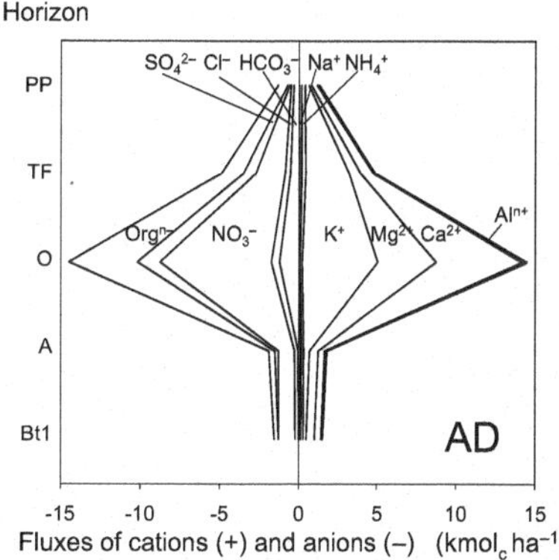

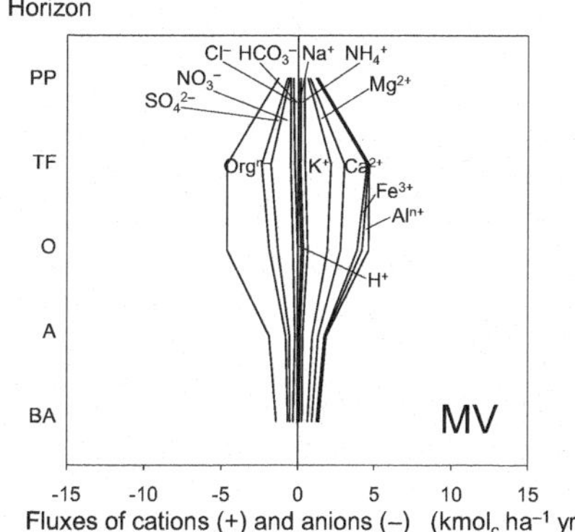

Fig. 9.7 Annual fluxes of cations and anions in each horizon. PP represents precipitation; TF represents throughfall; and O, A, Bt1, and BA represent soil horizons

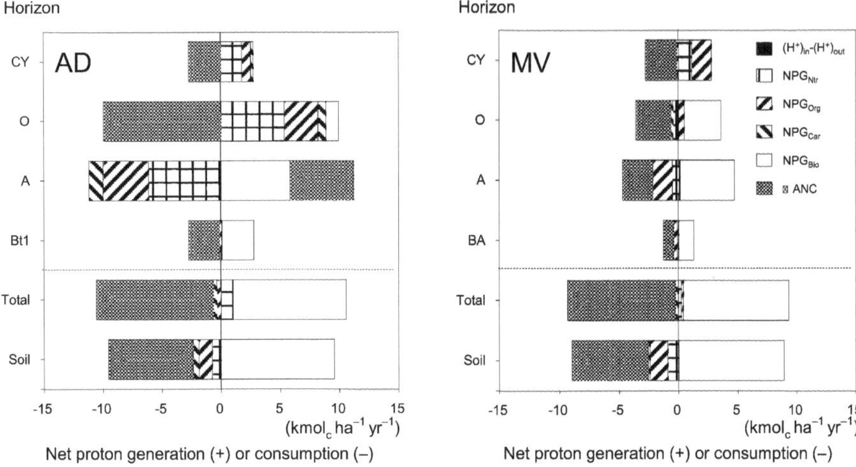

Fig. 9.8 Net proton generation and consumption in the soil profiles. CY represents the canopy; TF represents throughfall; O, A, Bt1, and BA represent soil horizons; and soil represents total proton generation and consumption minus those of the CY

contributions of NPG$_{Car}$, NPG$_{Org}$, and NPG$_{Ntr}$ in the B horizon. Judging from the relatively higher leaching fluxes of nitrate and basic cations from the 30-cm soil depth, as well as the positive contribution of NPG$_{Ntr}$ to the total ecosystem in AD, plants should also take up bases and nitrate from the deeper horizons of Ferralsols, as has been reported previously (Freycon et al. 2015; Nepstad et al. 1994). Accordingly, the results we observed in AD, which has Ferralsol soils, imply that plants "mine" and pump soil-derived nutrients up from the deeper B horizon, "retake" the O horizon-derived nutrients mainly from the A horizon, and then accumulate them in the canopy and O horizon. Our results also suggest that the acid load derived from nitrification and the dissociation of organic anions contribute to basic cations leaching from the canopy and the O horizon to the A horizon; then, proton generation by plant cation uptake in the A horizon is not sufficient to balance the net proton consumption, which contributes to the accumulation of base cations, and possibly C, in the "developing" A horizon. This is consistent with the deeper A horizon in AD compared with that of tropical forests in Southeast Asia (Fujii et al. 2009a, b).

In MV, because of the thick fine-root mat in the O horizon, 86 % of the proton generation resulted from plant excess cation uptake, while the remainder was due to NPG$_{Org}$. This is contrast with the minor contribution (11 %) of NPG$_{Bio}$ in the O horizon of AD, where nitrification and organic acid dissociation were the major proton-generating processes, although the function of the MV canopy was similar to that of AD. Judging from the similar tree family compositions of AD and MV, soil acidity could be the factor that controls proton budgets and fine-root distribution, as we discussed in the previous section and in Chap. 7. The low NPG$_{Ntr}$ in the O horizon of MV indicated that N that was produced by mineralization and nitrification seemed to be immediately taken up by plants within the O horizon. In the A and B horizons, protons generated by plant cation uptake were consumed by NPG_{Org} and NPG_{Ntr} in combination with basic cation release from the soil. In MV, which has Acrisol soils, the major proton-generating contribution of excess cation uptake by plants was due to the concentrated, thick root mat in the O horizon, which resulted in intensive acidification of the surface soil and, to some degree, the subsequent proton efflux to the subsoil, as shown in BS and BB in Chap. 7.

9.8 Conclusions

Our study suggests that N input can be very high in tropical forests of Central Africa because of the dominance of leguminous species. Nevertheless, the horizon in which plants take up N, as well as in which N transformation processes occur, differs depending on soil type, although the C balance is similar. In a tropical forest that has moderately acidic Ferralsols, plants take up N both in the O and mineral horizons, which is indicative of open N cycling. The high acid load is distributed in the O horizon via nitrification and organic anion dissociation, which subsequently results in nutrient leaching to the A horizon, where plants mainly absorb nutrients. In contrast, in a tropical forest that has strongly acidic Acrisols, almost all of the N uptake by plants occurs in the O horizon because of the presence of a dense root mat, which is indicative of tight N cycling. The high acid load is concentrated in the O and A horizons by plant cation uptake, which results in less nutrient leaching. Soil acidity might control the fine-root distribution, which causes heterogeneous soil acidification throughout the soil profile. Thus, the ecosystem processes and strategies used by plants to acquire nutrients differ depending on soil type (i.e., soil acidity).

References

Aber JD, Melillo JM, Nadelhoffer KJ, McClaugherty CA, Pastor J (1985) Fine root turnover in forest ecosystems in relation to quantity and form of nitrogen availability: a comparison of two methods. Oecologia 66:317–321

Alves LF, Vieira SA, Scaranello MA, Camargo PB, Santos FAM, Joly CA, Martinelli LA (2010) Forest structure and live aboveground biomass variation along an elevational gradient of tropical Atlantic moist forest (Brazil). For Ecol Manag 260:679–691

Asner GP, Martin RE, Ford AJ, Metcalfe DJ, Liddell MJ (2009) Leaf chemical and spectral diversity in Australian tropical forests. Ecol Appl 19(1):236–253

Asner G, Martin R, Suhaili A (2012) Sources of canopy chemical and spectral diversity in lowland Bornean forest. Ecosystems 15(3):504–517

Asner GP, Martin RE, Tupayachi R, Anderson CB, Sinca F, Carranza–Jiménez L, Martinez P (2014) Amazonian functional diversity from forest canopy chemical assembly. Proc Natl Acad Sci U S A 111(15):5604–5609

Binkley D, Richter D (1987) Nutrient cycles and H$^+$ budgets of forest ecosystems. Adv Ecol Res 16:2–51

Campbell JL, Hornbeck JW, McDowell WH, Buso DC, Shanley JB, Likens GE (2000) Dissolved organic nitrogen budgets for upland, forested ecosystems in New England. Biogeochemistry 49:123–142

Chandler G (1985) Mineralization and nitrification in three Malaysian forest soils. Soil Biol Biochem 17(3):347–353

Crews T (1999) The presence of nitrogen fixing legumes in terrestrial communities: evolutionary vs ecological considerations. Biogeochemistry 46:233–246

Davidson EA, Hart SC, Firestone MK (1992) Internal cycling of nitrate in soils of a mature coniferous forest. Ecology 73:1148–1156

De Graaf MCC, Bobbink R, Verbeek PJM, Roelofs JGM (1997) Aluminium toxicity and tolerance in three heathland species. Water Air Soil Pollut 98:229–239

Dezzeo N, Chacón N (2006) Nutrient fluxes in incident rainfall, throughfall, and stemflow in adjacent primary and secondary forests of the Gran Sabana, southern Venezuela. For Ecol Manag 234:218–226

Diabate M, Munive A, de Faria SM, Ba A, Dreyfus B, Galiana A (2005) Occurrence of nodulation in unexplored leguminous trees native to the West African tropical rainforest and inoculation response of native species useful in reforestation. New Phytol 166:231–239

Doyle JJ, Luckow MA (2003) The rest of the iceberg. Legume diversity and evolution in a phylogenetic context. Plant Physiol 131:900–910

Freycon V, Wonkam C, Fayolle A, Laclau JP, Lucot E, Jourdan C, Cornu G, Gourlet-Fleury S (2015) Tree roots can penetrate deeply in African semi-deciduous rain forests: evidence from two common soil types. J Trop Ecol 31:13–23

Fujii K, Funakawa S, Hayakawa C, Kosaki T (2008) Contribution of different proton sources to pedogenetic soil acidification in forested ecosystems in Japan. Geoderma 144:478–490

Fujii K, Funakawa S, Hayakawa C, Sukartiningsih, Kosaki T (2009a) Quantification of proton budgets in soils of cropland and adjacent forest in Thailand and Indonesia. Plant Soil 316:241–255

Fujii K, Uemura M, Funakawa S, Hayakawa C, Sukartiningsih, Kosaki T, Ohta S (2009b) Fluxes of dissolved organic carbon in two tropical forest ecosystems of East Kalimantan, Indonesia. Geoderma 152:127–136

Fujii K, Funakawa S, Shinjo H, Hayakawa C, Mori K, Kosaki T (2011a) Fluxes of dissolved organic carbon and nitrogen throughout Andisol, Spodosol and Inceptisol profiles under forest in Japan. Soil Sci Plant Nutr 57:855–866

Fujii K, Hartono A, Funakawa S, Uemura M, Kosaki T (2011b) Fluxes of dissolved organic carbon in three tropical secondary forests developed on serpentine and mudstone. Geoderma 163:119–126

Fyllas NM, Patiño S, Baker TR, Nardoto GB, Martinelli LA, Quesada CA, Paiva R, Schwarz M, Horna V, Mercado LM, Santos A, Arroyo L, Jiménez EM, Luizão FJ, Neill DA, Silva N, Prieto A, Rudas A, Silviera M, Vieira ICG, Lopez–Gonzalez G, Malhi Y, Phillips OL, Lloyd J (2009) Basin-wide variations in foliar properties of Amazonian forest: phylogeny, soils and climate. Biogeosciences 6:2677–2708

Gaston G, Brown S, Lorenzin M, Singh KD (1998) State and change in carbon pools in the forests of tropical Africa. Glob Chang Biol 4:97–114

Guggenberger G, Zech W (1994) Dissolved organic carbon control in forest floor leachate: simple degradation products or humic substances? Sci Total Environ 152:37–47

Homann PS, Grigal DF (1992) Molecular weight distribution of soluble organics from laboratory-manipulated surface soils. Soil Sci Soc Am J 56:1305–1310

IUSS Working Group WRB. 2014. World Reference Base for Soil Resources 2014. International soil classification system for naming soils and creating legends for soil maps. World Soil Resources Reports No. 106. FAO, Rome.

Jones A, Breuning-Madsen H, Brossard M, Dampha A, Deckers J, Dewitte O, Gallali T, Hallett S, Jones R, Kilasara M, Le Roux P, Micheli E, Montanarella L, Spaargaren O, Thiombiano L, Van Ranst E, Yemefack M, Zougmoré R. (eds.) (2013) Soil atlas of Africa. European commission, publications office of the European Union, Luxembourg, p 144

Jordan CF, Todd RL, Escalante G (1979) Nitrogen conservation in a tropical rain forest. Oecologia 39:123–128

Joslin JD, Henderson GS (1987) Organic matter and nutrients associated with fine root turnover in a white oak stand. For Sci 33:330–346

Kaiser K, Zech W (2000) Dissolved organic matter sorption by mineral constituents of subsoil clay fractions. J Plant Nutr Soil Sci 163:531–535

Kalbitz K, Solinger S, Park JH, Michalzik B, Matzner E (2000) Controls on the dynamics of dissolved organic matter in soils: a review. Soil Sci 165:277–304

Kingsbury N, Kellman M (1997) Root mat depths and surface soil chemistry in southeastern Venezuela. J Trop Ecol 13:475–479

Knops JMH, Bradley KL, Wedin DA (2002) Mechanisms of plant species impacts on ecosystem nitrogen cycling. Ecol Lett 5:454–466

Markewitz D, Davidson E, Moutinho P, Nepstad D (2004) Nutrient loss and redistribution after forest clearing on a highly weathered soil in Amazonia. Ecol Appl 14(4):177–199

McDowell WH (1998) Internal nutrient fluxes in a Puerto Rican rain forest. J Trop Ecol 14:521–536

Murach D, Ulrich B (1988) Destabilization of forest ecosystems by acid deposition. GeoJournal 17:253–260

Nakane K (1980) Comparative studies of cycling of soil organic carbon in three primeval moist forests. Jpn J Ecol 3:155–172

Neff JC, Asner GP (2001) Dissolved organic carbon in terrestrial ecosystems: synthesis and a model. Ecosystems 4:29–48

Nepstad DC, de Carvalho CR, Davidson EA, Jipp PH, Lefebvre PA, Negreiros GH, da Silva ED, Stone TA, Trumbore SE, Vieira S (1994) The role of deep roots in the hydrological and carbon cycles of Amazonian forests and pastures. Nature 372:666–669

Oliveira EP, Toteu SF, Araújo MNC, Carvalho MJ, Nascimento RS, Bueno JF, McNaughton N, Basilici G (2006) Geologic correlation between the Neoproterozoic Sergipano belt (NE Brazil) and the Yaounde´ belt (Cameroon, Africa). J Afr Earth Sci 44:470–478

Parton W, Silver WL, Burke IC, Grassens L, Harmon ME, Currie WS, King JY, Adair EC, Brandt LA, Hart SC, Fasth B (2007) Global-scale similarities in nitrogen release patterns during long-term decomposition. Science 315:361–364

Peh KSH, Sonké B, Taedoung H, Séné O, Lloyd J, Lewis SL (2012) Investigating diversity dependence of tropical forest litter decomposition: experiments and observations from Central Africa. J Veg Sci 23:223–235

Persson H (1980) Spatial distribution of fine-root growth, mortality, and decomposition in a young Scots pine stand in central Sweden. Oikos 34:77–87

Pons TL, Perreijn K, van Kessel C, Werger MJA (2007) Symbiotic nitrogen fixation in a tropical rainforest:^{15}N natural abundance measurements supported by experimental isotopic enrichment. New Phytol 173:154–167

Qualls RG, Haines BL (1992) Biodegradability of dissolved organic matter in forest throughfall, soil solution, and stream water. Soil Sci Soc Am J 56:578–586

Qualls RG, Haines BL, Swank WT (1991) Fluxes of dissolved organic nutrients and humic substances in a deciduous forest. Ecology 72:254–266

Sayer EJ, Tanner EVJ, Cheesman AW (2006) Increased litterfall changes fine root distribution in a moist tropical forest. Plant Soil 281:5–13

Schoenau JJ, Bettany JR (1987) Organic matter leaching as a component of carbon, nitrogen, phosphorus, and sulfur cycles in a forest, grassland, and gleyed soil. Soil Sci Soc Am J 51:646–651

Schwendenmann L, Veldkamp E (2005) The role of dissolved organic carbon, dissolved organic nitrogen, and dissolved inorganic nitrogen in a tropical wet forest ecosystem. Ecosystems 8:339–351

Sleutel S, Vandenbruwane J, De Schrijver A, Wuyts K, Moeskops B, Verheyen K, De Neve S (2009) Patterns of dissolved organic carbon and nitrogen fluxes in deciduous and coniferous forests under historic high nitrogen deposition. Biogeosciences 6:2743–2758

Soil Survey Staff (2014) Keys to soil taxonomy, 12th edn. United States Department of Agriculture Natural Resources Conservation Service, Washington, DC

Solinger S, Kalbitz K, Matzner E (2001) Controls on the dynamics of dissolved organic carbon and nitrogen in a Central European deciduous forest. Biogeochemistry 55:327–349

Spehn EM, Scherer–Lorenzen M, Schmid B, Hector A, MC C, PG D, JA F, Jumpponen A, O'Donnovan G, JS P, ED S, AY T, Körner C (2002) The role of legumes as a component of biodiversity in a cross–European study of grassland biomass nitrogen. Oikos 98:205–218

Stark NM, Jordan CF (1978) Nutrient retention by the root mat of Amazonian rain forest. Ecology 59(3):434–437

Toteu SF, Fouateu RY, Penaye J, Tchakounte J, Mouangue ACS, Van Schmus WR, Deloule E, Stendal H (2006) U–Pb dating of plutonic rocks involved in the nappe tectonic in southern Cameroon: consequence for the Pan-African orogenic evolution of the central African fold belt. J Afr Earth Sci 44:479–493

Van Wambeke A (1992) Oxisols. In: Van Wambeke A (ed) Soils of the tropics—properties and appraisal. McGraw-Hill, New York, pp 139–161.

Yahara T, Javadi F, Onoda Y, de Queiroz LP, Faith DP, Prado DE, Akasaka M, Kadoya T, Ishihama F, Davies S, Slik JWF, Yi T, Ma K, Bin C, Darnaedi D, Pennington RT, Tuda M, Shimada M, Ito M, Egan AN, Buerki S, Raes N, Kajita T, Vatanparast M, Mimura M, Tachida H, Iwasa Y, Smith GF, Victor JE, Nkonki T (2013) Global legume diversity assessment: concepts, key indicators, and strategies. Taxon 62(2):249–266

Chapter 10
Changes in Elemental Dynamics After Reclamation of Forest and Savanna in Cameroon and Comparison with the Case in Southeast Asia

Makoto Shibata

Abstract To understand the effects of original vegetation (forest or savanna) on changes of solute leaching and proton budgets after reclamation of Oxisols in Cameroon, we quantified the soil nutrient fluxes in forest, adjacent savanna, and each adjacent maize cropland and compared with the case in Southeast Asia. In forest plot, excess cation accumulation in wood has contributed to soil acidification in the entire soil profile, while soil acidification rates were much lower in savanna plot because of limited plant uptake. As a result of cultivation, NO_3^- fluxes were substantially increased and nitrification was the main process of soil acidification in both croplands. Reflecting the nutrient flux pattern of original vegetation, protons generated by nitrification in forest cropland plot (9.4–10.1 $kmol_c$ ha^{-1} $year^{-1}$) were significantly higher than that in savanna cropland (3.4–4.5 $kmol_c$ ha^{-1} $year^{-1}$). The rate in savanna cropland was comparable to that in Thailand Ultisol (5.0 $kmol_c$ ha^{-1} $year^{-1}$) with moderate soil pH and higher than that in Indonesian Ultisol (1.5 $kmol_c$ ha^{-1} $year^{-1}$) with low pH. Despite low pH of bulk Oxisols of Cameroon, they would provide favorable habitat for nitrifiers with physically well-structured microaggregates, allowing active nitrification in the plots of Cameroon. High rate of nitrification suggests the risk of nutrient deficiency in cropland is more serious in nutrient-poor Oxisols. The effects of reclamation on soil acidification processes would depend on the original vegetation and also on soil pH and physical structure, which affect the nitrification activity. Since the ratio of K^+ concentrations to sum of Mg^{2+} and Ca^{2+} concentrations increased with decreasing soil solution pH, the lower solution pH, which could stem from cultivation, might promote K leaching from cropland.

M. Shibata (✉)
Graduate School of Global Environmental Studies, Kyoto University, Kitashirakawaoiwake-cho, Sakyo-ku, Kyoto 606-8502, Japan
e-mail: makotosbt@gmail.com

© Springer Japan KK 2017
S. Funakawa (ed.), *Soils, Ecosystem Processes, and Agricultural Development*,
DOI 10.1007/978-4-431-56484-3_10

10.1 Introduction

In West and Central Africa, the transitory ecotone between Congo basin forest and Guinea savanna, called as "forest-savanna mosaic" is widely distributed to the north and south of the forest belt (Mitchard et al. 2011). In this region, there is sufficient rainfall (ca. 1500 mm year^{-1} with udic moisture regime) to support forest growth, and the savanna vegetation is maintained by anthropogenic disturbance such as fire (Sankaran et al. 2005; Staver et al. 2011). The area of forest-savanna mosaic on the continent of Africa reaches approximately 1.28 million km^2, comparable to 2.36 million km^2 dense forest (Mayaux et al. 2004), as was mentioned in Chap. 8. Because of the rapid population growth in sub-Saharan Africa—the populations will double over next two decades, being nearly equal to the China region (Lutz et al. 2001)—and large dependence on natural resources due to the low capita GNP (Khasa et al. 1995), agricultural pressure on a local nature (i.e., forest and savanna) is likely to increase (Gaston et al. 1998). The sustainable way of food production to keep up with rising population pressure is a major concern in the agricultural development of West Africa (Smith et al. 1994) where farmers cultivate both in forest and savanna.

Cultivation accelerates soil acidification: biological activity produces various acids, such as carbonic acid, nitric acid, and organic acid or acid (proton) generated by cation uptake in excess of anion uptake by plants, all of which could enhance mineral weathering, resulting in the release of basic cations into soil solution—the source of essential nutrients for plants. The effects of cultivation on the individual acid-generating processes are various (Fujii et al. 2009), and therefore, they are to be quantified to understand soil acidification processes owing to cultivation. Besides, these processes are driving forces of nutrient leaching from cropland. Fujii et al. (2009) concluded that the influence of cultivation on proton budget is different depending on soil pH and texture in Southeast Asia, where Ultisols (Alisols/Acrisols) are distributed. However, it is still unclear about the case of Oxisols (Ferralsols) in Africa, which are characterized by low available mineral nutrients and low retention capacity of nutrients owing to extreme weathering (van Wambeke 1992; Jones et al. 2013). Oxisols have the great macroporosity and hydraulic conductivity with physically well-structured microaggregates, which contributes to a high rate of solute downward movement (Anamosa et al. 1990; Melgar et al. 1992). Due to the combination of such high soil permeability with flat topography of the continent, leaching becomes the primary cause of soil nutrient losses from croplands in equatorial Africa; this is in contrast with Southeast Asia where erosion and runoff are the major risk with hilly topography with less permeable soils (Juo and Manu 1996). Hence, how farming activity will disturb the nutrient cycles, which possibly results in substantial soil nutrient loss by leaching, should be quantified on Oxisols in relation to soil acidification processes as a driving force of basic cation leaching. Besides, it is urgently indispensable for the establishment of sustainable land use management to clarify the effects of original vegetation (forest vs savanna) on nutrient leaching from croplands in forest-savanna transition zone, where soil nutrient stocks and microbial nutrient status are different according to the vegetation as was introduced in Chap. 8.

In this chapter, firstly, we quantified the soil nutrient fluxes, as well as C flow, in forest and adjacent savanna through in situ soil solution study for grasping the original situation. Secondly, we compared the nutrient leaching fluxes for 2 years in each adjacent maize cropland with original vegetation to reveal the effect of cultivation on soil nutrient losses both from forest and savanna. We then compared the case of Oxisols in Africa to that of Ultisols in Southeast Asia.

10.2 Description of Study Sites

This study was carried out in the farmer's field of village of Andom, East Region, Cameroon, where we named AD in Chap. 9 (Fig. 9.1). It was located in forest-savanna transition area. The field experiments were conducted from April 2010 to March 2012 for 2 years. Among the years in which the field experiments were carried out, the mean annual precipitation and mean annual air temperature at our monitoring site were recorded as 1512 mm and 23.1 °C, respectively, as usual year. The rainy season is bimodal, i.e., the minor rainy season (from March to July) and the major rainy season (from September to November). Experimental plots consisted of forested plot (ADf) and savanna plot (ADs), and each adjacent cropland plot (<50 m from each original vegetation plot) cleared in March 2010 (ADfc and ADsc, respectively). The results of the first year (April 2010–March 2011) and the second year (April 2011–March 2012) of croplands were shown separately. The ADf and ADs, ~5 km apart, were located on near-level topography and within the same soil color and texture. Most of forest soils were classified as Typic Kandiudox (Soil Survey Staff 2014) or Acric Ferralsols (Vetic) (IUSS 2014); while most of savanna soils were classified as Kandiudalfic Eutrudox (Soil Survey Staff 2014) or Lixic Ferralsols (Vetic) (IUSS 2014). The dominant vegetation in ADf was *Albizia zygia* (Fabaceae; N_2-fixing species), while that in ADs was *Chromolaena odorata* (Asteraceae). Maize (*Zea mays*) was cultivated in both croplands for each rainy season (twice a year) both in 2010 and in 2011, as practiced by the local farmers.

10.3 General Physicochemical Properties of the Soils and Environmental Factors

The surface A1 horizons at ADf and ADfc were more acidic, having pH (H_2O) values of 4.0–4.2, than those at ADs and ADsc with pH (H_2O) values of 5.0–5.5, consistent with lower base saturation and higher Al saturation at ADf and ADfc (Table 10.1). Clay contents in the soils were consistently high at all plots (49–74 %)

Table 10.1 Physicochemical properties of soils at studied plots

Plot	Horizon	Depth (cm)	pH H$_2$O	pH KCl	Sand (%)	Silt	Clay	Total C (g kg^{-1})	Total N	C/N (g g^{-1})	CEC (cmol$_c$ kg^{-1})	Na	K	Mg	Ca	Al	H	BS[b] (%)
ADf forest	A1	0–10	4.2	3.9	40	5	55	18.8	1.6	11	8.0	0.0	0.2	0.4	1.7	2.3	0.7	43
	A2	10–20	4.2	4.0	39	6	54	13.6	1.1	12	6.1	0.0	0.1	0.2	1.3	2.5	0.7	33
	Bt1	20–35	4.8	4.3	29	4	68	9.5	0.9	11	7.0	0.0	0.0	0.1	1.6	0.9	0.7	52
	Bt2	35–50	4.3	4.3	25	4	70	6.9	0.6	11	6.2	0.0	0.0	0.1	1.0	1.5	0.3	38
	Bt3	50–80+	4.9	4.7	23	3	73	5.7	0.6	9	6.4	0.3	0.1	0.1	1.1	0.4	0.4	69
ADfc cropland	Ap1	0–10	4.0	3.7	42	4	54	24.2	1.9	13	10.1	0.0	0.2	0.4	1.5	3.3	0.8	34
	Ap2	10–25	4.0	3.8	37	4	59	15.4	1.2	13	8.4	0.0	0.1	0.1	0.9	3.3	0.9	20
	Bt1	25–35	4.2	3.9	32	3	65	8.7	0.8	11	7.3	0.0	0.0	0.1	1.0	3.0	0.8	23
	Bt2	35–50	4.2	3.9	30	3	67	6.7	0.6	11	7.9	0.0	0.0	0.1	1.0	2.8	0.5	25
	Bt3	50–75+	4.4	4.1	27	4	69	5.0	0.5	10	9.7	0.0	0.0	0.1	0.9	2.2	0.6	27
ADs savanna	A1	0–8	5.5	4.4	39	8	53	29.1	1.8	16	10.3	0.0	0.2	1.2	3.1	0.8	0.2	81
	A2	8–19	4.7	4.1	37	6	57	18.0	1.2	16	7.6	0.0	0.1	0.5	1.9	2.0	0.4	51
	Bt1	19–38	4.6	4.1	27	4	69	9.4	0.8	12	5.6	0.0	0.0	0.3	1.4	1.4	0.7	46
	Bt2	38–65	4.7	4.4	23	3	74	6.9	0.6	12	6.8	0.1	0.1	0.2	1.5	0.9	0.4	58
	Bt3	65–90+	4.9	4.7	23	3	74	5.2	0.5	11	6.5	0.1	0.0	0.2	1.7	0.3	0.3	77
ADsc cropland	Ap1	0–9	5.0	4.2	45	6	49	21.5	1.3	17	7.8	0.0	0.2	1.1	2.1	1.0	0.4	71
	Ap2	9–15	6.5	5.2	46	5	49	13.6	1.1	12	6.0	0.1	0.1	0.3	3.1	0.2	0.3	86
	Bt1	15–30	4.6	4.2	39	4	57	10.7	0.8	13	5.6	0.0	0.0	0.1	1.1	1.5	0.5	39
	Bt2	30–60	6.0	5.4	30	3	68	6.6	0.7	9	3.9	0.0	0.1	0.2	2.1	0.2	0.2	86
	Bt3	60–90+	4.8	4.5	27	4	69	5.3	0.5	11	6.2	0.0	0.0	0.1	1.2	0.6	0.4	56

The *Particle size distribution*[a] columns are Sand, Silt, Clay. The *Exchangeable cation* columns are Na, K, Mg, Ca, Al, H.

[a]Clay($<$0.002 mm), silt(0.002–0.05 mm), sand(0.05–2 mm)

[b]Base saturation: $(Na^{+}+K^{+}+Ca^{2+}+Mg^{2+})/(Na^{+}+K^{+}+Ca^{2+}+Mg^{2+}+Al^{3+}+H^{+})$

and increased with soil depth. Silt contents in the soils were consistently low ranging 3–8 % through the horizons at all plots, resulting from a long period of weathering. C/N ratios in the surface A horizon (0–15 cm) were lower at ADf (12) and ADfc (13) than at ADs (16) and ADsc (15) (N availability in ADf and ADfc was higher), whereas this ratio in the subsurface horizon were equally low consistently at all plots. Bulk density in ADf and ADfc was lower than in ADs and ADsc, respectively. All of these characteristics of soil physicochemical properties for each vegetation were in accordance with the previous report (Sugihara et al. 2014). Kaolin minerals were dominant and weak diffraction peak of mica was detected at all plots according to XRD analysis (data not shown).

Figure 9.4 in Chap. 9 shows daily precipitation and half-hourly soil temperature at 5 cm (a) and seasonal fluctuations of volumetric water content (VWC) and daily water fluxes at 0 cm (b), at 15 cm (c), and at 30 cm (d) for ADf. Calculated evapotranspiration (the differences between the water fluxes of precipitation and soil water percolated from 100 cm) in ADf (2390 mm for 2 years) was in accordance with the previous reports of equatorial Africa near the sites (Riou 1984; van der Ent et al. 2010). Evapotranspiration in ADs (1872 mm for 2 years) was 22 % smaller than that in ADf because ADf vegetation extracts more water of soil profile due to higher demand for transpiration (Table 10.2). This is comparable with the previous study by Jipp et al. (1998), which compared water balance of forest and adjacent pasture in Amazonia and reported that pasture evapotranspiration was 14–18 % lower than forest with annual rainfall of 1485–1755 mm. Land conversion to cropland resulted in a considerable increase of drainage owing to limited plant water uptake (evapotranspiration in ADfc: 1478 mm for 2 years; in ADsc: 1637 mm for 2 years). This is also in agreement with the repot of Williams and Melack (1997) in a mixture of crops and bush fallow converted from forest in Amazonia, who estimated evapotranspiration was 667 mm with annual rainfall of 2754 mm.

10.4 Carbon Stock and Flow

Stock and annual flow of carbon are shown in Table 10.3. Aboveground biomass in ADf (122.4 Mg C ha^{-1}) was 20 times higher than that in ADs (6.1 Mg C ha^{-1}), although soil organic matter in ADs (61.7 Mg C ha^{-1}) was slightly higher than that in ADf (44.6 Mg C ha^{-1}) due to the higher bulk density and C concentration of surface horizon in ADs. In cropland plots, soil organic matter was comparable (53.2 and 56.1 Mg C ha^{-1} in ADfc and ADsc, respectively). Aboveground biomass in ADfc (3.9 and 2.9 Mg C ha^{-1} in 1st year and 2nd year, respectively) was higher than that in ADsc (1.9 and 1.7 Mg C ha^{-1} in 1st year and 2nd year, respectively) since more nutrients were available in soil solution of ADfc, as will be discussed in the next section.

Seasonal fluctuations of the rates of organic matter decomposition are presented in Fig. 10.1. The annual rates of organic matter decomposition were calculated using the average rates of CO$_2$ efflux measured. In the natural vegetation plots, the annual rates

Table 10.2 Estimated water fluxes during 2 years (*PP* precipitation, *TF* throughfall, *ET* evapotranspiration)

	ADf	ADfc	ADs	ADsc
	(mm for 2 years)			
PP	3024	3024	3024	3024
15 cm	1401	1930	1807	1693
30 cm	1027	1643	1575	1416
100 cm	634	1546	1152	1387
ET	2390	1478	1872	1637

of organic matter decomposition were 4.1 and 3.5 Mg C ha^{-1} year^{-1} in ADf and ADs, respectively. Assuming that root litter turnover was 0.20 year^{-1} in ADf (Nakane, 1980), organic matter decomposition rates were comparable with C inputs (4.7 Mg C ha^{-1} year^{-1} in ADf). Likewise in ADs, assuming that root litter turnover was 0.73 year^{-1} (Chen et al. 2004) and that the aboveground turnover was 0.29 year^{-1} (Scholes and Hall 1996), organic matter decomposition rates were also comparable with C inputs (3.9 Mg C ha^{-1} year^{-1} in ADs). In the cropland plots, surprisingly, the annual rates of organic matter decomposition did not increase as compared to each original vegetation, but still they were higher than C inputs due to smaller biomass of maize in each cropland plot. This resulted in the loss of soil organic matter in both cropland plots (1.7–2.8 and 2.2–2.6 Mg C ha^{-1} year^{-1} in ADfc and ADsc, respectively). These values are smaller than the previous reports by Funakawa et al. (2006) in which tropical forests were just cultivated in Thailand (5.0 Mg C ha^{-1} year^{-1}). Such low rates of decomposition in our plots can be explained by higher clay contents and sesquioxides of Oxisols in this region, which would contribute to the strong bonding with soil organic matter.

10.5 Soil Solution Composition and Soil Acidification Processes in Forest and Savanna

We found that a driving force of basic cation leaching was clearly different between vegetation: NO_3^- in ADf and Org^{n-} in ADs (Table 10.4; Fig.10.2). In ADf, NO_3^- dominated anions (0.16–0.19 mmol$_c$ L^{-1}, 60–61 % of the anionic charge of soil solution), meanwhile in ADs organic anion was the majority (0.10–0.15 mmol$_c$ L^{-1}, 52–62 %) and NO_3^- was double-digit lower (2.7–3.4 µmol$_c$ L^{-1}, 1–2 %). Many previous studies reported that NO_3^- seldom accounts for a large proportion of the anionic composition of forest soil solutions neither in the temperate zone (Johnson and Cole 1980; Fujii et al. 2008; Perakis and Sinkhorn 2011) nor the tropical zone (Jordan et al. 1979; Wickle and Lilienfein 2005; Vernimmen et al. 2007; Fujii et al. 2011), although only some cases were found with a high contribution of NO_3^- in tropical forest (Markewitz et al. 2004; Schwendenmann and Veldkamp 2005), where leguminous trees are dominant on Oxisols. Therefore, NO_3^- dominance in soil solution in ADf could result from low C/N ratio of litterfall (high N concentration) owing to the N_2-fixing species (Table 9.5). Soil solution in ADf was approximately

Table 10.3 Annual flow and stock of carbon

Location	Cameroon Forest		Cropland				Savanna		Cropland				Indonesia[f] Forest		Cropland		Thailand[f] Forest		Cropland	
	ADf		ADfc (1st year)		ADfc (2nd year)		ADs		ADsc (1st year)		ADsc (2nd year)		BSf		BSc		RPf		RPc	
C stock (Mg C ha^{-1})																				
Above-ground biomass	122.4	(–)	3.9	(0.7)	2.9	(0.5)	6.1	(0.9)	1.9	(0.2)	1.7	(0.2)	292.6	(–)	1.2	(0.2)	169.1	(–)	5.8	(0.8)
Fine-root biomass																				
O horizon	0.1	(0.0)	–		–		–		–		–		1.0	(0.4)	0.2	(0.1)	2.4	(0.3)	0.1	(0.0)
0–15 cm	0.7	(0.0)	0.1	(0.0)	0.1	(0.0)	0.4	(0.0)	0.1	(0.0)	0.1	(0.0)	1.2	(0.3)	0.1	(0.0)	2.2	(0.2)	0.1	(0.0)
15–30 cm	0.3	(0.1)	0.1	(0.0)	0.0	(0.0)	0.1	(0.0)	0.0	(0.0)	0.0	(0.0)	2.1	(0.4)			1.7	(0.1)	–	
Soil organic matter																				
O horizon	2.0	(0.4)	–		–		0.2	(0.0)	–		–		3.5	(0.3)			2.6	(0.1)	–	
Mineral soil horizons[a]	44.6	(–)	53.2	(–)	–		61.5	(–)	56.1	(–)	–		26.6	(0.8)	24.6	(0.3)	65.2	(5.0)	55.8	(2.3)

(continued)

Table 10.3 (continued)

Location		Cameroon												Indonesia[f]				Thailand[f]			
		Forest		Cropland				Savanna		Cropland				Forest		Cropland		Forest		Cropland	
		ADf		ADfc (1st year)		ADfc (2nd year)		ADs		ADsc (1st year)		ADsc (2nd year)		BSf		BSc		RPf		RPc	
C flow (Mg C ha^{-1} year^{-1})																					
Organic matter decomposition	(a)	4.1	(0.6)	3.6	(0.2)	4.4	(0.3)	3.5	(0.1)	3.7	(0.2)	3.2	(0.3)	5.4	(0.5)	3.6	(0.3)	5.5	(0.3)	8.2	(0.3)
Organic matter production																					
C input to soil[b]	(b)	4.7	(0.5)	1.9	(0.1)	1.6	(0.1)	3.9	(0.4)	1.1	(0.1)	1.0	(0.2)	5.0	(0.5)	1.4	(0.1)	5.2	(0.2)	4.3	(0.7)
Litterfall[c]	(c)	4.5	(0.5)	1.8	(0.1)	1.5	(0.1)	3.6	(0.4)	1.0	(0.1)	0.9	(0.1)	4.1	(0.5)	1.1	(0.1)	4.0	(0.2)	4.1	(0.7)
Root litter[d]	(d)	0.2	(0.1)	0.1	(0.0)	0.1	(0.0)	0.3	(0.0)	0.1	(0.0)	0.1	(0.0)	0.9	(0.1)	0.3	(0.1)	1.2	(0.1)	0.2	(0.1)
Product removal		–		2.1	(0.7)	1.5	(0.5)	–		0.9	(0.1)	0.8	(0.1)	–		0.2	(0.0)	–		1.7	(0.3)
Wood increment		8.1	(–)	–		–		1.6	(0.3)	–		–		10.6	(–)	–		5.8	(–)	–	
C budget in soil[e]	(e)	0.6	(0.8)	–1.7	(0.3)	–2.8	(0.3)	0.4	(0.4)	–2.6	(0.3)	–2.2	(0.3)	–0.4	(0.8)	–2.2	(0.3)	–0.2	(0.4)	–3.9	(0.8)

The figures in parenthesis represent standard errors

[a]Organic carbon in soil at 0–30 cm depth was counted

[b]C input was calculated as the sum of litterfall and root litter (b = c + d)

[c]The annual rates of litterfall production in savanna were assumed to be aboveground biomass with the turnover rate of 0.29 year^{-1} (Scholes and Hall 1996)

[d]The annual rates of root litter production in forest and in savanna were assumed to be 20 % and 73 % of the fine-root biomass, respectively (Nakane 1980; Chen et al. 2004)

[e]C budget in soil was calculated as the difference between organic matter decomposition and C input (e = b − a)

[f]Data were cited from Fujii et al. (2009), Fujii (2009), and Fujii et al. (2013)

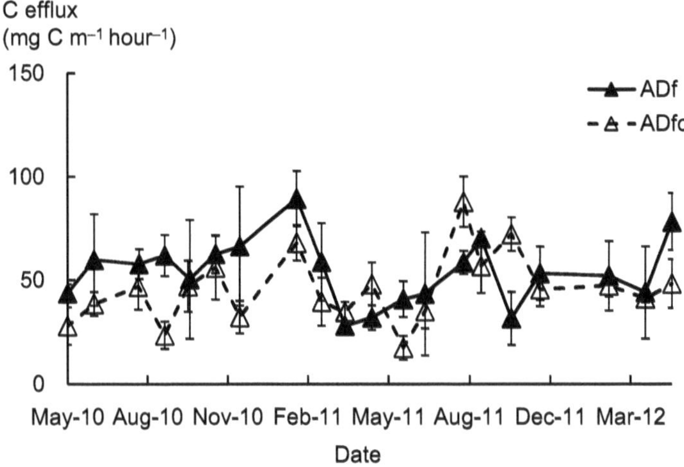

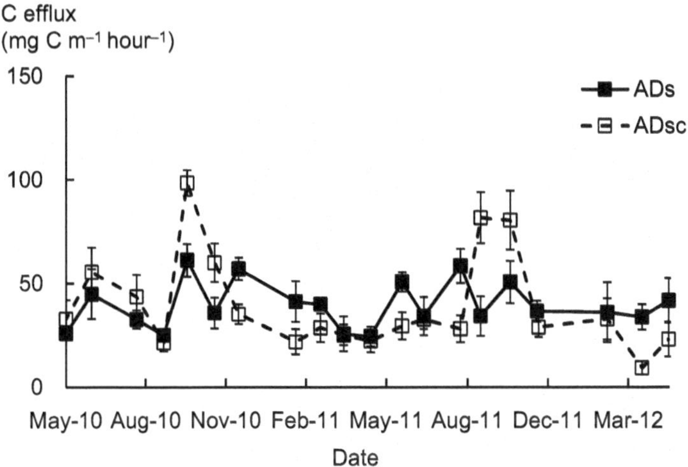

Fig. 10.1 Seasonal fluctuations of the rates of organic matter decomposition in each plot. Bars indicate standard errors ($n = 5$)

one unit more acidic (Fig. 10.3; median value: 15 cm, 4.9; 30 cm, 5.5) than that in ADs (median value: 15 cm, 6.3; 30 cm, 6.4). Lower soil solution pH in ADf compared to ADs, owing to cation excess uptake by trees in ADf, is consistent with the previous reports in forest-savanna transition zone (Dezzeo et al. 2004).

Based on the ion fluxes in each horizon (Fig. 10.2; precipitation, A and B horizons) and the excess cation uptake by vegetation, we calculated carbonic acid dissociation and protonation (NPG_{Car}); organic anion dissociation, mineralization, and adsorption to soil (NPG_{Org}); N mineralization, nitrification, and N uptake by plants (NPG_{Ntr}); plant excess cation uptake (NPG_{Bio}); and cation release and accumulation (ΔANC) (Fig. 10.4), as described in Chap. 7. In ADf, NPG_{Bio}

Table 10.4 Annual volume-weighed mean concentrations of ions in precipitation, throughfall, and soil solution

Site	Horizon	pH	HCO_3^-	Cl^-	NO_3^-	SO_4^{2-}	Org^{n-} [b]	H^+	Na^+	NH_4^+	K^+	Mg^{2+}	Ca^{2+}	Σcation	Σanion
			$mmol_c\ L^{-1}$												
ADf forest	TF[a]	6.4	0.06	0.03	0.16	0.07	0.12	0.00	0.01	0.03	0.24	0.06	0.07	0.43	0.43
	15 cm	5.1	0.00	0.03	0.16	0.03	0.04	0.01	0.01	0.02	0.06	0.08	0.06	0.25	0.27
	30 cm	5.4	0.02	0.04	0.19	0.01	0.06	0.01	0.02	0.03	0.04	0.12	0.10	0.31	0.32
ADfc cropland 1st year	PP[a]	6.1	0.01	0.01	0.01	0.01	0.03	0.00	0.01	0.01	0.01	0.01	0.02	0.07	0.07
	15 cm	4.7	0.01	0.05	1.03	0.11	0.12	0.04	0.02	0.06	0.28	0.35	0.52	1.27	1.31
	30 cm	4.5	0.00	0.05	0.78	0.04	0.05	0.06	0.02	0.03	0.24	0.25	0.31	0.90	0.92
ADfc cropland 2nd year	PP[a]	6.5	0.05	0.02	0.00	0.01	0.03	0.00	0.02	0.03	0.02	0.01	0.03	0.12	0.12
	15 cm	4.9	0.01	0.02	1.06	0.10	0.07	0.04	0.02	0.04	0.12	0.37	0.59	1.17	1.26
	30 cm	5.4	0.02	0.02	0.32	0.12	0.08	0.01	0.03	0.02	0.10	0.17	0.27	0.60	0.56
ADs savanna	PP[a]	6.3	0.03	0.01	0.01	0.01	0.03	0.00	0.02	0.02	0.02	0.01	0.03	0.09	0.09
	15 cm	6.2	0.03	0.01	0.00	0.02	0.10	0.00	0.02	0.02	0.01	0.05	0.06	0.17	0.17
	30 cm	6.3	0.05	0.01	0.00	0.01	0.15	0.00	0.02	0.02	0.02	0.07	0.09	0.23	0.23
ADsc cropland 1st year	PP[a]	6.1	0.01	0.01	0.01	0.01	0.03	0.00	0.01	0.01	0.01	0.01	0.02	0.07	0.07
	15 cm	5.5	0.00	0.04	0.42	0.10	0.10	0.01	0.02	0.02	0.13	0.23	0.24	0.66	0.66
	30 cm	6.0	0.01	0.04	0.46	0.02	0.22	0.00	0.03	0.04	0.08	0.23	0.36	0.74	0.74
ADsc cropland 2nd year	PP[a]	6.5	0.05	0.02	0.00	0.01	0.03	0.00	0.02	0.03	0.02	0.01	0.03	0.12	0.12
	15 cm	6.3	0.04	0.03	0.51	0.06	0.08	0.00	0.03	0.03	0.10	0.28	0.27	0.71	0.72
	30 cm	6.5	0.06	0.02	0.19	0.04	0.10	0.00	0.03	0.03	0.05	0.13	0.16	0.41	0.41

[a] PP and TF represent precipitation and throughfall, respectively
[b] Org^{n-} represents anion deficit, the negative charge of organic acids

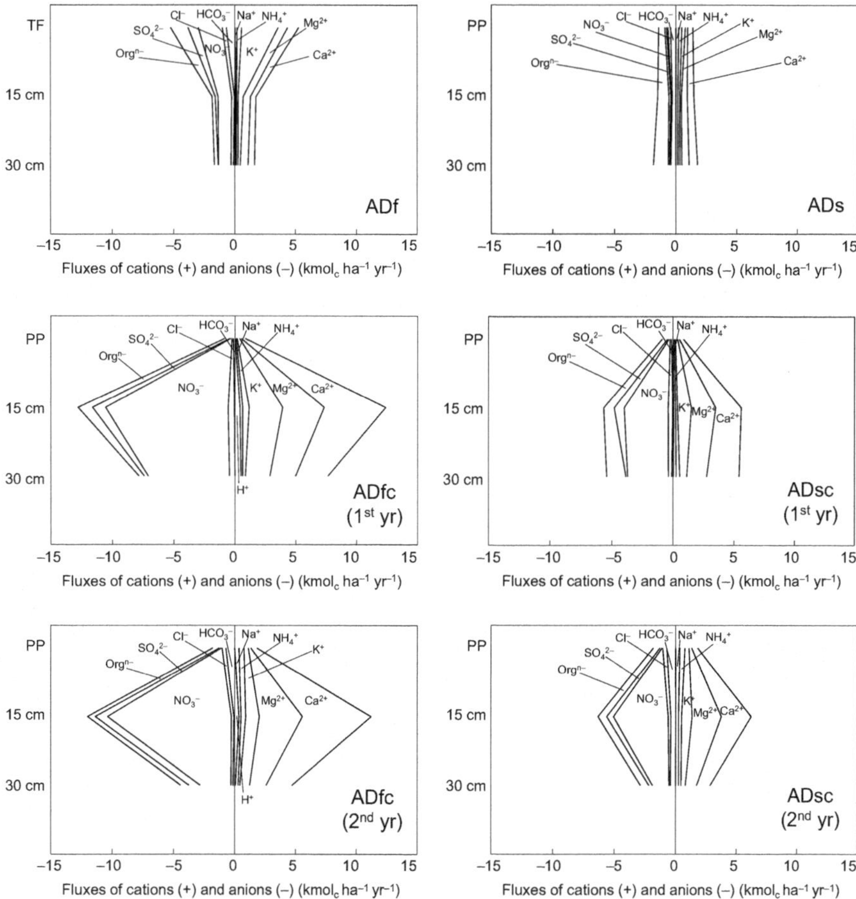

Fig. 10.2 Fluxes of solutes at each stratum at studied plots. PP, precipitation; TF, throughfall; 15 cm, soil solution at 15 cm; 30 cm, soil solution at 30 cm

contributed to net proton generation in the entire soil profile, which is in accordance with the previous reports in the tropical forests of Southeast Asia (Fujii et al. 2009, 2011). In ADs, on the other hand, soil acidification rates were much lower than in ADf because of limited plant uptake. Focusing on the process in detail, NPG_{Bio} and NPG_{Org} contributed comparably in the entire soil profile in ADs because of relatively higher production of Org^{n-}. As such, soil acidification processes proceed much more rapidly under forest vegetation due to the growing large biomass of forest. Therefore, current or the history of forest vegetation is likely to play an essential role in development of highly weathered soils, such as Oxisols. In other words, savanna vegetation alone tends not to be enough for Oxisols development as was shown by almost neutral pH of soil solution of ADs (Fig. 10.3).

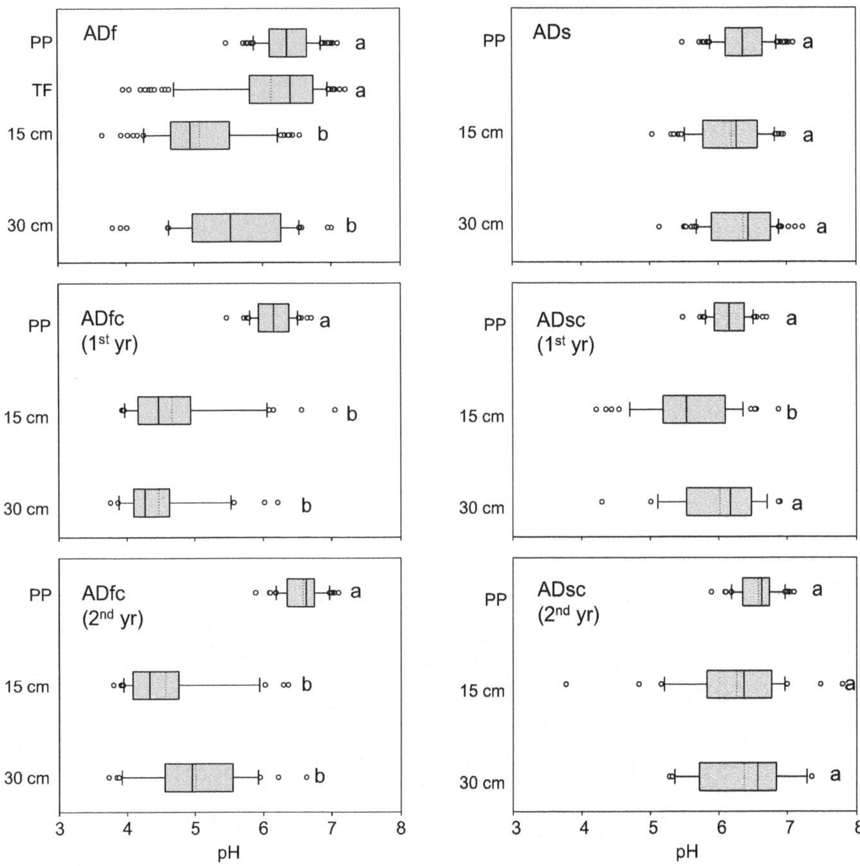

Fig. 10.3 Box-and-whisker plots of solution pH leached from each stratum at studied plots. PP, precipitation; TF, throughfall; 15 cm, soil solution at 15 cm; 30 cm, soil solution at 30 cm; the inner bar, the median; the inner dotted bar, the mean. The box boundaries indicate the upper and lower quartiles, the whisker caps indicate 90th and 10th percentiles, and the circles indicate the outliers. The boxes are preceded by letters which, if different, indicate that the values between strata are significantly different, $p < 0.05$, Dunn's post-hoc tests

10.6 Effect of Cultivation on Soil Solution Composition and Soil Acidification Processes

As a result of cultivation, NO_3^- concentrations in soil solution and leaching fluxes were substantially increased both in ADfc and ADsc, as compared to each original vegetation, and it caused much loss of basic cations beyond the crop-rooting zone (>30 cm depth) (Figs. 10.2 and 10.4). In the 2nd year of ADfc and ADsc, total ionic concentrations of 30 cm solution were significantly smaller than that of the 1st year of ADfc and ADsc, respectively (t-test; $p < 0.01$), indicating that the impact of cultivation was reduced in the 2nd year after reclamation. Total ionic

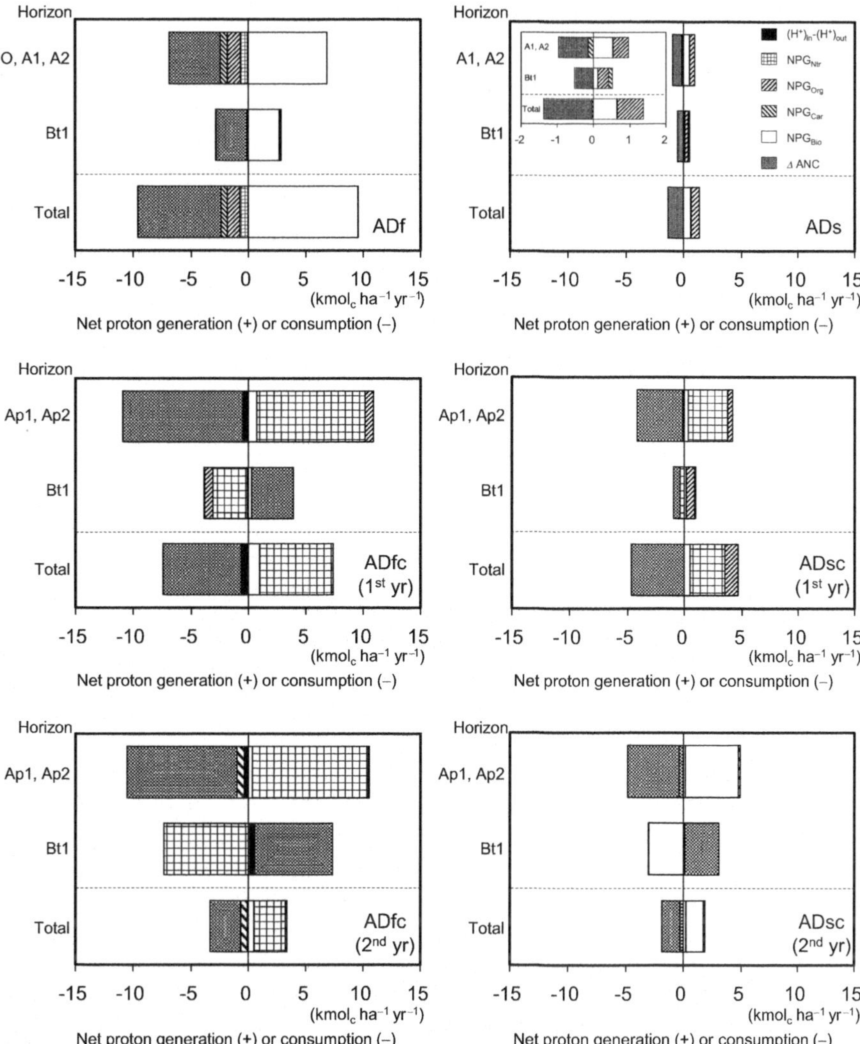

Fig. 10.4 Net proton generation and consumption and soil acidification in the soil profiles at studied plots. O, A1, A2, Ap1, Ap2, and Bt1 represent soil horizons

concentrations in ADfc were significantly higher than those in ADsc for both years (t-test; $p < 0.01$). In both 1st and 2nd year of ADfc, the pH values of soil solution were significantly lower than that in ADf for 15 and 30 cm, respectively (Dunn's test). In ADsc, on the other hand, the pH values of soil solution were significantly lower than that in ADs only for 15 cm of the 1st year. Those results suggest that the impact of cultivation on soil nutrient cycles was larger in ADfc than in ADsc during 2 years after reclamation. Reflecting the nutrient flux pattern of original vegetation, NO_3^- leaching flux in ADfc was significantly higher than that in ADsc, showing nitrification was more active in ADfc. This means bulk soil acidity did not suppress nitrification (Sierra 2006), since we observed soil and solution pH in ADfc was lower than in ADsc.

Calculating the proton budgets, contribution of NPG_{Ntr} was more than 90 % of proton generation in Ap horizon for both ADfc and ADsc, showing that nitrification was the main process of soil acidification in cropland. Product removal contributed very little to acidification because of limited plant (maize) biomass. Most of protons produced by nitrification were neutralized by basic cations, resulting in intensive basic cation leaching from Ap horizon (Fig. 10.2). Although NPG_{Ntr} was enhanced by cultivation, its degree depends on the original vegetation before reclamation. NPG_{Ntr} in Ap horizon of ADfc was high (Fig. 10.2; 9.4–10.1 $kmol_c$ ha^{-1} $year^{-1}$), and those values even ranged higher end of the reported ones in the fertilized croplands (up to 13.5 $kmol_c$ ha^{-1} $year^{-1}$) (Poss et al. 1995; Lesturgez et al. 2006; Noble et al. 2008; Meng et al. 2013). Considering that the loss rates of soil organic matter following cultivation was relatively low in ADfc, as compared to other croplands converted from tropical forests, labile organic matter with low C/N ratio in the litter layer of this legume-dominated forests would be the source of NO_3^- and the reason of such a high value of NPG_{Ntr}. In B horizon, NO_3^- produced in Ap horizon was partly taken up by plants in both cropland plots. Although NPG_{Ntr} in Ap horizon showed similar value both in 1st and 2nd year for both ADfc and ADsc, the contribution of plant uptake in B horizon was larger in the 2nd year possibly due to enhanced weed and/or secondary vegetation. As a result, soil acidification rates in the entire soil profiles, which was mainly caused by nitrification, decreased in the 2nd year after reclamation for both ADfc and ADsc. In the 1st year of reclamation of ADsc, the contribution of NPG_{Org} in the entire soil profiles was relatively high with approximately 25 % of net proton generation, which reflected on the nature of original vegetation. Overall, soil acidification in cropland was enhanced mainly by nitrification at the expense of the loss of SOM but not by product removal in both plots. The effects of reclamation on soil acidification processes would depend on the original vegetation (forest vs savanna), possibly due to the different quality (e.g., C/N ratio) of organic matter supplied from each vegetation as a substrate for soil microbes in croplands.

10.7 Comparison of Soil Acidification Processes in Cameroon with the Case in Indonesia and Thailand

We will compare the cultivation-induced soil acidification in Cameroon (ADf, ADfc, ADs, and ADsc) to the case in tropical forests of Indonesia (BSf and BSc) and Thailand (RPf and RPc), where proton budget of forest plots was evaluated in Chap. 7 and the influence of cultivation on it was quantified by Fujii et al. (2009) (Figs. 10.5 and 10.6).

In all forested plots, NPG_{Bio} (vegetation uptake) was the dominant proton source in the entire profile, although NPG_{Org} (dissociation of organic acid) also accounted

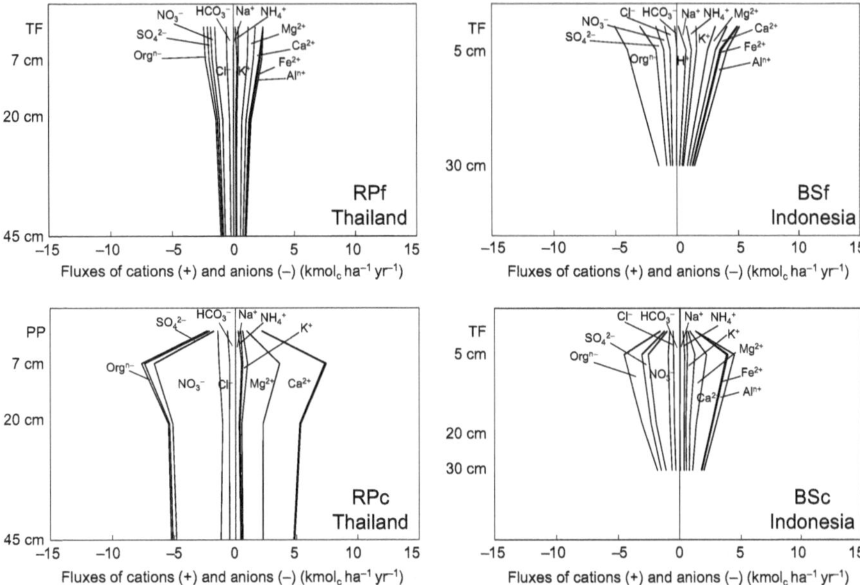

Fig. 10.5 Fluxes of solutes at each stratum in Thailand and Indonesia. PP, precipitation; TF, throughfall; 5 cm, soil solution at 5 cm; 7 cm, soil solution at 7 cm; 20 cm, soil solution at 20 cm; 30 cm, soil solution at 30 cm; 45 cm, soil solution at 45 cm (Data was modified from Fujii et al. 2009)

for the soil acidification in ADs because of small vegetation increment in savanna vegetation. On the other hand, in all cropland plots, generally saying, it was nitrification that caused soil acidification in the Ap horizon. The degree of NPG_{Ntr} (nitrification) in Ap horizon was, however, various: in descending order, ADfc (9.4–10.1 $kmol_c$ ha^{-1} $year^{-1}$) > RPc (5.0 $kmol_c$ ha^{-1} $year^{-1}$) $\simeq$ ADsc (3.4–4.5 $kmol_c$ ha^{-1} $year^{-1}$) > BSc (1.1 $kmol_c$ ha^{-1} $year^{-1}$). Dividing these values by the loss amount of SOC, as organic matter decomposition exceeded the inputs of C in cropland plots, the rates of NPG_{Ntr} were in descending order, ADfc (43–66 $mmol_c$ mol^{-1} C) > ADsc (16–25 $mmol_c$ mol^{-1} C) $\simeq$ RPc (15 $mmol_c$ mol^{-1} C) > BSc (6 $mmol_c$ mol^{-1} C). As such, in terms of proton generation process in Ap horizon of cropland, ADfc, ADsc, and RPc could be categorized into nitrification-dominated type, while dissociation of organic acid, as well as nitrification, contributed significantly to proton generation in BSc. Fujii et al. (2009) pointed out that higher nitrification activity in RPc would be allowed by moderately acidic soils. Although the bulk soils of ADfc and ADsc were more acidic and even close to the low pH level of BSc, clayey Oxisols of ADfc and ADsc, which have physically well-structured microaggregates, would provide favorable habitat for microbes in micropores where pH was not as low as the bulk soil. This can be the reason for active nitrification in the plots of Cameroon. Such a high nitrification rate might be the characteristics of Oxisols, which might accommodate the nitrifiers with suitable environment. In the case of entire soil profile, protons generated by nitrification still highly contributed to the soil acidification for ADfc,

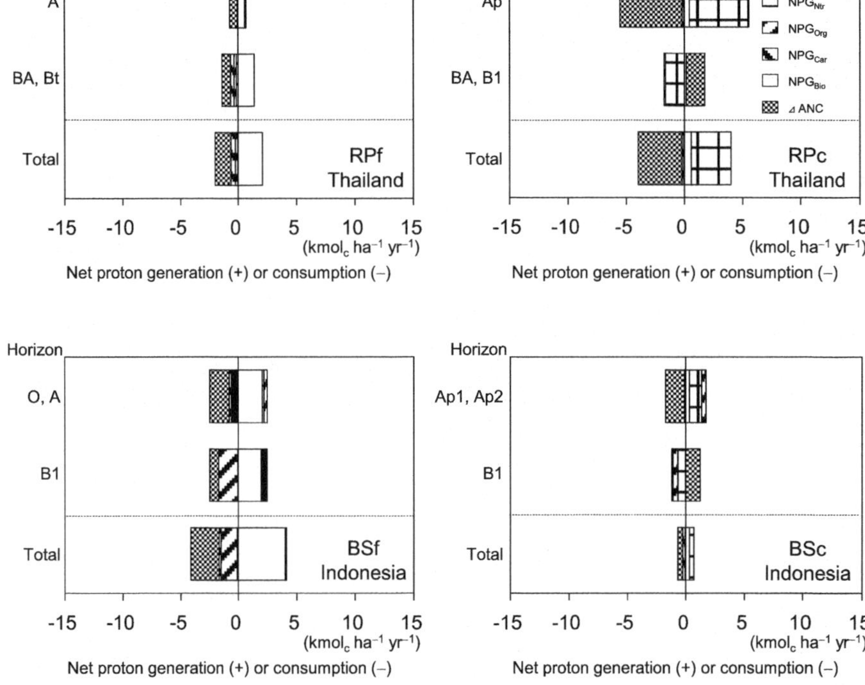

Fig. 10.6 Net proton generation and consumption and soil acidification in the soil profiles in Thailand and Indonesia. O, A, Ap, Ap1, Ap2, BA, B1, and Bt represent soil horizons (Data was modified from Fujii et al. 2009)

ADsc, and RPc. On the other hand in BSc, cation excess accumulation in products (grains) and nitrification comparably contributed to the proton generation due to the almost balanced proton budget in the whole profile. Accordingly, the effects of cultivation on soil acidification processes would depend on soil pH and physical structure, which affect the nitrification activity.

10.8   Cation Leaching Pattern Can Depend on pH of Soil Solution

The protons, or the Al cations, may displace exchangeable bases, such as K^+, Mg^{2+}, and Ca^{2+}, into the soil solution, which could cause basic cation leaching (Cahn et al. 1993). We have discussed NO_3^- and, to some degree, Org^{n-} can be the driving force of cation leaching in the previous sections.

How about the behavior of those basic cations? In nine natural vegetation soils and four cropland soils from Cameroon, Indonesia, and Thailand, the molar charge ratio of K^+ concentrations to sum of Mg^{2+} and Ca^{2+} concentrations $(K^+/(Mg^{2+}+Ca^{2+}))$ increased with decreasing soil solution pH in the mineral horizon (Fig. 10.7). This indicates that the low pH of soil solutions could enhance the dissolution of K-bearing minerals of highly weathered tropical soils. Hence, the lower solution pH, which could stem from cultivation, might promote K leaching following K-bearing mineral dissolution. RPf was unlikely to follow this pattern possibly because of higher content of K-bearing minerals (K_2O content in RP was 3.4–3.9 %, as compared to 0.1–0.7 % in the remaining plots). Soils that are originally high in K-bearing minerals generally have greater K availability than those with low K content. The values of MV were also odd partly because the pH of gravitational soil solution was much higher (almost 2 unit) than that of the bulk soil. Even though Oxisols are highly weathered soils that rarely have K-bearing minerals (Buol 2006), such an intensive soil acidification is possibly the reason for large K^+ leaching in ADfc and lower soil K stock in ADf as compared in ADs in spite of same climatic and geologic condition, as Melo et al. (2004) suggested the potential of K-bearing minerals to supply K to the plants from Oxisols. Our results indicate that soil acidification, which may be caused by continuous cultivation (Fujii et al. 2009), could have the potential to accelerate K^+ leaching through mineral weathering especially under udic moisture regime with intense leaching condition. Nutrient leaching fluxes from Oxisols of Cameroon were comparable or even higher than the case of Ultisols in Thailand and Indonesia, which generally stocks more basic cations than Oxisols (Funakawa et al. 2006; Fujii et al.

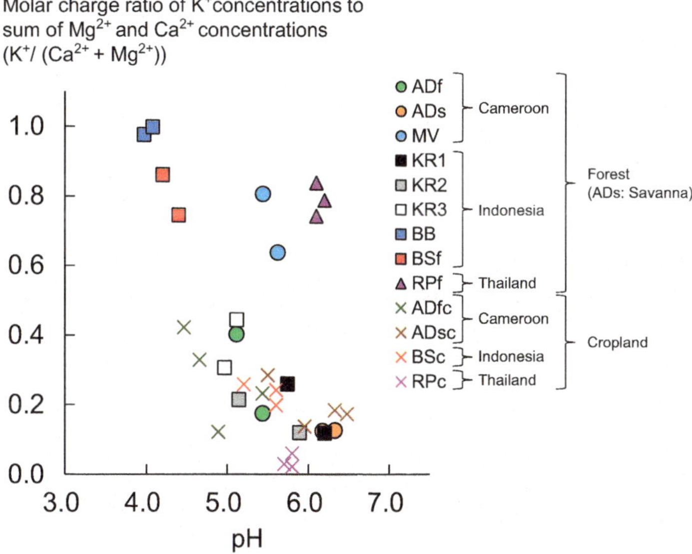

Fig. 10.7 Relationship between solution pH and the molar charge ratio of K^+ concentrations to sum of Mg^{2+} and Ca^{2+} concentrations. Data sources include five plots in Cameroon from the present study and Chap. 9, six plots in Indonesia from Fujii et al. (2011) and Chap. 7, and two plots in Thailand from Fujii et al. (2009) and Chap. 7

2011). This suggests the risk of nutrient deficiency in cropland is more serious in Oxisols.

10.9 Conclusions

Our study revealed that NO_3^- flux was high and contributes to the cation movement through the soil profile in forest, while in savanna, NO_3^- was almost negligible in soil solution and organic anion was the driving force of cation leaching. Cation excess uptake by vegetation was the main process for soil acidification in forest, while dissociation of organic acid also contributed to net proton generation in savanna due to the limited growing biomass. Soil acidification in cropland was enhanced by nitrification in both forest cropland and savanna cropland, although cultivation in forest caused higher rates of NO_3^- leaching than that in savanna, which was influenced by the pattern of nutrient dynamics in original vegetation. Hence, the effects of reclamation on soil acidification processes would depend on the original vegetation (forest vs savanna) under Oxisols in Cameroon. High rates of nitrification in our plots in spite of low bulk soil pH could be the characteristics of well-structured Oxisols, which provide the favorable habitat for the nitrifiers. This indicates the importance of soil physical structure, as well as soil pH, for the effects of cultivation on soil acidification processes in relation to nitrification. Our results also indicate that soil acidification could have the potential to accelerate K^+ leaching from cropland. Since nutrient leaching fluxes from Oxisols of Cameroon were comparable or even higher than the case of Ultisols of Thailand and Indonesia, which generally stocks more basic cations, the risk of nutrient deficiency from cropland is more serious in Oxisols.

References

Anamosa PR, Nkedi-Kizza P, Blue WG, Sartain JB (1990) Water movement through an aggregated, Gravelly Oxisol from Cameroon. Geoderma 46:263–281

Buol SW (2006) Oxisols. In: Lal R (ed) Encyclopedia of soil science, 2nd edn, vol 2. Taylor & Francis, New York, pp 1233–1235

Cahn MD, Bouldin DR, Cravo MS, Bowen WT (1993) Cation and Nitrate leaching in an Oxisol of the Brazilian Amazon. Agron J 85:334–340

Chen X, Derek E, Lindsay BH (2004) Seasonal patterns of fine-root productivity and turnover in a tropical savanna of northern Australia. J Trop Ecol 20:221–224

Dezzeo N, Chacon N, Sanoja E, Picon G (2004) Changes in soil properties and vegetation characteristics along a forest-savanna gradient in southern Venezuela. For Ecol Manag 200:183–193

Fujii K (2009) Quantitative analysis of soil acidification under forests developed in different geological and climatic conditions in humid Asia. Doctoral dissertation

Fujii K, Funakawa S, Hayakawa C, Kosaki T (2008) Contribution of different proton sources to pedogenetic soil acidification in forested ecosystems in Japan. Geoderma 144:478–490

Fujii K, Funakawa S, Hayakawa C, Sukartiningsih KT (2009) Quantification of proton budgets in soils of cropland and adjacent forest in Thailand and Indonesia. Plant Soil 316:241–255

Fujii K, Funakawa S, Hayakawa C, Sukartiningsih HC, Kosaki T (2013) Fluxes of dissolved organic carbon and nitrogen in cropland and adjacent forests in a clay-rich Ultisol of Thailand and a sandy Ultisol of Indonesia. Soil Till Res 126:267–275

Fujii K, Hartono A, Funakawa S, Uemura M, Sukartiningsih KT (2011) Acidification of tropical forest soils derived from serpentine and sedimentary rocks in East Kalimantan, Indonesia. Geoderma 160:311–323

Funakawa S, Hayashi Y, Tazaki I, Sawada K, Kosaki T (2006) The main functions of the fallow phase in shifting cultivation by Karen people in northern Thailand—a quantitative analysis of soil organic matter dynamics. Tropics 15(1):1–27. doi:10.3759/tropics.15.1

Gaston G, Brown S, Lorenzini M, Singh KD (1998) State and change in carbon pools in the forests of tropical Africa. Glob Change Biol 4:97–114

IUSS Working Group WRB. (2014). World reference base for soil resources 2014. International soil classification system for naming soils and creating legends for soil maps. World Soil Resources Reports No. 106. FAO, Rome

Jipp PH, Nepstad DC, Cassel DK, de Carvalho CR (1998) Deep soil moisture storage and transpiration in forests and pastures of secondary-dry Amazonia. Clim Chang 39:395–412

Johnson DW, Cole DW (1980) Anion mobility in soils: relevance to nutrient transport from forest ecosystems. Environ Int 3:79–90

Jones A, Breuning-Madsen H, Brossard M, Dampha A, Deckers J, Dewitte O, Gallali T, Hallett S, Jones R, Kilasara M, Le Roux P, Micheli E, Montanarella L, Spaargaren O, Thiombiano L, Van Ranst E, Yemefack M, Zougmoré R (2013) Soil Atlas of Africa. European Commission, Publications Office of the European Union, Luxembourg City

Jordan CF, Todd RL, Escalante G (1979) Nitrogen conservation in a tropical rain forest. Oecologia 39:123–128

Juo ASR, Manu A (1996) Chemical dynamics in slash-and-burn agriculture. Agric Ecosyst Environ 58:49–60

Khasa PD, Valle G, Belanger J, Bousquet J (1995) Utilization and management of forest resources in Zaire. For Chron 71:479–488

Lesturgez G, Poss R, Noble A, Grünberger O, Chintachao W, Tessier D (2006) Soil acidification without pH drop under intensive cropping systems in Northeast Thailand. Agric Ecosyst Environ 114:239–248

Lutz W, Sanderson W, Scherbov S (2001) The end of world population growth. Nature 412:543–545

Markewitz D, Davidson E, Moutinho P, Nepsad D (2004) Nutrient loss and redistribution after forest clearing on a highly weathered soil in Amazonia. Ecol Appl 14:177–199

Mayaux P, Bartholomé E, Fritz S, Belward A (2004) A new land-cover map of Africa for the year 2000. J Biogeogr 31:861–877

Melgar RJ, Smyth TJ, Sanchez PA, Cravo MS (1992) Fertilizer nitrogen movement in a Central Amazon Oxisol and Entisol cropped to corn. Fert Res 31:241–252

Melo GW, Meurer EJ, Pinto LFS (2004) Potassium-bearing minerals in two oxisols of Rio Grande do Sul state, Brazil. R Bras Ci Solo 28:597–603

Meng HQ, Xu MG, Lü JL, He ZH, Li JW, Shi XJ, Peng C, Wang BR, Zhang HM (2013) Soil pH Dynamics and Nitrogen transformations under long-term chemical fertilization in four typical Chinese croplands. J Integr Agric 12(11):2092–2102

Mitchard ETA, Saatchi SS, Lewis SL, Feldpausch TR, Woodhouse IH, Sonke B, Rowland C, Meir P (2011) Measuring biomass changes due to woody encroachment and deforestation /degradation in a forest–savanna boundary region of central Africa using multi-temporal L-band radar backscatter. Remote Sens Environ 115:2861–2873

Nakane K (1980) Comparative studies of cycling of soil organic carbon in three primeval moist forests. Jap J Ecol 30:155–172

Noble AD, Suzuki S, Soda A, Ruaysoongnern S, Berthelsen S (2008) Soil acidification and carbon storage in fertilized pastures of Northeast Thailand. Geoderma 144:248–255

Perakis SS, Sinkhorn ER (2011) Biogeochemistry of a temperate forest nitrogen gradient. Ecology 92(7):1481–1491

Poss R, Smith CJ, Dunin FX, Angus JF (1995) Rate of soil acidification under wheat in a semi-arid environment. Plant Soil 177:85–100

Riou C (1984) Experimental study of potential evapotranspiration (PET) in Central Africa. J Hydrol 72:275–288

Sankaran M, Hanan NP, Scholes RJ, Ratnam J, Augustine DJ, Cade BS et al (2005) Determinants of woody cover in African savannas. Nature 438:846–849

Scholes RJ, Hall DO (1996) The carbon budget of tropical savannas, woodlands and grasslands. In: Breymeyer AI, Hall DO, Melillo JM, Ågren GI (eds) Global change: effects on coniferous forests and grassland. Wiley, New York, pp 69–100

Schwendenmann L, Veldkamp E (2005) The role of dissolved organic carbon, dissolved organic nitrogen, and dissolved inorganic nitrogen in a tropical wet forest ecosystem. Ecosystems 8:339–351

Sierra J (2006) A hot-spot approach applied to nitrification in tropical acid soils. Soil Biol Biochem 38:644–652

Smith J, Barau AD, Goldman A, Mareck JH (1994) The role of technology in agricultural intensification: the evolution of maize production in the northern Guinea savanna of Nigeria. Econ Dev Cult Chang 42:537–554

Soil Survey Staff (2014) Keys to soil taxonomy, 12th edn. United States Department of Agriculture Natural Resources Conservation Service, Washington, DC

Staver AC, Archibald S, Levin S (2011) Tree cover in sub-Saharan Africa: rainfall and fire constrain forest and savanna as alternative stable states. Ecology 92(5):1063–1072

Sugihara S, Shibata M, Mvondo Ze AD, Araki S, Funakawa S (2014) Effect of vegetation on soil C, N, P and other minerals in oxisols at the forest–savanna transition zone of Central Africa. Soil Sci Plant Nutr 60:45–59

van der Ent RJ, Savenije HHG, Schaefli B, Steele-Dunne SC (2010) Origin and fate of atmospheric moisture over continents. Water Resour Res 46:W09525. doi:10.1029/2010WR009127

van Wambeke A (1992) Soils of the tropics—properties and appraisal. McGraw-Hill, New York, pp 139–161.

Vernimmen RRE, Verhoef HA, Verstraten JM, Bruijnzeel LA, Klomp NS, Zoomer HR, Wartenbergh PE (2007) Nitrogen mineralization, nitrification and denitrification potential in contrasting lowland rain forest types in Central Kalimantan, Indonesia. Soil Biol Biochem 39:2992–3003

Wilcke W, Lilienfein J (2005) Nutrient leaching in Oxisols under native and managed vegetation in Brazil. Soil Sci Soc Am J 69:1152–1161

Williams MR, Melack JM (1997) Solute export from forested and partially deforested catchments in the central Amazon. Biogeochemsitry 38(1):67–102

Chapter 11
Shifting Cultivation in Northern Thailand with Special Reference to the Function of the Fallow Phase

Shinya Funakawa

Abstract To clarify the various functions of the fallow phase in the shifting cultivation system in northern Thailand, the fluctuation of fertility-related properties of soils throughout land-use stages was analyzed and the soil organic matter (SOM) budget was quantitatively evaluated, with special reference to soil microbial activities. The factors that have ensured the long-term sustainability of the shifting cultivation system can be summarized as follows: (1) Some soil properties relating to soil acidity improve when SOM increases in the late stages of fallow. The litter input may be supplying bases that are obtained via tree roots from further down the soil profile to the surface soil. This simultaneous increase in SOM and bases in the surface soil through forest-litter deposition in the late stages of fallow has an increasing effect on nutritional elements. (2) The decline in soil organic C during the cropping phase may be compensated by litter input during 6–7 years of fallow. With regard to the overall budget, the organic matter input through incorporation of initial herbaceous biomass into the soil system after establishment of tree vegetation (approximately in the fourth year) was indispensable for maintaining the SOM level. (3) The succession of the soil microbial community from rapid consumers of resources to stable and slow utilizers, along with the establishment of secondary forest, retards further leaching loss of nitrogen (N) and enhances N accumulation in the forestlike ecosystems. The functions of the fallow phase listed above can be considered essential to the maintenance of this forest-fallow system. Agricultural production can therefore be maintained with a fallow period of around 10 years, which is somewhat shorter than widely believed. Traditional shifting cultivation in the study village is shown to be well adapted to its soil-ecological condition.

Keywords Function of fallow • Karen people • Shifting cultivation • Soil organic matter budget • Thailand

S. Funakawa (✉)
Graduate School of Global Environmental Studies, Kyoto University, Yoshida-Honmachi, Sakyo-ku, Kyoto 606-8501, Japan
e-mail: funakawa@kais.kyoto-u.ac.jp

© Springer Japan KK 2017
S. Funakawa (ed.), *Soils, Ecosystem Processes, and Agricultural Development*,
DOI 10.1007/978-4-431-56484-3_11

11.1 Introduction

Traditional shifting cultivation systems consist of two distinct phases: a cropping phase that occurs after the slash and burn of the forest and the fallow phase. The main effects of burning on crop production are considered to be the supply of bases and phosphorus via ash input (Alegre et al. 1988; Ewel et al. 1981; Nye and Greenland 1960; Raison et al. 1985; Tanaka et al. 1997). In addition, N supply will be increased by enhanced mineralization after burning (Knoepp and Swank 1993; Marion et al. 1991; Nye and Greenland 1960; Sakamoto et al. 1991; Tanaka et al. 2001). On the other hand, the function of the fallow phase is said to guarantee enough accumulation of nutritional elements into the forest vegetation (Richter et al. 2000) and SOM via litter input (Ruark 1993) for the next cropping phase to be successful. The effects of burning and subsequent ash input during shifting culti- vation systems have been intensively studied (e.g., Leonard et al. 1998). Although quantitative analysis of the fallow phase is still limited, this information is thought to be imperative for a comprehensive understanding of shifting cultivation in relation to its historical sustainability.

When considering possible functions of the fallow phase, we could ask the question: How many years of fallow period are needed to maintain the productivity of the shifting cultivation in northern Thailand? In order to answer this question, several field experiments were conducted. In the present chapter, in situ SOM budget under shifting cultivation in northern Thailand is analyzed with special reference to functions of soil microbial community.

11.2 Description of Study Sites

The study was carried out in a farmer's field in the village of Ban Du La Poe, located in the Mae La Noi District, Mae Hong Son Province, northern Thailand (DP village). The village is inhabited by Karen people. The cropping fields of the villagers are mostly on steep slopes (usually exceeding 20°) of hilly landscape, with an elevation of 1100–1300 m above sea level. There is a distinct dry season in the area, which is under the influence of a monsoon climate, and precipitation is concentrated in the rainy season from April to October. In the year in which the field experiments were carried out (2001), the annual precipitation and mean annual air temperature at our monitoring site were recorded as 1222 mm and 20.2 °C, respectively. Parent materials of the soils are derived from partly metamorphosed fine-textured sedimentary rocks and granite. Most of the soils studied were classi- fied into Ustic Haplohumults or Humic/Typic Dystrustepts in the USDA classifi- cation system (Soil Survey Staff 2014). The shifting cultivation system of the village was similar to a traditional land rotation system, in which cropping for upland rice is limited to only 1 year, followed by at least 7 years of fallow. According to our field observation, this fallow period seems to be the longest one in this area. In fact, at many of the nearby villages, the fallow period has recently been shortened to around 4 years. Only a limited area is used for the cultivation of cash crops, such as cabbage.

11.3 Dynamics of Soil Fertility Status Throughout a Land Rotation System of Shifting Cultivation in Northern Thailand

Table 11.1 shows the average values of general physicochemical properties of the surface soils studied, with respect to the stage of land use. The soils studied were moderately acidic, having $pH(H_2O)$ values of 4.79–6.41. Total C content ranged from 31 to 79 g kg^{-1}. Most of the soils showed medium to heavy textures, with clay contents ranging between 23 and 42 %. Although data are not presented here, the dominant clay minerals were mica and kaolin minerals. These soils showed relatively high fertility that had been formed in highland areas under a monsoon climate. In Table 11.1, the amounts of readily mineralizable C (C_0) and N (N_0), which were determined by the long-term incubation experiment, were also provided. The C_0 and N_0 values ranged from 2090 to 14070 and from 100 to 436 mg kg^{-1}, respectively, which is equivalent to 3.9–19.5 % and 3.5–10.1 % of total C and N, respectively.

To summarize several soil parameters relating to soil fertility and to assess their fluctuation throughout a land rotation system, principal component analysis was conducted. Variables used included $pH(H_2O)$; $pH(KCl)$; exchangeable K^+, Mg^{2+}, Ca^{2+}, and Al^{3+}; available phosphate; total C and N; C/N ratio; and contents of sand, silt, and clay fractions. Table 11.2 shows the factor pattern for the first four principal components after varimax rotation, which accounted for 82 % of the total variance.

High coefficients, positive or negative, were given for $pH(H_2O)$, $pH(KCl)$, and exchangeable K^+, Mg^{2+}, Ca^{2+}, and Al^{3+} for the first component. These variables corresponded to the properties derived from soil acidity and, hence, the first component was referred to as the "acidity" factor. The second component exhibited high positive coefficients of total C and N, indicating that a close relationship exists between these and SOM accumulation and that this relationship could be referred to as the "SOM" factor. The third component showed high positive coefficients with available P, NC ratio, and sand content and was considered to be a "P and NC ratio" factor. The forth was correlated with silt and clay contents and was characterized by a "texture" factor.

Figure 11.1 shows the dynamics of the first, second, and third factors throughout different land-use stages of shifting cultivation. In the initial stage of fallow, the scores in the "acidity" factor were kept low, presumably reflecting the ash input at the beginning of cultivation (Fig. 11.1a). It then sharply increased in the middle stage of fallow (3–6 years) and then decreased again in the late stage of fallow. As a similar, but more prolonged, acidification was reported for subsoils under fallow in

Table 11.1 Selected chemical properties of the soils from surface 5-cm layers and parameters determined for the long-term incubation experiment ($n = 27$)

Land-use stage	Number of samples	pH (H_2O)	Clay content (%)	Total C (g kg^{-1})	Total N (g kg^{-1})	Base saturation (%)	Sum of exch. bases (cmol$_c$ kg^{-1})	Exch. Al (cmol$_c$ kg^{-1})	Available P by Truog method (mg P_2O_5 kg^{-1})	C_0 (mg kg^{-1})	N_0 (mg kg^{-1})
0–1 year of fallow	6	5.98 a	28.5 ab	45.8 a	3.38 a	99 a	11.78 a	0.09 a	61.5 a	2980 a	199 ab
2–5 years of fallow	9	5.41 ab	28.5 a	47.5 a	3.14 a	75 a	5.86 a	1.35 ab	45.0 ab	3618 a	159 a
6–8 years of fallow	7	5.70 ab	32.3 ab	56.7 ab	3.84 a	94 a	9.33 a	0.41 a	36.8 b	4957 ab	264 ab
Seminatural forest	5	5.13 b	35.8 b	71.5 b	5.16 b	68 a	6.60 a	2.21 b	39.7 ab	11003 b	318 b

The values with the same letters are not significantly different ($p < 0.05$)

Table 11.2 Factor pattern for the first four principal components determined for the surface soil samples from northern Thailand ($n = 27$)

Variable	PC1	PC2	PC3	PC4
pH(H_2O)	−0.78 **	−0.48	0.17	−0.08
pH(KCl)	−0.90 **	−0.27	0.20	0.15
Exch. K	−0.68 **	0.25	0.38	−0.07
Exch. Mg	−0.91 **	0.22	0.03	−0.14
Exch. Ca	−0.94 **	0.07	0.02	0.10
Exch. Al	0.83 **	0.38	0.12	0.04
Avail. P	−0.18	−0.05	0.80 **	−0.01
Total C	0.14	0.94 **	−0.13	−0.10
Total N	0.04	0.98 **	0.07	−0.09
NC ratio	−0.32	0.41	0.70 *	0.02
Sand	0.26	−0.37	0.70 *	−0.08
Silt	−0.16	−0.13	0.13	0.82 **
Clay	0.18	0.01	−0.19	0.82 **
Proportion (%)	35	21	15	11
	Acidity	SOM	P and NC ratio	Texture

Significant correlation coefficients at **$p < 0.01$ and *$p < 0.05$

the area (Tanaka et al. 1997), soil acidification may commonly occur as a result of cation uptake by fallow vegetation.

The values of factor 2 scores, the "SOM" factor, were low throughout the cropping year and the initial stage of fallow (0–6 years) and then slightly increased along with the establishment of secondary forest and an increasing supply of forest litter (7–8 years) (Fig. 11.1b). A similar trend observed for the amounts of readily decomposable fractions of organic C and N (C_0 and N_0, respectively) supported an accumulation of organic materials through litter incorporation into the surface soils (Fig. 11.2). As the "acidity" factor showed a decreasing trend at the same time in the late stage (Fig. 11.1a), this litter may be supplying bases (obtained by tree roots from further down the soil profile) to the surface soil. This simultaneous change in factors 1 and 2 in the surface soil through forest-litter deposition in a late stage of fallow has an increasing effect on nutritional elements and can be considered essential to the maintenance of this forest-fallow system. However, since in the long fallow the factor 1 scores were often high, cumulative cation uptake by forest vegetation may occasionally cause soil acidification. It is necessary to investigate in more detail whether prolonged fallow can accumulate bases in surface soil layers. In the long fallow, the values of factor 2 scores were usually high due to cumulative litter input.

The "P and NC ratio" factor did not exhibit any clear trend throughout the land-use stages, including long fallow (Fig. 11.1c). The fallow practice does not seem to improve the soil properties relating to this factor. The slash-and-burn practice was, therefore, considered to be indispensable for improving the soil properties relating to this factor for the next cropping. This is in contrast to the acidity and/or SOM-related properties of the soils, which already showed improvements in the late stage of fallow.

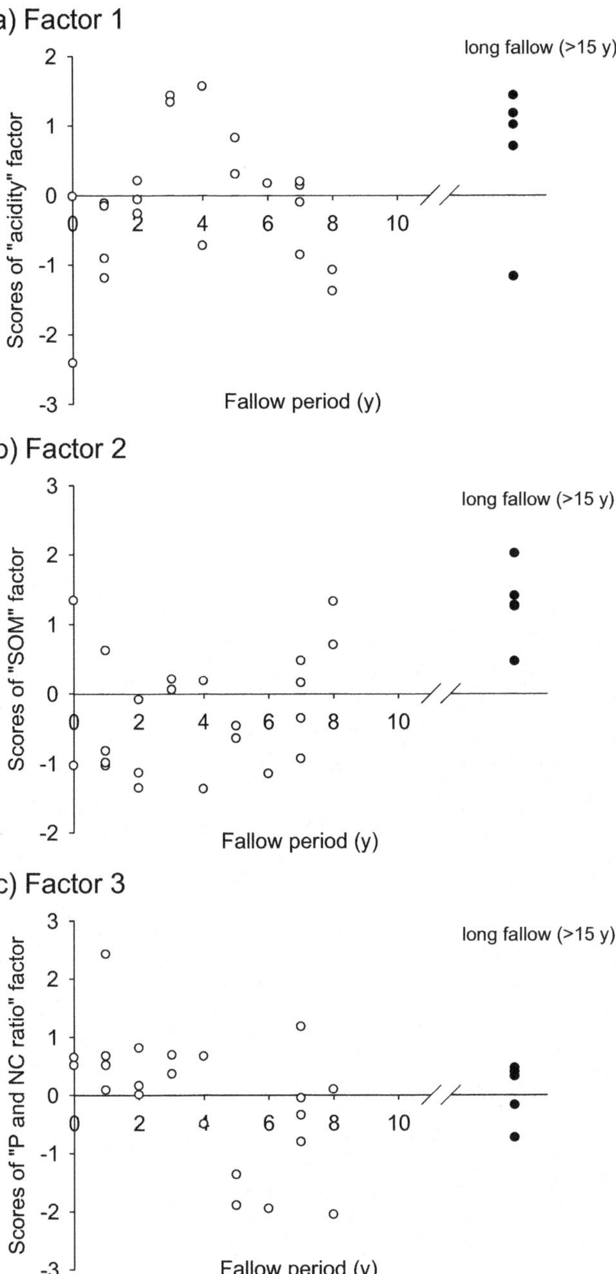

Fig. 11.1 Distribution of factor scores in relation to land-use stages of shifting cultivation in northern Thailand

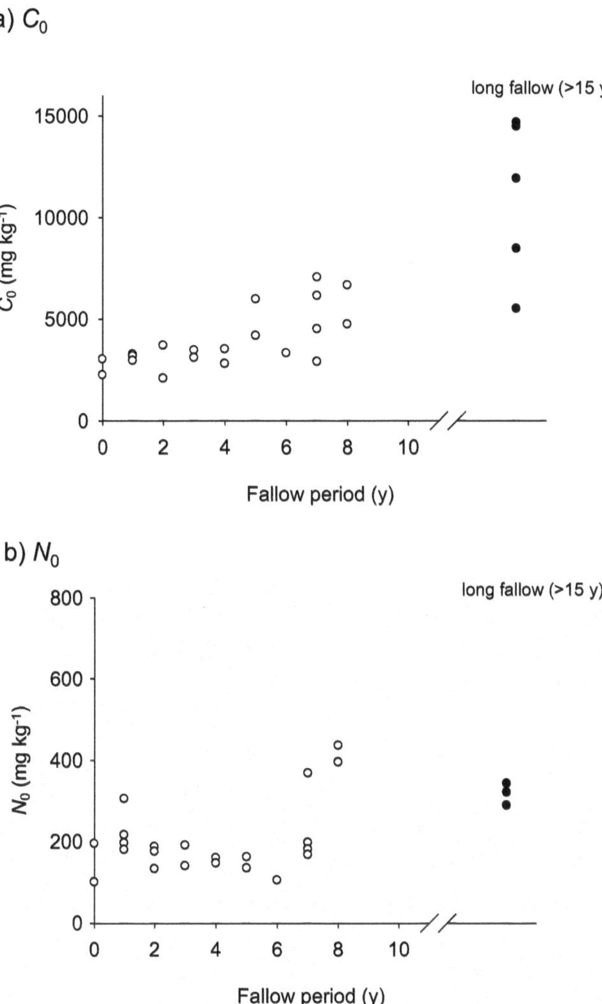

Fig. 11.2 Distribution of C_0 and N_0 in relation to land-use stages of shifting cultivation in northern Thailand

11.4   Field Measurements of SOM Budget in Different Stages of Land Use Under Shifting Cultivation

The field experiments were conducted from April 2001 to April 2002. To compare SOM dynamics under different stages of land use (cropped field after burning, fallow forest, and natural forest), six plots were set up as follows: cropland in 2001 (*CR01*); first, second, fourth, and sixth years of fallow forest (*F1*, *F2*, *F4*, and *F6*, respectively); and unburned seminatural forest stand of about 30 years old (*NF*) for comparison.

To estimate the amount of C input into the soils, we measured litter input in *CR01*, *F1*, *F2*, *F4*, *F6*, and *NF* from 19 April 2001 to 19 April 2002, as follows. First, litter input in cropland is assumed to be equal to the amount of rice residues after harvest, which were collected for 1-m^2 plots in triplicate just after harvest. Second, in the early stage of fallow (first to third year), fallowed fields are occupied by perennial herbaceous species, among which *Eupatorium odoratum* is dominant. Tree species gradually increase, but their biomass and litter supply are still limited at this stage. As it was difficult to collect litter fall in these fields, we postulate that in the first and second year, the amount of litter input is equivalent to the amount of leaves at the end of the rainy season, and in the third year, all the herbaceous biomass and half of the litter input of the establishing tree vegetation are returned to the soils. The biomass of herbaceous species was measured in *F1* and *F2* at the end of the rainy season (in 1-m^2 plots in triplicate). Third, in the later stage of fallow (fourth to seventh year), secondary forest is successively established and litter fall from tree vegetation comprises a major part of litter input. The amount of litter fall was collected every month in *F4*, *F6*, and *NF* (in approximately 0.4-m^2 plots in five replications). In addition, the amounts of litter stock were measured for 1-m^2 plots in triplicate in *F4*, *F6*, and *NF* on 16 May 2001 and 19 April 2002. The results on litter input are given in Table 11.3. Annual litter fall was calculated to be 6.87, 5.37, and 9.00 Mg ha^{-1} year^{-1} in *F4*, *F6*, and *NF*, respectively. These values can be converted to carbon: 3.19, 2.61, and 4.48 Mg C ha^{-1} year^{-1}, respectively. For these forest stands, as it is difficult to measure the amount of annual supply of dead root litter, we used the ratio of root litter to litter fall of 0.32, according to the simulation of Nakane (1980), which was proposed for matured forests under different climates. In *F1* and *F2*, the total biomass of herbaceous species was 12.0 and 14.3 Mg ha^{-1} year^{-1}, respectively, of which approximately 47 % consisted of leaves. In *CR01*, only 1.31 Mg C ha^{-1} year^{-1} of rice residues remained on the soil after harvest. As a result, total litter input is calculated to be 4.3–9.0 Mg ha^{-1} year^{-1} in the fallowed plots, which equals to 2.0–4.2 Mg C ha^{-1} year^{-1}. It is noteworthy that especially high concentrations of N or P could be returned to the soils via the forest litter compared with via the herbaceous litter (Table 11.3).

On the other hand, the soil C output was estimated by measuring the in situ field soil respiration rate using a closed chamber method for several times during the experimental period. At the same time, soil temperature at a depth of 5 cm and volumetric moisture content of the soil at depths of 0–15 cm were measured at all sites and also monitored continuously using data loggers at two sites (*CR01* and *NF*). Figure 11.3 shows the seasonal fluctuations of soil temperature (5 cm), volumetric water content of the soils (0–15 cm), and soil respiration rate (mol C ha^{-1} h^{-1}) with or without root respiration (C_{em+R} and C_{em-R}) measured at the plots. Soil temperature at *CR01* was the highest, within the range of 23–32 °C, whereas that at *NF* was the lowest, within the range of 18–21 °C. In the fallowed plots, soil temperature was between *CR01* and *NF*. Volumetric water content of the *CR01* soil was lowest among all the plots throughout the year. The temperature and moisture data monitored by the data loggers also exhibited the same trend. The values of C_{em-R} were usually lower than C_{em+R}, indicating that whole-soil respiration is also

Table 11.3 Litter layer and annual litter input in the cropland and fallow fields under shifting cultivation in northern Thailand

	Litter layer in May 2001				Litter layer in April 2002				Annual litter input					
									Herbaceous biomass at the end of rainy season	Annual litter-fall input (C content)	Total litter input estimated[a]			
Site	Amount	C content	N content	P content	Amount	C content	N content	P content			Amount	C content	N content	P content
	$(Mg\ ha^{-1})$	$(Mg\ C\ ha^{-1})$	$(kg\ N\ ha^{-1})$	$(kg\ P\ ha^{-1})$	$(Mg\ ha^{-1})$	$(Mg\ C\ ha^{-1})$	$(kg\ N\ ha^{-1})$	$(kg\ P\ ha^{-1})$	$(Mg\ ha^{-1}\ year^{-1})$	$(Mg\ ha^{-1}\ year^{-1})$	$(Mg\ ha^{-1}\ year^{-1})$	$(Mg\ C\ ha^{-1}\ year^{-1})$	$(kg\ N\ ha^{-1}\ year^{-1})$	$(kg\ P\ ha^{-1}\ year^{-1})$
CR01									3.07		3.07	1.31	17.8	0.355
F1									11.95		5.56	2.59	21.4	0.428
F2									9.18		4.27	1.99	22.3	0.446
F4	5.29	2.66	70.4	1.41	4.69	2.35	54.9	1.10		6.87 (3.19)	9.04	4.19	99.2	3.48
F6	3.67	1.50	29.4	0.588	5.93	2.88	39.3	0.532		5.37 (2.61)	7.06	3.44	49.3	1.51
NF	7.52	3.63	93.0	1.86	6.81	3.43	97.8	1.96		9.00 (4.08)	11.84	5.89	143.1	4.21

[a]For the herbaceous vegetation in *F1* and *F2* , only leaves are assumed to be incorporated into soils. The coefficient is 0.47. On the other hand, the amount of root litter under the forest stand (*F4* , *F6*, and *NF*) is estimated to be litter fall $\times$ 0.12/0.38 according to simulation of Nakane (1980)

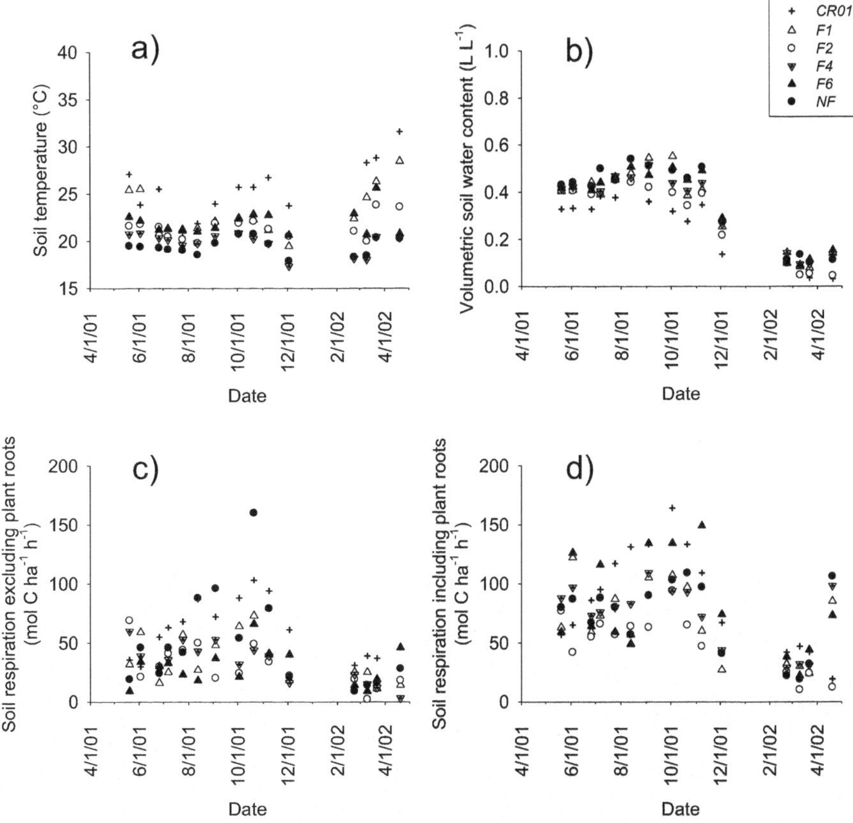

Fig. 11.3 Fluctuation of soil temperature, soil moisture, and soil respiration rates measured from April 2001 to April 2002

contributed by plant root respiration. Both the C_{em+R} and C_{em-R} at all the plots exhibited the highest values in the late rainy season (September to October). The C_{em+R} and C_{em-R} values were low throughout the dry season (December to March).

In order to estimate total soil respiration throughout the year, we first established an equation that represented a relationship of the in situ hourly soil respiration rate and environmental factors (such as soil temperature and moisture) by multiple regression analysis. We then calculated hourly soil respiration rate by substituting each parameter of the equation using monitored data and summed up hourly soil respiration rates for a given period. In the first step, we assume the Arrhenius-type relationship between soil temperature and respiration rate, as follows:

$$C_{em} = a\,\theta^{b}\exp(-E/RT)\exp(cD) \tag{11.1}$$

where C_{em} is an hourly soil respiration rate with or without root respiration (mol C ha^{-1} year.$^{-1}$), θ is a volumetric soil moisture content (L L^{-1}), E is the activation energy (J mol^{-1}), R is the gas constant (8.31 J mol^{-1} K^{-1}), T is an absolute soil temperature (K), D is days after the rainy season started on 14 May (until 13 November, on which θ was falling below 0.3), and a, b, and c are coefficients. As soil respiration rate and microbial biomass N showed an increasing trend in some of the plots during the rainy season, a parameter D was introduced. For the samples in the dry season (after 13 November, $\theta < 0.3$), the value of D was set to 0. The equation was then rewritten in the logarithm form:

$$\ln C_{em} = \ln a + b\ln\theta - E/RT + cD \tag{11.2}$$

Following this, a series of coefficients (a, b, c, and E) are calculated by stepwise multiple regression analysis ($p < 0.15$) using the measured data, C_{em}, θ, T, and D (SPSS Inc. 1998).

The results of regression analysis are summarized in Table 11.4. The coefficient relating to soil temperature (E) was often rejected ($p < 0.15$). For example, this occurred in $F1$ and $F4$ for C_{em-R} and in $CR01$ for C_{em+R}, indicating that under the subtropical conditions in this study, the effects of seasonal fluctuation of temperature on soil respiration rate were limited. In contrast, coefficient b usually caused a fluctuation of the soil respiration rate due to the presence of a distinct dry season. In $CR01$, the influence of both soil temperature and moisture was uncertain, and only the date factor, c, was strongly related to fluctuations of soil respiration. A drastic change in microbial community after slash and burn may cause such a unique behavior of soil respiration, which continuously increased throughout the rainy season. Using these regression equations, and monitoring soil temperature and moisture data in $CR01$ and NF, cumulative soil respiration during the year was calculated. As the monitored data was available for only two of the six plots, fluctuations in soil temperature and moisture for the remaining four plots were assumed based on the relationship between the actual data and either monitored data of $CR01$ or of NF with a higher correlation. We used monitoring data of $CR01$ for simulation in $CR01$, $F1$, and $F2$ and that of NF for simulation in NF, $F4$, and $F6$. The results of these calculations are also given in Table 11.4. Annual soil respiration without root respiration was highest in $CR01$, followed by NF and then $F2$, $F1$, $F4$ and $F6$, ranging from 2.15 to 6.34 Mg C ha^{-1} year^{-1}. The contribution of root respiration on whole-soil respiration was calculated to be 23, 48, 29, 57, 69, and 38 % in $CR01$, $F1$, $F2$, $F4$, $F6$, and NF, respectively (an average of 55 % in forest plots $F4$, $F6$, and NF), which was similar to the estimation (50 %) by Nakane (1980).

Table 11.5 summarizes the amounts of SOM stock (up to 15-cm depth) and C budget in the study plots. Total SOM stock, including litter layer, in the soil profiles ranges from 37.3 to 66.7 Mg C ha^{-1}. Compared with the total stock, annual output of soil C under the fallow forest is small (2.15–3.35 Mg C ha^{-1} year^{-1}), as is that of cropland (6.34 Mg C ha^{-1} year^{-1}); this is equivalent to only 4.5–10.3 % of total SOM. This may be one of the reasons why soil degradation did not become pronounced after 1 year of cropping in this area. Considering the fact that, even in NF, the SOM budget is still positive, the SOM level after the forest ecosystem

Table 11.4 Parameters determined by stepwise multiple regression analysis

Site	Coefficients				R^2	n	$C_{em}=a\theta^b\ exp(\text{-}EIRT)\ exp(cD)$ (at $T = 293K$, $\theta=0.4LL^{-1}$, D=0) (mol C ha^{-1} h^{-1})	Annual soil respiration (Mg C ha^{-1} year^{-1})
	$\ln a$	b	E (kJ mol^{-1})	c				
Excluding roots (C_{em-R})								
CR01	−9.39 *	–	−32.3 *	5.96E-03 ***	0.81 ***	15	49.1	6.34
F1	3 73 ***	0.394 *	–	3.41E-03 *	0.52 ***	15	29.1	3.02
F2	157 ***	0.769 ***	398 ***	3.01E-03 *	0.75 ***	15	18.1	3.35
F4	4.50 ***	0.993 ***	–	–	0.55 ***	15	36.3	2.99
F6	59.0 ***	0.612 ***	135 **	–	0.45 **	15	19.2	2.15
NF	58.6 *	0.484 *	134*	6.24E-03 **	0.77 ***	15	28.2	4.14
Whole respiration (C_{em+R})								
CR01	457 ***	0.360 ***	–	4.27E-03 ***	0.81 ***	15	76.4	8.23
F1	45.5 ***	0.931 ***	99.3 ***	–	0.81 ***	15	47.2	5.76
F2	53.3 ***	0.732 ***	118 **	–	0.92 ***	15	52.2	4.72
F4	62.5 ***	0.393 ***	141 ***	–	0.73 ***	15	74.1	6.96
F6	49 2 ***	0.708 ***	109 ***	2.69E-03 *	0.74 ***	15	55.6	7.03
NF	85.3 ***	0.621 ***	196 ***	–	0.74 ***	15	74.2	6.64

*, **, ***: Singnificant at 25%, 5%, and 1% levels, respectively

$C_{em} = a\theta^b\ exp(\text{-}E/RT)\ exp(cD)$

C_{em}: Rate of CO_2 emission (mol C ha^{-1} h^{-1})

T: Temperature (K)

θ: Volumetric moisture content of soil (L L^{-1})

D: Days after rain season started on May 14. The value of D is 0 for the samples in dry season (after November 13, $\theta<0.3$)

$R = 8.315$ (J K^{-1} mol^{-1})

E: Activation energy (J mol^{-1})

a, b, c: Coefficients

This equation is converted to; $\ln C_{em} = \ln a + b\ \ln\theta - E/RT + cD$

Table 11.5 Stock and budget of soil organic carbon from 19 April 2001 to 19 April 2002

Site	C stock in litter layer (Mg C ha^{-1})	C stock in 0–15-cm depth (Mg C ha^{-1})	Total litter input estimated (a) (from Table 11.4) (Mg C ha^{-1} year^{-1})	Annual C output by decomposition (b) (from Table 11.5) (Mg C ha^{-1} year^{-1})	Annual budget of SOM (a-b) (Mg C ha^{-1} year^{-1})
CRO1		61.9	1.31	6.34	−5.03
F1		64.4	2.59	3.02	−0.43
F2		61.8	1.99	3.35	−1.36
F4	2.7	64.0	4.19	2.99	1.20
F6	1.5	35.8	3.44	2.15	1.29
NF	3.6	50.8	5.89	4.14	1.75

reaches its equilibrium would possibly be higher. In turn, the SOM level under the current shifting cultivation practice can be said to be a secondary one, in which SOM stock had already been decreasing during repeating cultivation, compared with natural forest. Figure 11.4 summarizes annual input and output of organic matter in the soils with a cumulative budget based on the data in Table 11.5. For the fifth year, the average value of *F4* and *F6* was used for the calculation, and for the third year, the average value of *F1* and *F2* and half of the litter input of *F4* were assumed to consist of litter input. During the cropping phase and the initial stage of fallow, the annual C budget was negative. It became positive when the initial herbaceous vegetation was succeeded by tree vegetation (after the third or fourth year of fallow), and a cumulative budget became positive after the sixth year of fallow. As the litter fall supplied may require another 6 months to 1 year to be completely incorporated into the soils after initial decomposition (based on the C stock in litter layer and the annual C input from litter fall (see Table 11.5)), approximately 6–7 years of fallow is needed to balance the SOM budget under the current fallow/cropping practice. A possible error can derive, in this study, from uncertainty regarding the estimation of dead plant roots in the burning stage or during the later stage of fallow. Even so, the minimum fallow period required would not fall below 4 years, judging from the fact that the SOM budget in the system strongly depends on incorporation of initial herbaceous biomass into the soil system after establishment of tree vegetation (in approximately the fourth year).

It was possible to establish a multiple regression equation between the annual soil respiration and soil parameters for the five noncultivated plots (excluding *CR01*), as follows:

$$\text{Annual soil respiration } (\text{Mg C ha}^{-1}\text{year}^{-1}) = 1.291 + 0.483 \times C_0 (\text{g kg}^{-1})$$
$$+0.643 \times \text{pH}(\text{H}_2\text{O}) - 0.116 \times \text{clay content } (\%); \ (r^2 = 0.93, p < 0.25, n = 5)$$

$$(11.3)$$

These parameters were chosen from variables that correlated with the different principal components determined previously. Although the reason for the negative

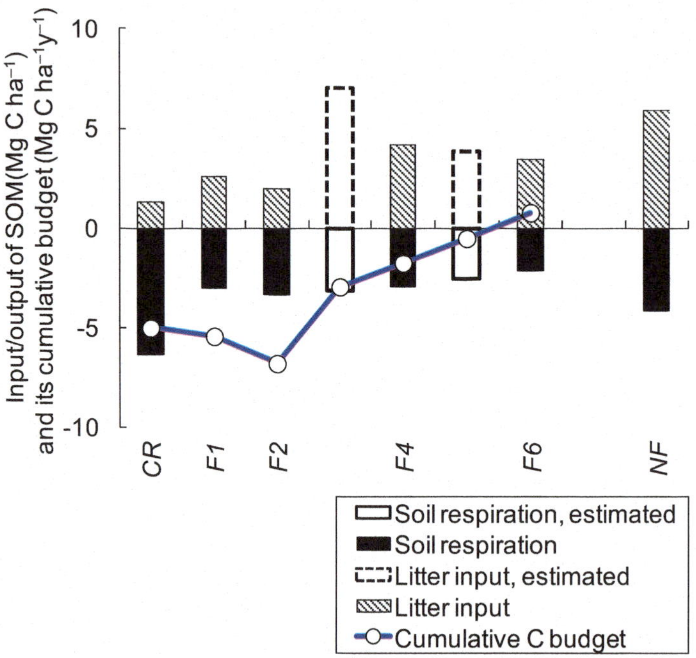

Fig. 11.4 Annual input and output of organic matter of the soils with a cumulative budget

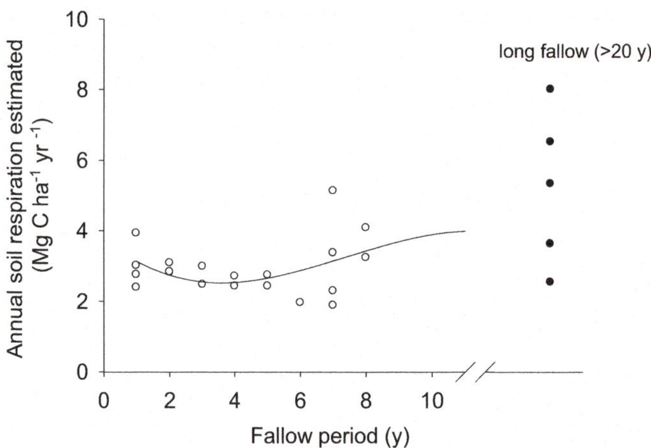

Fig. 11.5 Annual soil respiration estimates based on the regression equation obtained

contribution of soil acidity is uncertain, it seemed to be worthwhile to be further investigated. Using this equation, the annual soil respiration was estimated for all 27 soils analyzed in Table 11.1 (excluding those in the cropping year) and plotted in Fig. 11.5. In most cases, the annual soil respiration in the fallowed plots ranged from 2 to 4 Mg C ha^{-1} year^{-1} and were almost consistent with the measured values. Therefore, it is possible to conclude that the annual soil respiration or the annual decay of SOM in the studied village falls in this range and that amounts of organic material should be incorporated into soils annually to keep the present level of SOM.

11.5 In Situ Soil Solution Composition Under Shifting Cultivation

As decomposed products of SOM, such as NO_3^-, were expected to be present in the soil solution, we collected the soil solution by a porous-cup method (Funakawa et al. 1992) to analyze the process of N translocation in soil profiles.

Figure 11.6 shows that the NO_3^- concentration in the soil solution was high in May and sharply decreased in early June at all the plots, except for *CR01*, in which the NO_3^- concentration remained as high as 0.5 mg N L^{-1} at a depth of 45 cm even on June 22, suggesting that a higher amount of NO_3^- could have been leached out from the top 45 cm of the soil layer in *CR01* along with rainfall events. In contrast, a low concentration of NO_3^- was detected in the soil solution at a depth of 45 cm in *NF* throughout the rainy season, indicating that almost no N leaching from the forested ecosystem occurred. Figures for fallow forest were between that of *CR01* and *NF*, in that the high concentrations of NO_3^- were observed only at the initial stage of the rainy season (May), and they soon dropped to a low level.

11.6 Fluctuation of Microbial Biomass and the Substrate-Induced Microbial Activities (Respiration and N Assimilation/Nitrification)

To estimate the dynamics of microbial biomass and its activity in relation to the dynamics of C and N, the amounts of microbial biomass C and N were determined for several times during the experimental period by the chloroform fumigation-extraction method (Brookes et al. 1985; Vance et al. 1987). The metabolic quotient, $q\mathrm{CO}_2$, was calculated from the soil respiration rate divided by microbial biomass C.

Figure 11.7a shows that microbial biomass C in *NF* was the highest, ranging from 530 to 1849 mg kg^{-1}, whereas that in *CR01* was usually the lowest throughout the year (85–614 mg kg^{-1}). Salt extractable C in *CR01* increased remarkably up to 600 mg kg^{-1} in the dry season (Fig. 11.7b), in which microbial biomass C dropped. Such an inverse fluctuation of biomass C and salt extractable C indicates an increase of microbial debris during the dry season due to exposure to a strongly fluctuating environment (such as in the soil temperature), with a limited vegetation cover. According to Fig. 11.7c, microbial biomass N was the lowest in *CR01*,

Fig. 11.6 Concentration of NO_3^- in soil solution

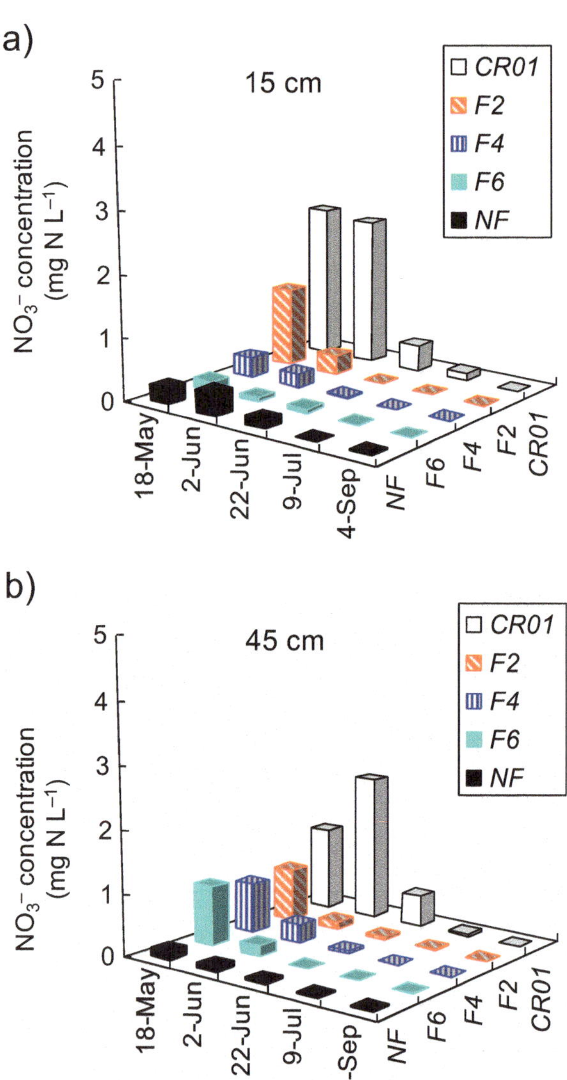

whereas the highest was seen in *NF*. Microbial biomass N exhibited an increasing trend throughout the rainy season, probably due to continuous assimilation of N. Similarly, the value of qCO_2 was the highest in the late rainy season (Fig. 11.7d) and then sharply dropped at the start of the dry season. Soil respiration also decreased, whereas the microbial biomass C level remained high, except for in *CR01* (Fig. 11.7a). It is considered that a large proportion of the soil microbes could have survived in the forested soils without a drastic decrease of biomass during the dry season, presumably by suppressing their activity.

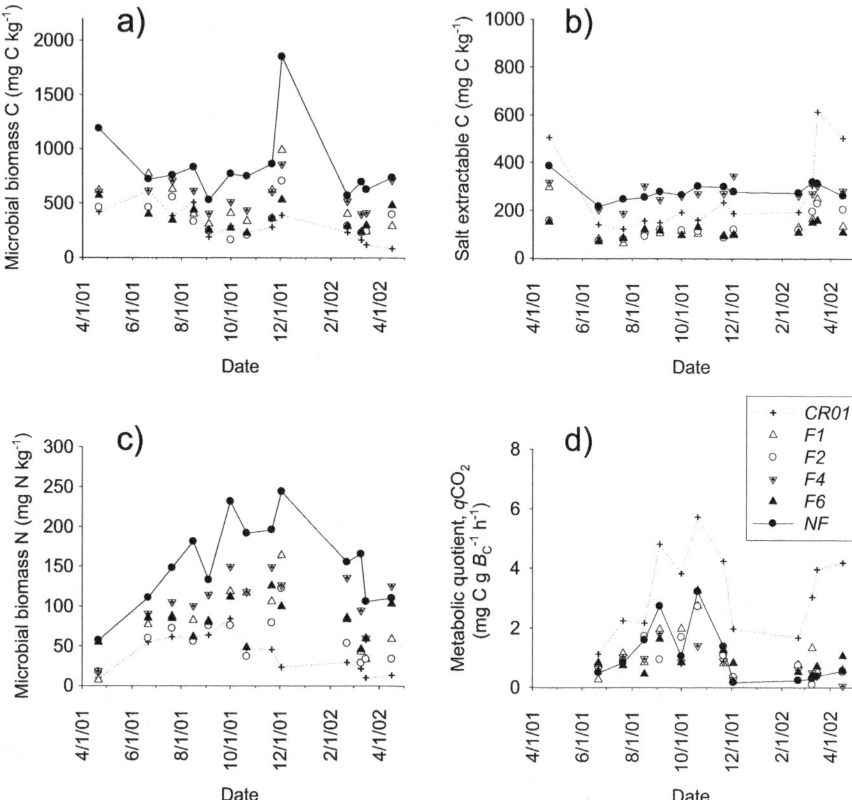

Fig. 11.7 Seasonal fluctuation of microbial biomass C (**a**), salt extractable C (**b**), microbial biomass N (**c**), and metabolic quotient (**d**) during the experiments

To analyze the behavior of the microbial community as a controlling factor of N dynamics in shifting cultivation in more detail, short-term responses of soil microbes after the addition of C and N substrates were comparatively investigated for forest and cropland soils from the study village using fresh soil samples collected in March 2003 (late dry season), early June and July 2003 (rainy season), from a field that was used for rice cropping in 2003 (CR_{03}), a field in its second year of fallow ($CR01_{03}$), a field in its sixth year of fallow ($F4_{03}$), and a seminatural forest stand (NF_{03}). The latter three are the same plots as those used for the field experiments in 2001. Then substrate-induced microbial activities were traced using a fresh soil after addition of glucose (equivalent to 4000 mg C kg^{-1} soil) with or without 0.0168 g of NH$_4$NO$_3$ (equivalent to approximately 400 mg N kg^{-1} soil). In the treatment with NH$_4{}^+$, enough N was added to eliminate the microbial activity limitations that occur as a result of N shortage. Then the mineralized C was measured continuously up to 94 h at 25 °C. Similarly short-term N transformation after addition of NH$_4{}^+$–N as an N source was traced up to 168 h at 25 °C under aerobic conditions.

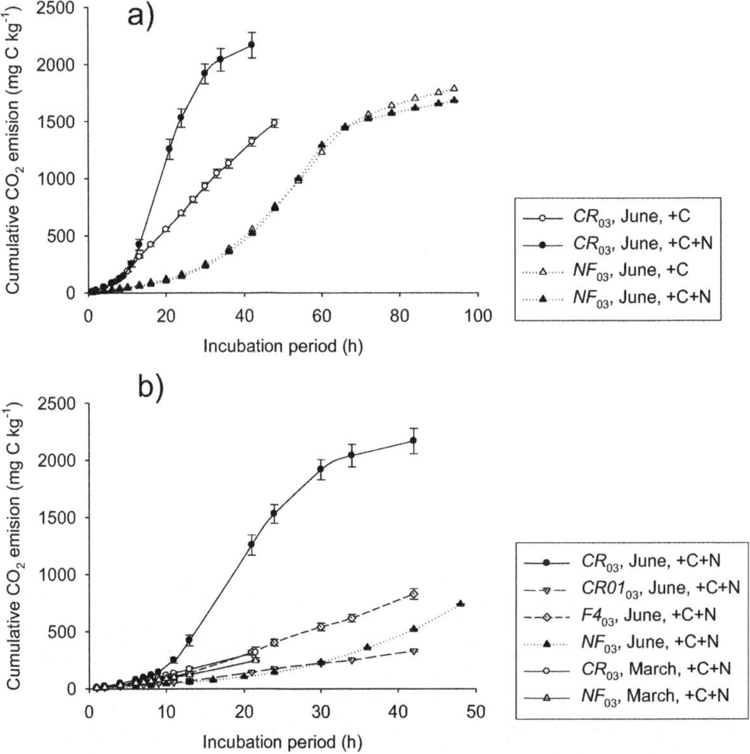

Fig. 11.8 Short-term CO_2 emission (cumulative) after glucose addition with or without NH_4^+ to the fresh soils; (**a**) effect of NH_4^+ addition to cropped (CR_{03}) and forest (NF_{03}) soils and (**b**) comparison of the soils in different land-use stages

Figure 11.8a compares cumulative CO_2 emissions after adjustment of glucose levels with or without N in the cropland soil (CR_{03}) and the forest soil (NF_{03}). Irrespective of N addition, glucose-induced respiration in NF_{03} was significantly delayed compared with that in CR_{03}. In the case of CR_{03}, N addition resulted in further acceleration of respiration, suggesting that the soils of CR_{03} originally did not contain enough N for the microbial community to demonstrate its maximum potential or the microbial community could utilize additional N efficiently for multiplication of the community. Such a trend was, however, scarcely observed for the forest soil, NF_{03}. A higher and efficient utilization of additional C and N resources is obvious in CR_{03} compared with NF_{03}. It was notable that such a response of the microbial community was observed only in CR_{03} during the rainy season (after burning and moistening) (Fig. 11.8b). The soils from the second and sixth years of fallow forests ($CR01_{03}$ and $F4_{03}$) showed a similar trend as NF_{03}. Even in CR_{03}, the soil collected before the burning event (but after forest clearcutting) did not exhibit active respiration after glucose addition with N. Therefore, the unique property of the microbial community in CR_{03} was probably introduced after slash and burn of fallow forest.

At the end of the incubation period (42 h), 2170 mg kg^{-1} of soil C was mineralized in CR_{03} (in the +C+N treatment), which amounted to 54 % of added C (Fig. 11.8a). Coody et al. (1986) reported that, using a ^{14}C-labeling technique, 27–44 % of added glucose was mineralized and most of the remaining glucose was already assimilated by soil microbes within 96 h at 25 °C. It is highly probable in this experiment that the added glucose was almost completely consumed, either through respiration or via assimilation by soil microbes — even considering that our results may involve some degree of priming effect and, therefore, an overestimation of substrate decomposition. On the contrary, the soils from fallowed ($CR01_{03}$ and $F4_{03}$) or matured (NF_{03}) forests exhibited slow decomposition rates compared with CR_{03}, and the amounts of mineralized C at 42 h were only 8, 21, and 13 % of added glucose, respectively. Since in this experiment the possibility of a deficiency in N was practically eliminated, such a slow utilization of the added C could be attributed to a specific property of the microbial community in these forest plots.

Figure 11.9 demonstrates the fluctuation of levels of inorganic nitrogen (NH$_4^+$ and NO$_3^-$) after addition of NH$_4^+$ in the incubation experiment. The nitrifying

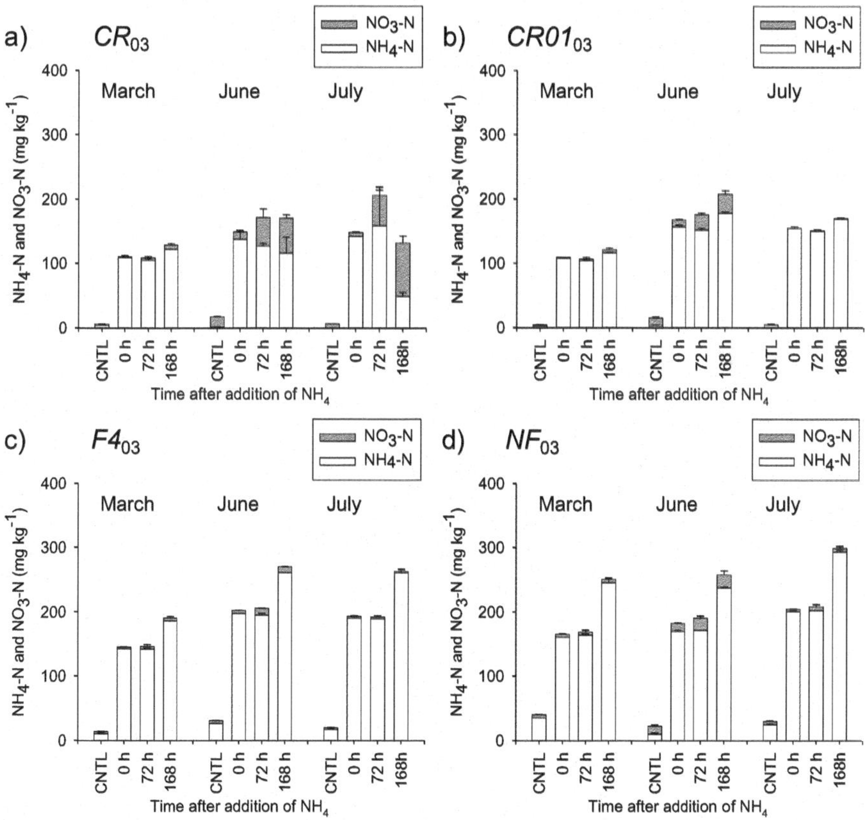

Fig. 11.9 Short-term transformation of NH$_4$-N added to the fresh soils

activity was detected only in the CR_{03} samples collected in the rainy season (June and July). In the soils from fallowed ($CR01_{03}$ and $F4_{03}$) or matured forest (NF_{03}), minimal nitrifying activity was traced, if any. As the concentration of NH_4^+ increased within the period of incubation (168 h), gross mineralization of organic N was superior to gross immobilization of NH_4^+. So, in the forested plots, including in the initial stage of fallow ($CR01_{03}$ in the second year), both the nitrifying activity and the NH_4^+ assimilation are noticeably low. In contrast, judging from the fact that the sum of the remaining NH_4^+–N and NO_3^-–N was decreasing during the experiment, N immobilization also occurred at the same time as active nitrification in cropland (CR_{03}). These results were generally consistent with the fact that, in most cropped ecosystems, NH_4^+ is mineralized and then readily nitrified to form NO_3^- if it is not immobilized again rapidly. In contrast, in forest soils, such an active nitrification is sometimes retarded due to low activity of nitrifying bacteria (Robertson 1982).

11.7 Discussion on the Dynamics of Microbial Activities During Different Stages of Shifting Cultivation with Special Reference to the Function of the Fallow Phase

Figure 11.7d shows that the values of qCO_2 in $CR01$ were much higher than those in the other plots, indicating a higher activity of soil microbes per unit weight of biomass despite apparent lower biomass there (Fig. 11.7a). This is partly because the method used in this study (the chloroform fumigation-extraction method) could evaluate the whole biomass including inactive biomass that may be dominant in the forest soils. Two explanations are possible for this apparent high microbial activity in $CR01$. First, there may be a possible effect of disturbance, as well as slash and burn, in $CR01$. Mamilov and Dilly (2002) reported that stress, such as drying and re-wetting, makes microbial activity high, which is expected to be more pronounced in the field just after forest reclamation (as in $CR01$). According to Wardle and Ghani (1995), who described how qCO_2 could be used as a bioindicator of disturbance, the cropland ecosystem in the present study ($CR01$) is considered to be a more disturbed one than the forested ecosystems. Second, the soil microbial community in $CR01$ drastically changes after slash and burn of the forest due to a sharp increase in soil pH along with ash addition. As was observed in Table 11.1 and Fig. 11.1, a few years after a burning event, it generally produces higher pH values in soils than in fallow or matured forest in the same area. In any case, the distinct difference in the values of qCO_2 between the cropland ($CR01$) and the others suggests some essential difference in composition and/or function of soil microbial communities.

On the other hand, soil solution composition revealed that, in the cropland ($CR01$) soil, a fairly high concentration of NO_3^- was released into the soil solution, unlike in forest soil (*NF*) (Fig. 11.6). The values for fallow forests were between them. A small stand of upland rice with possibly poor nutrient-uptake ability in *CR01* during the early rainy season may further increase the chance of NO_3^- leaching. Ellingson et al. (2000) and Neill et al. (1999) also reported that deforestation and burning increased the concentration of NO_3^- in the soils in Mexican tropical dry forests and Amazonian forests, respectively. There are two explanations in which a higher amount of NO_3^- could be released into the solution in *CR01* in the present study: either there was higher nitrification activity of soil microbes in *CR01* or a higher N immobilization (assimilation) activity of soil microbes in NF.

As shown in Fig. 11.8, the laboratory incubation experiment for analyzing the substrate-induced respiration clearly showed that the soil microbes in the cropland (*CR03*) soil responded very quickly to addition of glucose, unlike fallowed or matured forest soils (*CR01*$_{03}$, *F4*$_{03}$, and *NF*$_{03}$). In the same way, soil microbes rapidly nitrified NH_4^+ added to NO_3^- in *CR03* while partially assimilating the NH_4^+ at the same time (Fig. 11.9). In the remaining plots, neither accelerated glucose consumption nor active NH_4^+ utilization was observed. Therefore, all the microbial activities that were tested in *CR03* were induced quite rapidly after the addition of the substrates. These properties of the microbial community in *CR03* were probably introduced after slash and burn of fallow forest. This, together with the fact that a higher qCO_2 was observed despite an apparent lower microbial biomass in *CR01* (Fig. 11.7), indicates that the microbial community in the cropped soils consists of a smaller number of, but more active, microbes compared with that in the fallowed or matured forests, in which a larger number of soil microbes coexisted with a low activity. The main reason for suppressing NO_3^- release into the soil solution is therefore a lower nitrifying activity of the microbial community in the forest soils than in the cropped soils. Such a high nitrifying activity in cropped soils under shifting cultivation is also reported by Tulaphitak et al. (1985a) and Tanaka et al. (2001). On the other hand, the contribution of re-immobilization of NH_4^+ once mineralized is secondary to suppressing NO_3^- release in the forest soil, as in *NF03* and the fallowed plots (*CR01*$_{03}$ and *F4*$_{03}$), where the microbial response for immobilization of added NH_4^+ was slow in the incubation experiment.

Conversely, the seasonal fluctuation of microbial biomass C and salt extractable C suggests that the microbial community in *CR01* was easily destroyed after sharp drought conditions during the dry season. With a succession of secondary vegetation, such a community seemed to be replaced by a more stable one, which did not show a clear decrease of microbial biomass C, even in the dry season. At the same time, it does not increase, or multiply, rapidly even under favorable conditions, such as the addition of substrates, as demonstrated in the short-term incubation experiment. The increasing trend of microbial biomass N that is observed during the rainy season (Fig. 11.7c) is considered to be a result of slow utilization of N by the microbial community and may contribute to accumulation of N into the SOM pool.

11.8 Conclusion: The Main Functions of the Fallow Phase in Shifting Cultivation by Karen People in Northern Thailand

Based on the results obtained in this study, the functions of the fallow phase in shifting cultivation that have ensured the long-term sustainability of the system can be summarized as follows: First, some soil properties relating to soil acidity improve at the same time as the SOM-related properties increase in the late stage of fallow. The litter input may be supplying bases (obtained by tree roots from further down the soil profile) to the surface soil. This simultaneous increase in SOM and bases in the surface soil, through forest-litter deposition in the late stage of fallow, has an increasing effect on nutritional elements. Second, the decline in soil organic C during the cropping phase could be compensated by litter input during 6–7 years of fallow. With regard to the overall budget, the organic matter input through incorporation of initial herbaceous biomass into the soil system after establishment of tree vegetation (approximately in the fourth year) was indispensable for maintaining the SOM level. Third, the succession of the soil microbial community from rapid consumers of resources to stable and slow utilizers, along with establishment of secondary forest, retards further N loss through leaching and enhances N accumulation into the forestlike ecosystems. It is noteworthy that, during the fallow period, nitrifying activity of soil microbes, which was once activated in the cropping phase, is apparently suppressed. As a result, $NO_3{}^-$ effluent from soil layers was remarkably low, even in the initial stage of fallow.

The functions of the fallow phase listed above can be considered essential to the maintenance of this forest-fallow system. Agricultural production can therefore be maintained with a relatively short fallow period of around 10 years. Traditional shifting cultivation in the study village can be seen to be well adapted to the respective soil-ecological conditions. Socioeconomic conditions are, however, drastically changing, making it difficult to sustain this system. Under such conditions, we should search for alternative technical tools that could maintain the SOM level, suppress the nitrifying activity of soil microbes, and avoid depletion of bases while mitigating soil acidification. This is imperative if the subsistence agriculture seen in this village is intended to continue in the near future.

Acknowledgement This chapter is derived in part from an article published in Tropics, 2006, copyright. The Japan Society of Tropical Ecology, available online: http://doi.org/10.3759/tropics.15.1.

References

Alegre JC, Cassel DK, Bandy DE (1988) Effect of land clearing method on chemical properties of an Ultisol in the Amazon. Soil Sci Soc Am J 52:1283–1288

Brookes PC, Landman A, Pruden G, Jenkinson DS (1985) Chloroform fumigation and the release of soil nitrogen: a rapid direct extraction method to measure microbial biomass nitrogen in soil. Soil Biol Biochem 17:837–842

Coody PN, Sommers LE, Nelson DW (1986) Kinetics of glucose uptake by soil microorganisms. Soil Biol Biochem 18:283–289

Ellingson LJ, Kauffman JB, Cummings DL, Sanford RL, Jaramillo VJ (2000) Soil N dynamics associated with deforestation, biomass burning, and pasture conversion in a Mexican tropical dry forest. For Ecol Manag 137:41–51

Ewel J, Berish C, Brown B, Price N, Raich J (1981) Slash and burn impacts on a Costa Rican wet forest site. Ecology 62:816–829

Funakawa S, Hirai H, Kyuma K (1992) Soil-forming processes under natural forest north of Kyoto in relation to soil solution composition. Soil Sci Plant Nutr 38:101–112

Knoepp JD, Swank WT (1993) Site preparation burning to improve southern Appalachian pine-hardwood stands: nitrogen responses in soil, water, and streams. Can J For Res 23:2263–2270

Leonard FD, Daniel CN, Peter FF (1998) Fire effects on ecosystems. Wiley, New York

Mamilov A-S, Dilly OM (2002) Soil microbial eco-physiology as affected by short-term variations in environmental conditions. Soil Biol Biochem 34:1283–1290

Marion GM, Moreno JM, Oechel WC (1991) Fire severity, ash deposition, and clipping effects on soil nutrients in chaparral. Soil Sci Soc Am J 55:235–240

Nakane K (1980) Comparative studies of cycling of soil organic carbon in three primeval moist forests. Jpn J Ecol 3:155–172

Neill C, Piccolo CM, Mellio MJ, Steudler AP, Cerri CC (1999) Nitrogen dynamics in Amazon forest and pasture soils measured by ^{15}N pool dilution. Soil Biol Biochem 31:567–572

Nye PH, Greenland DJ (1960) The soil under shifting cultivation, pp 1–12. CBS Tech. Commun. No. 51, Harpenden, UK

Raison RJ, Khanna PK, Woods PV (1985) Mechanisms of element transfer to the atmosphere during vegetation fires. Can J For Res 15:132–140

Richter DD, Markewitz D, Heine PR, Jin V, Raikes J, Tian K, Wells CG (2000) Legacies of agriculture and forest regrowth in the nitrogen of old-field soils. For Ecol Manag 138:233–248

Robertson GP (1982) Nitrification in forested ecosystems. Philos Trans R Soc Lond B 296:445–457

Ruark GA (1993) Modeling soil temperature effects on in situ decomposition rates for fine roots of loblolly pine. For Sci 39:118–129

Sakamoto K, Yoshida T, Satoh M (1991) Comparison of carbon and nitrogen mineralization between fumigation and heating treatments. Soil Sci Plant Nutr 38:133–140

Soil Survey Staff (2014) Keys to soil taxonomy, Twelfth edn. United States Department of Agriculture, National Resources Conservation Survice. Washington, DC

SPSS (1998) SYSTAT 8.0. Statistics. SPSS Inc, Chicago, pp 1086

Tanaka S, Funakawa S, Kaewkhongkha T, Hattori T, Yonebayashi K (1997) Soil ecological study on dynamics of K, Mg, and Ca, and soil acidity in shifting cultivation in northern Thailand. Soil Sci Plant Nutr 43:695–708

Tanaka S, Ando T, Funakawa S, Sukhrun C, Kaewkhongka T, Sakurai K (2001) Effect of burning on soil organic matter content and N mineralization under shifting cultivation system of Karen people in northern Thailand. Soil Sci Plant Nutr 47:547–558

Tulaphitak T, Pairintra C, Kyuma K (1985a) Changes in soil fertility and tilth under shifting cultivation II. Changes in soil nutrient status. Soil Sci Plant Nutr 31:239–249

Vance ED, Brookes PC, Jenkinson DS (1987) An extraction method for measuring soil microbial biomass C. Soil Biol Biochem 19:703–707

Wardle DA, Ghani A (1995) A critique of the microbial metabolic quotient (qCO_2) as a bioindicator of disturbance and ecosystem development. Soil Biol Biochem 27:1601–1610

Chapter 12
Slash-and-Burn Agriculture in Zambia

Kaori Ando and Hitoshi Shinjo

Abstract Soil organic matter (SOM) is important to sustain crop productivity for famers in semiarid tropics of Africa who use little fertilizer and rely on the nutrient released from SOM. It is, therefore, important to elucidate the effect of the conversion to cropland and shortened fallow period on SOM for evaluation and improvement of soil fertility. The results of our field study indicated the following: (1) More than 10-year cropping exhausted SOM and woody biomass; SOM and woody biomass were not restored during short fallow. (2) At spots burned with emergent and bush trees, SOM decreased only by burning and could be restored during short cropping, although N contents of SOM and woody biomass were not restored even during fallow. (3) Without burning, 2-year fallow after 3-year cropping is relatively useful management in terms of maintenance of POM, SOM, and grain yield, although woody biomass decreased compared to 2-year fallow after 1-year cropping. The C and N stock could not be fully recovered, especially burned spots. Thus, under rainfed agriculture without fertilizer in semiarid region, this type of slash and burn was relatively suitable to maintain grain yield and SOM contents although the yield and the SOM were low level compared to other regions where slash and burn is practiced. In other regions such as Central Africa, Southeast Asia, and South America, soil C and N stock decreased rapidly even during short cropping, while grain yield was higher than this semiarid region. However, in the semiarid region, the decrease in soil C and N stock during cropping was small with return of plant residue because long dry season constrains loss of SOM through leaching and decomposition and N_2 fixation by free-living N_2 fixation. Therefore, SOM will not decrease drastically under short cropping and short fallow rotation because recent decrease in emergent trees and increase in bush trees brought small loss of SOM during burning.

Keywords C and N budget • Fallow • Miombo • Soil burning effect

K. Ando
Graduate School of Agriculture, Kyoto University, Kitashirakawa-Oiwake, Sakyo, Kyoto 606-8502, Japan

H. Shinjo (⊠)
Graduate School of Global Environmental Studies, Kyoto University, Yoshida-Honmachi, Sakyo, Kyoto 606-8501, Japan
e-mail: shinhit@kais.kyoto-u.ac.jp

© Springer Japan KK 2017

S. Funakawa (ed.), *Soils, Ecosystem Processes, and Agricultural Development*,
DOI 10.1007/978-4-431-56484-3_12

12.1 Introduction

Soil organic matter (SOM) is important to sustain crop productivity for famers in semiarid tropics of Africa who use little fertilizer and rely on the nutrient released from SOM. The SOM which decreases during clearing and cropping has been restored during fallow in slash-and-burn agriculture in the Africa (Nandwa 2001). However, conversion of woodland to cropland followed by shortening fallow length has been increasing due to high land use pressure, which may cause the decrease in SOM (Nandwa 2001; Chirwa et al. 2004). Therefore, the effect of the conversion to cropland and shortened fallow period on SOM is needed to be elucidated for evaluation and improvement of soil fertility (Nandwa 2001).

When woodland is cleared for cropland, various slash-and-burn practices are used, based on the aboveground biomass present in an area, resulting in varied responses. In tropical humid regions such as Southeast Asia, South America, and Central Africa, with aboveground biomass of about 150–350 Mg ha^{-1} (Hertel et al. 2009), slashed trees cover an entire cleared area to be burned (Rasul and Thapa 2003). In contrast, the aboveground biomass in the semiarid tropics covered by miombo woodland composed of emergent tree and bush trees was only 69 Mg ha^{-1} (Chidumayo 1997). In Northern Province, Zambia, not only trees slashed in a cleared field but also branches collected from surrounding woodland are burned to compensate for the low levels of biomass. Then they are piled to cover the entire cleared field to be burned (Strømgaard 1984). In Eastern Province, trees are cut only within the field being opened, and the cut trees are collected and piled only in part of the field because of the low levels of biomass. Therefore, only the spots with the tree piles are burned, and the remaining cleared spots without piles are not burned.

Traditionally, most of the piles were composed of the emergent trees. However, shorter fallow periods caused decrease in emergent tree biomass recently, and the tree piles composed of bush trees in turn have increased. Therefore, two kinds of tree piles, such as emergent and bush tree piles, could be present simultaneously within the cleared field. This recent practice may increase the spatial variability of the extents of burning with these mosaic spots of unburned and burned with two levels of biomass. The extents of burning or fire intensity, which were determined by the maximum temperature and the duration of burning (Hatten and Zabowski 2009), affects the extent of SOM degradation and nutrient release (Tanaka et al., 2004). Therefore, it is essential to reveal the spatial variability of SOM and nutrient release immediately after burning.

The decline of SOM during cropping after clearing and burning has been attributed to various factors: the increase in mineralization of SOM via high microbial activity (Jobbagy and Jackson 2000), incorporation of plant materials and breakdown of soil aggregation by plowing (Six et al. 2002), and decrease in the amount of input returned to soil (Norton et al. 2012). Particularly, the amount of input may decrease with increase in cropping period, but the rate of the decrease may be different among the spots with different fire intensities. Additionally, the composition of input may also change in response to the decrease in tree coppicing ability with increase in cropping period and/or fire intensity (Luoga et al. 2004), which may affect decomposition rate. Therefore, the decrease process of SOM during cropping may be different among the spots unburned and burned with emergent and bush tree piles.

It should be concerned that those different extents of decrease in SOM during cropping with different fire intensity could be restored during the recent shortened fallow after cropping. Restoration of SOM during fallow is attributed to increase in litter returned to soil (Funakawa et al. 2006) and decrease in decomposition of SOM with leaving plant litter on soil surface (Ouattara et al. 2006). Several studies have claimed that the short fallow was not enough to recover the decrease in SOM by cropping or burning (Hauser et al. 2006). However, it depends on the extent of decrease in SOM and vegetation during cropping before fallow (Mobbs and Cannell 2006; Hauser et al. 2006). Hence, evaluation of restoration of SOM during fallow is needed to consider not only fallow period but also the cropping period and fire intensity before fallow. However, little has been reported on the evaluation of changes in SOM stock during different cropping period followed by short fallow under the slash-and-burn agriculture, with the changes in composition and amount of input.

The C and N budgets are a time-consuming method but could detect even small changes of C and N which could not be found by direct measurement of soil stock of C and N because of heterogeneity in soil itself. Therefore, the validation of changes in C and N contents of SOM by input to soil and output from soil is useful for the establishment of well-balanced cropping and short fallow rotation.

The objectives of this study were (1) to evaluate the effects of fire intensity on SOM and nutrient release immediately after burning, (2) to evaluate the changes of C and N contents in COM and POM under different cropping period followed by short fallow with reference to the changes of the composition and the amount of input, and (3) to estimate C and N budget during cropping and short fallow rotation from input and output flow.

12.2 Materials and Methods

12.2.1 Site Description

The study site is located in a village in Eastern Province of Zambia in southern Africa (Fig. 12.1: 14°08′S, 31°43′E; 890 m above sea level). The climate has a unimodal distribution of annual rainfall with a rainy season from November to April and a dry season in the remaining months. During the experimental period, the mean annual air temperature was 24 °C from 2008 to 2012, and annual rainfall in the rainy season was 762 mm in 2008/2009, 986 mm in 2009/2010, 1019 mm in 2010/2011, and 806 mm in 2011/2012. In the study site covered by miombo woodland composed of emergent trees and bush trees, aboveground biomass was

Fig. 12.1 Location of study
site

38.3 Mg ha^{-1}, much lower than that reported by Chidumayo (1997). This suggests that emergent trees have been lost under serious land use pressure in the region. The vegetation type of this woodland is classified as the eastern dry miombo (Chidumayo 1997), dominated by *Brachystegia manga*, *Julbernardia globiflora*, and *Diplorhynchus condylocarpon*. The soil was classified as Typic Plinthustalfs (Soil Survey Staff 2006). The soil at the depth of 0–15 cm in long fallow had a pH of 6.8 with water to soil ratio of 5; soil texture of sandy loam containing 66.5 % sand, 19.6 % silt, and 13.6 % clay; total C of 1.30 %; and total N of 0.082 %.

Villagers in Eastern Province, Zambia, clear and prepare land for cropping conventionally as follows. Trees are cut only within the field being opened, and the cut trees are collected and piled only in a part of the field because of the low biomass. Therefore, only the spots with the tree piles are burned; the remaining cleared spots without piles are not burned. The piles are left to dry during the dry season and burned at the end of dry season. The burned spots are not plowed during the first year following traditional local farming practices, while cleared spots without piles are plowed prior to burning to prevent them from burning.

12.2.2 Experimental Design

The woodland site (100 × 230 m) was selected in a flat and uniform place in terms of C and N contents of SOM and vegetation. Land clearing and preparation for the

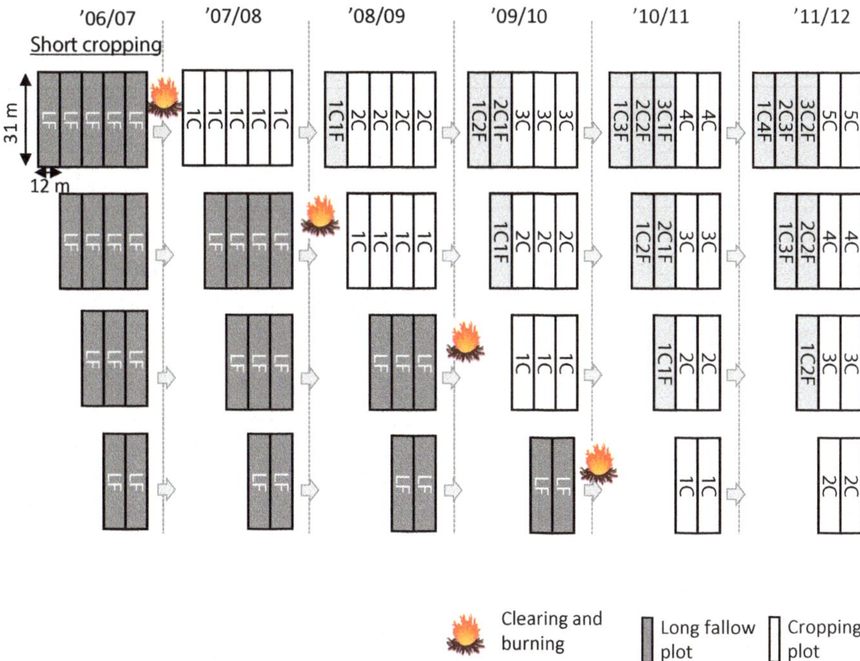

Fig. 12.2 Schematic diagram on history of experimental treatments (LF represents long fallow, C represents cropping, and F represents fallow. Values followed by C or F show the period of cropping or fallow, respectively. For example, 1C1F represents 1-year cropping followed by 1-year fallow. In the experimental site, the plots with these treatments were arranged with three replicates)

experiment was carried out according to the conventional practice as already described. A part of the experiment site was opened by slash-and-burn practice every year from October 2007 to October 2010 (Fig. 12.2). After land clearing, three seeds of maize (*Zea mays*) were planted in each hole with plant density of 1 × 1 m grid at the beginning of December. Weeds were removed about 3 and 6 weeks after planting. Then some of the plots were returned to fallow after cropping for 1–3 years as shown in Fig. 12.2. Each plot with three replication had spots unburned and burned with emergent and bush tree piles.

After trees were cut within each field being opened, bush trees and emergent trees were piled separately in different parts of the experimental field. Trees with a diameter at breast height (DBH) smaller than 27 cm are referred to as "small trees" or bush trees and trees with a DBH larger than 27 cm as "large trees" or emergent trees, hereinafter. The remaining spots without piles were plowed prior to burning. Then, the plots were divided into three treatment areas.

1. Unburned: consisting of unburned spots after clearing (85.6 % of total field area)
2. Bur S: spots burned with piles of small trees (7.5 % of total field area)
3. Bur L: spots burned with piles of large trees (6.9 % of total field area)

To establish plots of cropping and short fallow rotation for the experiment, the site opened by the slash-and-burn practice was divided into plots (12 × 31 m) with three replicates. After the 1-year cropping, some of the plots continued cropping, and the others returned to fallow every year according to Fig. 12.2. The treatments of 1C, 2C, 3C, 4C, and 5C represent cropping for 1, 2, 3, 4, and 5 years after clearing long fallow (LF) to cropland (C). The treatments of 1C1F, 1C2F, 1C3F, and 1C4F mean short fallow (F) for 1, 2, 3, and 4 years after 1-year cropping, respectively. The treatments of 2C1F, 2C2F, and 2C3F represent fallow for 1, 2, and 3 years, respectively, after 2-year cropping. The plots of 3C1F and 3C2F represent fallow for 1 and 2 years, respectively, after 3-year cropping. In 2011/2012, no plot was returned to fallow. A 20 × 30 m LF plot was also marked inside the experimental site.

12.2.3 Analytical Methods

Soil bulk density was determined by oven-drying undisturbed soils for 24 h. The dry weight of ash was measured after oven-drying for 24 h. Soil pH (H_2O) was measured with a soil to water ratio of 1:5. The total C and total N content of the soil and ash were analyzed by the dry combustion method with a NC analyzer (Vario Max CHN; Elementar, Germany). Available P was extracted by the Bray-1 method (Bray and Kurtz, 1945). Exchangeable Ca, Mg, K, and Na were extracted with 1 mol L^{-1} ammonium acetate at pH 7. Total P, Mg, Ca, K, and Na in ash were determined by wet digestion with nitric and sulfuric acid.

Carbon mineralization from the soil was determined by the aerobic incubation method (Kendawang et al. 2004). The samples were incubated at 30 °C for 56 days. The CO_2 collected by the alkali trap was sampled after 3, 7, 14, 28, 42, and 56 days of incubation and measured by an automatic titrimetric analyzer (COM-1600, Hiranuma Sangyo Co., Ltd., Japan). Nitrogen mineralization from the soil was determined by the aerobic incubation method (Funakawa et al. 2006). After incubation at 30 °C for 3, 7, 14, 28, 42, and 56 days, the samples were extracted with 2 mol L^{-1} KCl to determine the amount of NH_4–N and NO_3–N.

Maize stover and grain were collected at three hills for each treatment at the time of harvest. At weeding and harvesting times, aboveground parts of woody and herbaceous weeds were collected separately from three 1 × 1 m quadrates for each treatment. The samples were weighed after drying at 70 °C for 48 h.

12.3 Results and Discussion

12.3.1 Immediate Effects of Slash and Burn on Soil Organic Matter and Vegetation

Fire intensity increased with an increase in burned biomass per unit of burned area (Table 12.1 and Fig. 12.3). Soil was heated more deeply and for longer periods at

Table 12.1 Maximum soil temperature during burning measured by Thermo Crayon in 2008 (Ando et al. 2014b)

	Maximum soil temperature (°C)	
Depth (cm)	Bur L	Bur S
0	460–600	460
5	460–300	100–50
10	200–100	50
15	50	50

Values show the temperature of Thermo Crayon melting point ($n = 3$)

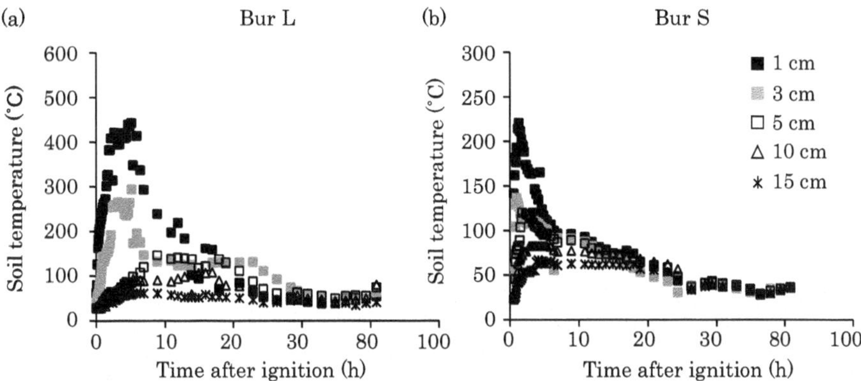

Fig. 12.3 Soil temperature during burning at depths of 1, 3, 5, 10, and 15 cm at Bur L and Bur S in 2010 (Ando et al. 2014b)

higher temperatures in Bur L than in Bur S. Kendawang et al. (2004) also found that the maximum soil temperature increased with an increase in burned biomass. The maximum temperature with the same burned biomass in the above study in Southeast Asia, however, was lower than in this study, 100–150 °C burning with 200 Mg ha^{-1} and 40–50 °C burning with 100 Mg ha^{-1} at the depth of 5 cm. High soil moisture content in Southeast Asia (around 20 %) may cause the lower maximum soil temperature observed there. The increase in fire intensity led to a further decrease in total C and N at a depth of 0–5 cm (Table 12.2). Some of the loss of C caused by soil heating will be compensated in second cropping year because ash will be incorporated with soil by plowing (Table 12.2).

Carbon mineralization during the first 3 days of incubation and NH$_4$–N (In–N$_0$) at a depth of 0–15 cm increased with an increase in fire intensity (Figs. 12.4 and 12.5). Those increases might be derived from the mortality of microbes; microbe mortality started when temperatures reached 80–120 °C (DeBano et al. 1998). Therefore, C mineralization and NH$_4$–N (In–N$_0$) were higher in Bur L than in Bur S because microbes were killed in deeper soil in Bur L than that in Bur S because of the deeper heating in Bur L (Table 12.1, Fig. 12.3).

An increase in C mineralization caused by soil heating during the early stage of incubation was also observed in Southeast Asia (Kendawang et al. 2004). Despite the documented increase in C mineralization with an increase in fire intensity, C

Table 12.2 Physicochemical properties of soil and ash in burned and unburned spots

Treatment	Depth	Bulk density	Total C	Total N	pH	Total P	Total Ca	Total Mg	Total K	Total Na
	(cm)	$(Mg\ m^{-3})$	$(Mg\ ha^{-1})$			$(kg\ ha^{-1})$	$(kg\ ha^{-1})$			
Bur S	ash	–	1.48 a	0.03 a	–	36 a	2120 a	239 a	681 a	17 a
Bur L	ash	–	4.14 b	0.01 b	–	132 b	6395 b	612 b	2186 b	39 b
						Avairable P	Exchangeable base			
							Ca	Mg	K	Na
						$(kg\ ha^{-1})$	$(kg\ ha^{-1})$			
Unburned	0–15	1.26 a	18.35 a	1.27 a	6.8 a	107 a	1381 a	254 a	305 a	29 a
Bur S	0–15	1.18 a	15.64 b	1.15 a	7.6 b	338 b	1354 a	263 a	372 ab	30 a
Bur L	0–15	1.16 a	13.73 b	1.08 b	8.1 b	920 c	2151 b	252 a	408 b	16 a
Bur S	0–5	1.15 a	6.73 a	0.49 a	8.0 a	230 a	687 a	109 a	167 a	10 a
Bur L	0–5	1.14 a	5.08 b	0.39 a	8.5 a	450 b	1240 b	116 a	184 a	4 a
Bur S	5–10	1.17 a	5.13 a	0.37 a	7.0 a	71 a	378 a	80 a	103 a	12 a
Bur L	5–10	1.18 a	4.66 a	0.37 a	7.5 a	394 b	530 a	69 a	114 a	6 a
Bur S	10–15	1.23 a	3.78 a	0.29 a	7.3 a	38 a	289 a	73 a	102 a	9 a
Bur L	10–15	1.23 a	3.99 a	0.32 a	7.2 a	76 a	381 a	67 a	110 a	6 a

Ando et al. (2014b)

Values are mean ($n = 3$)

Value followed by different letters vertically indicates that the means are significantly different ($P < 0.05$) at each depth

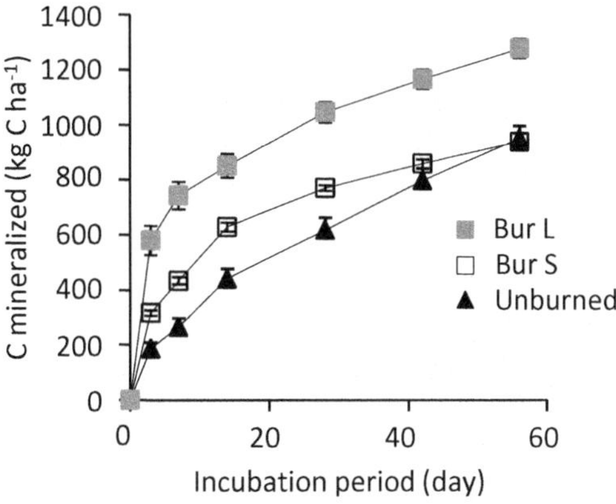

Fig. 12.4 Cumulative C mineralization at a depth of 0–15 cm during the incubation experiment. Bars indicate standard error ($n = 3$) (Ando et al. 2014b)

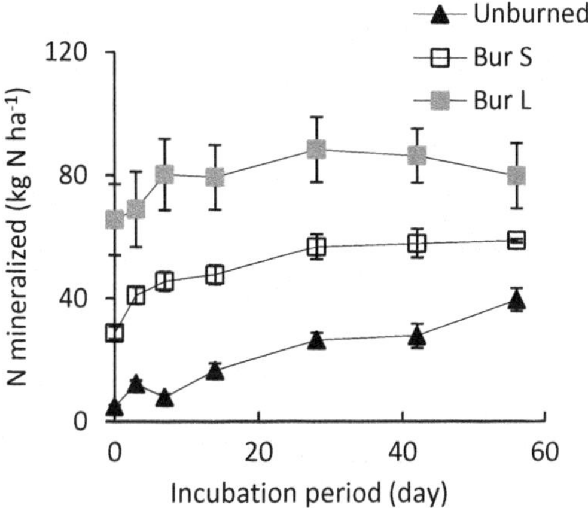

Fig. 12.5 Net N mineralization at a depth of 0–15 cm during the incubation experiment. Bars indicate standard error ($n = 3$) (Ando et al. 2014b)

mineralization at the end of incubation was higher only in Bur L compared with that in the Unburned (Fig. 12.4). Presumably, soil heating in Bur S might produce not only mineralizable organic material through volatilization of SOM or mortality of microbial biomass but also recalcitrant forms of C such as char or black C by imperfect combustion of SOM (González-Pérez et al. 2004).

An increased release of NH_4–N caused by soil heating based on an increase in fire intensity was also found in Southeast Asia (Tanaka et al. 2004). Although the amount of inorganic N was highest in Bur L during the incubation at a depth of 0–15 cm, net N mineralization was apparently the lowest in Bur L among the three treatments (Fig. 12.5). Although the rate of C mineralization and In–N_0 were the same between samples at a depth of 0–5 cm and 5–10 cm in Bur L, N was significantly immobilized at a depth of 0–5 cm in Bur L during the first 3 days of incubation. This might be related to the decrease in the N content of labile organic matter at a depth of 0–5 cm, which was suggested by the significant decrease in total N only at a depth of 0–5 cm.

The amount of available P increased with fire intensity (Table 12.2). Heat-induced microbial mortality may have been the primary factor leading to the increase in available P (Giardina et al. 2000). In Bur L, available P and exchangeable Ca and K increased because degradation of SOM was promoted by the high maximum temperature and long duration of burning (Giardina et al. 2000). The exchangeable bases and available P also seemed to increase depending on fire intensity in Southeast Asia (Tanaka et al. 2004). Total P and bases in ash will gradually be incorporated into the soil through the percolation of rain and by plowing during the second cropping year.

Herbaceous weeds had a low resistance to soil heating (Table 12.3); weed seeds were killed at about 90 °C resulting in low herbaceous biomass (Martin et al. 1975). In addition, biomass of woody weeds decreased with strong fire intensity (Table 12.3) because their large roots could survive only low-intensity fires.

In unburned spots, accounting for 85 % of the total field, maize yield was 0.7 Mg ha^{-1} because of the low available nutrient (Table 12.3). In Bur S and Bur L, covering 7.5 % and 6.9 % of the total field, maize yield was 2.3 Mg ha^{-1} and 3.2 Mg ha^{-1}, respectively (Table 12.3). Those increases in maize yield are due to the increase in available nutrient by the degradation of SOM and leachate from ash and the decrease in weeds. Those extents were more pronounced in Bur L than in Bur S. However, the content of C and N in SOM decreased more in Bur L than in Bur S (Table 12.2). Thus, the recent slash-and-burn practice, with the presence of both emergent and bush tree piles, increased the spatial variability in SOM, available nutrients, and weeds, which affected maize grain. Because more areas are being burned with bush trees owing to the decrease in emergent trees, grain yield may decrease, although the severe decrease in SOM may be alleviated.

Table 12.3 Aboveground biomass of weed and maize in burned and unburned spots

	Maize stover (Mg ha^{-1})	Maize grain (Mg ha^{-1})	Herbaceous weed (Mg ha^{-1})	Woody weed (Mg ha^{-1})
2008/2009				
Unburned	1.06 ± 0.10 a	0.65 ± 0.09 a	0.94 ± 0.01 a	0.93 ± 0.01 a
Bur S	4.75 ± 0.42 b	2.27 ± 0.26 b	0.19 ± 0.01 b	0.34 ± 0.02 a
Bur L	6.58 ± 0.47 b	3.17 ± 0.34 c	0.02 ± 0.00 b	0.00 ± 0.00 b
2009/2010				
Unburned	3.37 ± 0.23 a	1.18 ± 0.19 a	0.26 ± 0.04 a	0.39 ± 0.17 a
Bur S	7.71 ± 0.63 b	2.51 ± 0.43 b	0.16 ± 0.05 b	0.18 ± 0.04 a
Bur L	8.25 ± 0.72 b	3.85 ± 0.38 c	0.10 ± 0.05 b	0.00 ± 0.00 b

Ando et al. (2014b)
Values are mean ±standard error ($n = 3$). Value followed by different letters vertically indicates that the means are significantly different ($P < 0.05$) in each year

12.3.2   Effects of Cropping and Short Fallow Rotation on Stock and Budget of Soil Carbon and Nitrogen

The soil C and N stocks did not decrease at Unburned by cropping for 1–5 years apparently (Fig. 12.6). The soil C stock decreased by burning was restored partly during cropping, while soil N stock decreased by burning did not change. After returned to fallow, the soil C and N stock did not change at Unburned and Bur L significantly.

The annual C budget during short cropping was estimated by the input (plant materials and rainfall) and output (decomposition of organic matter and leaching) (Tables 12.4 and 12.5). The major flow of C loss was CO_2 efflux as a result of decomposition of SOM which was restricted by soil moisture (Fig. 12.7) and the amount and composition of input. In the semiarid tropics of Zambia, long dry season for 6 months constrained the rapid decomposition of SOM during cropping. The changes of quantity of input were partly due to the different amount of available nutrients by burning and cropping, while composition of input changed

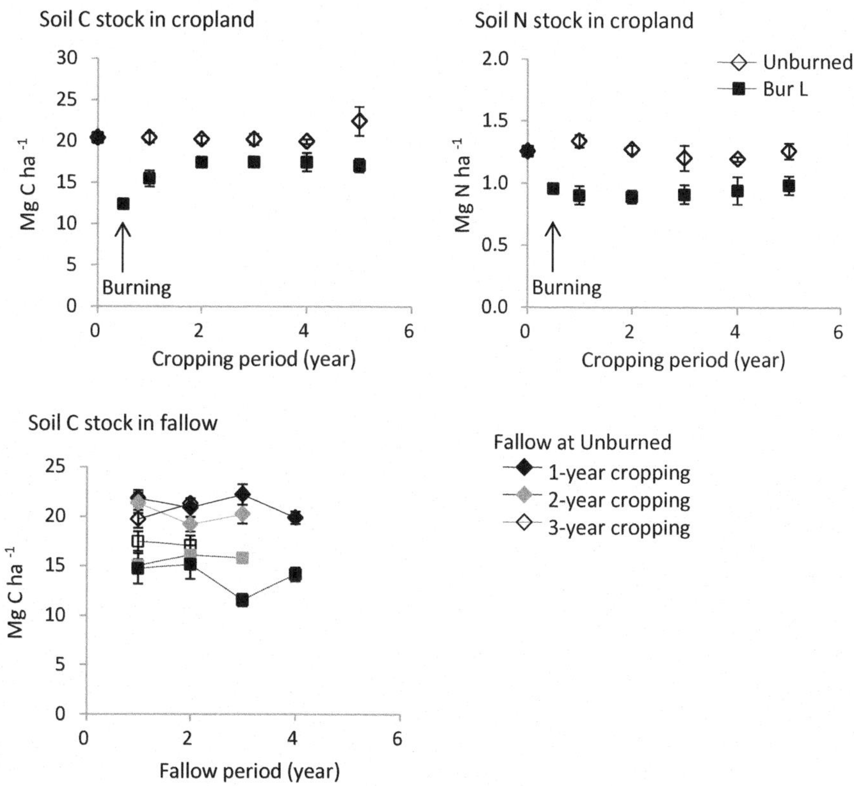

Fig. 12.6 Soil C and N stock during cropping and fallow at Unburned and Bur L (Bars indicate the standard error (5C, 1C4F, 2C3F, and 3C2F for $n = 3$, others for $n = 6$) (Ando et al. 2014a))

Table 12.4 Annual C budget during short cropping at Bur L

	2010/2011				2011/2012			
	1C	2C	3C	4C	2C	3C	4C	5C
Output (Mg C ha^{-1} year^{-1})								
Decomposition	3.3	3.1	3.3	3.1	2.9	2.9	2.9	2.8
Leachate	0.1	0.0	0.0	0.0	0.0	0.0	0.0	0.0
Sum	3.3	3.1	3.3	3.1	2.9	2.9	2.9	2.8
Input (Mg C ha^{-1} year^{-1})								
Plant litter or residue	5.5	3.7	2.7	2.4	3.4	2.3	2.3	2.3
Rainfall	0.1	0.1	0.1	0.1	0.1	0.1	0.1	0.1
Sum	5.6	3.8	2.8	2.5	3.5	2.4	2.3	2.4
Budget (input-output) (Mg C ha^{-1} year^{-1})	2.3	0.8	−0.4	−0.5	0.6	−0.6	−0.5	−0.4
SC change[a] (Mg C ha^{-1} year^{-1})	3.1	1.9	0.0	0.6	1.8	−0.3	1.2	−0.5

Ando (2014)

[a]Annual SC (soil C stock) change was calculated by subtracting soil C stock in the previous year from that in the respective year. For example, SC change in 1C was the difference between soil C stock in long fallow and that in 1C

Table 12.5 Annual C budget during short cropping at Unburned

	2010/2011				2011/2012			
	1C	2C	3C	4C	2C	3C	4C	5C
Output (Mg C ha^{-1} year^{-1})								
Decomposition	4.0	3.8	3.4	2.9	3.6	3.2	2.8	2.5
Leachate	0.0	0.0	0.0	0.0	0.0	0.0	0.0	0.0
Sum	4.0	3.8	3.4	2.9	3.6	3.2	2.8	2.5
Input (Mg C ha^{-1} year^{-1})								
Plant litter or residue	6.0	2.8	2.8	2.3	2.6	2.2	2.2	2.3
Rainfall	0.1	0.1	0.1	0.1	0.1	0.1	0.1	0.1
Sum	6.1	2.9	2.9	2.4	2.7	2.3	2.2	2.4
Budget (input-output) (Mg C ha^{-1} year^{-1})	2.1	−0.8	−0.5	−0.5	−0.9	−0.9	−0.5	−0.1
SC change (Mg C ha^{-1} year^{-1})	0.4	0.3	−0.6	0.2	0.4	−0.7	0.4	0.6

Ando (2014)

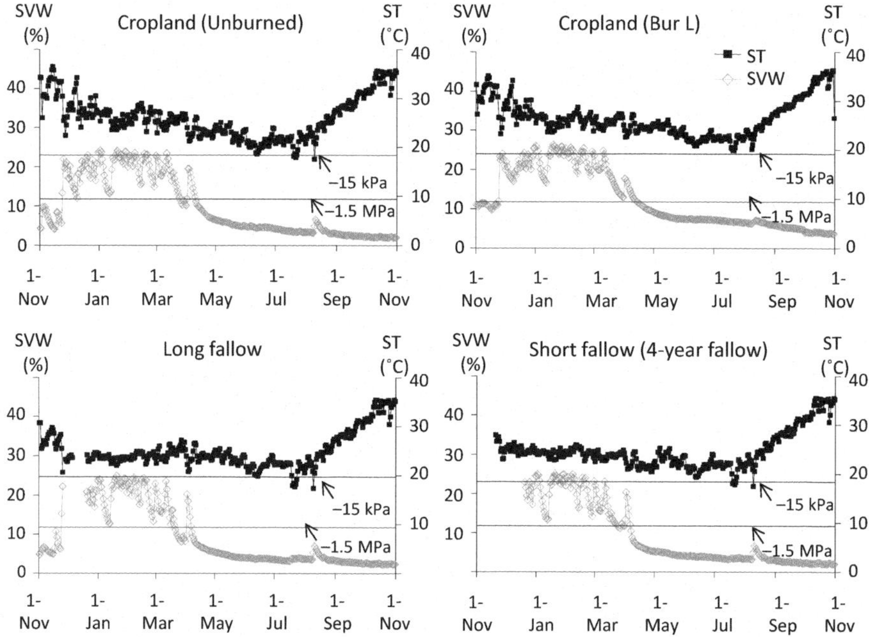

Fig. 12.7 Fluctuation of mean daily soil volumetric water (SVW) content and mean daily soil temperature (ST) in 2011/2012 (Ando et al. 2014a)

by cropping and burning. For instance, decrease in tree coppicing ability by increase in cropping period and fire intensity brought decrease in the proportion of woody weed or litter in input (Figs. 12.8 and 12.9).

Large amount of input in 1-year cropping at the unburned spots could compensate the negative soil C budget during cropping for 2–4 years partly. In spots burned with emergent trees and bush trees, soil C stock and labile fraction C decreased by burning (Fig. 12.6) could be restored by the large amount of input of ash, hard stem of maize to sustain large maize stover which was decomposed slowly.

After returned to fallow at Unburned, the decomposition of organic matter became lower than in cropland due to decrease in the input (Table 12.6) and plant materials left on surface. Therefore, soil C stock was not restored during short fallow for 1–4 years because the input of plant materials could not exceed the decomposition of organic matter, which was also reported during short fallow in tropical dry region (Aweto 1981).

Soil N stock decreased during burning via volatilization of N, during cropping via high leachate and high production in 1-year cropping at spots burned with emergent trees, and during cropping for more than 10 years via continuous grain harvesting (Fig. 12.6). The small N loss from soil during short cropping at unburned and burned spots (except for 1-year cropping) (Tables 12.7, 12.8 and 12.9) was attributed to low leachate and low grain yield with restricted mineralization of SOM

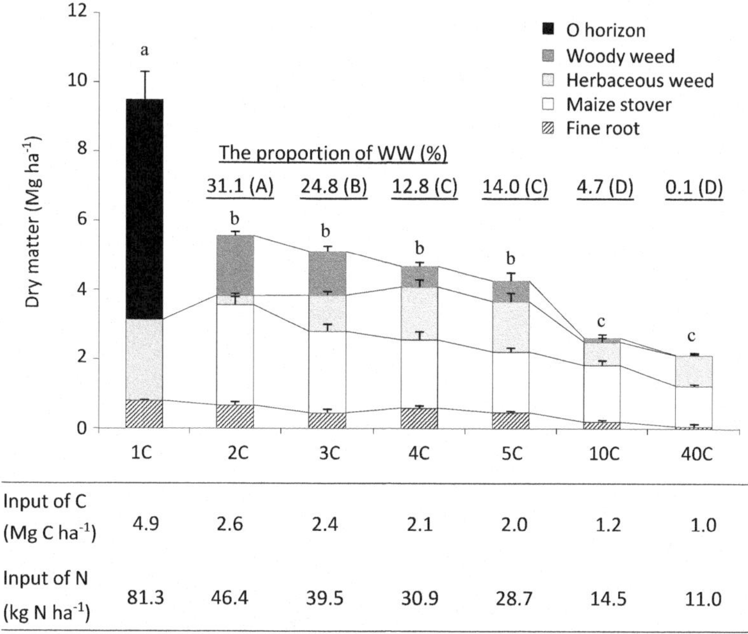

	1C	2C	3C	4C	5C	10C	40C
Input of C (Mg C ha⁻¹)	4.9	2.6	2.4	2.1	2.0	1.2	1.0
Input of N (kg N ha⁻¹)	81.3	46.4	39.5	30.9	28.7	14.5	11.0

Fig. 12.8 The composition and amount of input of plant materials during cropping in Unburned. Bars indicate the standard error (1C, 4C, and 10C were $n = 6$, 2C and 3C were $n = 9$, 5C was $n = 3$, 40C was $n = 2$). The ratio of WW shows the ratio of woody weed to total input, and different uppercase letters (A, B, C, D) indicate that the means of the proportion of woody weed are significantly different ($P < 0.05$). Different lowercase letters (a, b, c) indicate that the means of total input are significantly different ($P < 0.05$) (Ando et al. 2014a)

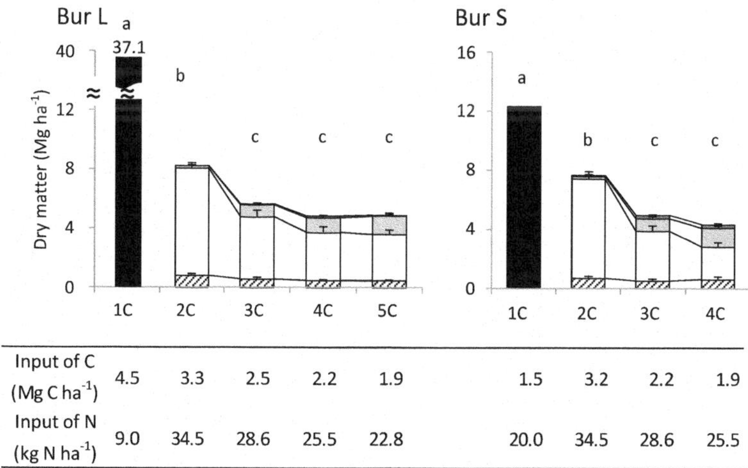

	1C	2C	3C	4C	5C		1C	2C	3C	4C
Input of C (Mg C ha⁻¹)	4.5	3.3	2.5	2.2	1.9		1.5	3.2	2.2	1.9
Input of N (kg N ha⁻¹)	9.0	34.5	28.6	25.5	22.8		20.0	34.5	28.6	25.5

Fig. 12.9 The composition and amount of input during cropping in Bur L and Bur S (C represents cropping and value followed by C shows the cropping period. Bars indicate the standard error ($n = 3$). Different letters indicate that the means of total input are significantly different among the cropping period ($P < 0.05$) (Ando 2014))

Table 12.6 Annual C budget during short fallow after different cropping periods at Unburned

	2010/2011						2011/2012					
	1C1F	LC2F	1C3F	2C1F	2C2F	3C1F	1C2F	1C3F	1C4F	2C2F	2C3F	3C2F
Output (Mg C ha^{-1} year^{-1})												
Decomposition	2.6	2.6	2.4	2.8	2.2	2.7	2.4	1.9	1.9	2.1	2.0	2.2
Leachate	0.0	0.0	0.0	0.0	0.0	0.0	0.0	0.0	0.0	0.0	0.0	0.0
Sum	2.6	2.6	2.4	2.8	2.2	2.7	2.5	1.9	1.9	2.1	2.0	2.2
Input (Mg C ha^{-1} year^{-1})												
Plant litter or residue	1.9	1.2	1.4	2.1	1.5	2.0	1.6	1.2	1.5	1.8	1.5	2.5
Rainfall	0.2	0.2	0.2	0.2	0.2	0.2	0.1	0.1	0.1	0.1	0.1	0.1
Sum	2.0	1.4	1.6	2.3	1.7	2.2	1.7	1.3	1.6	1.9	1.5	2.6
Budget (input-output) (Mg C ha^{-1} year^{-1})	−0.6	−1.2	−0.8	−0.6	−0.5	−0.6	−0.7	−0.7	−0.3	−0.2	−0.4	0.4
SC change[*] (Mg C ha^{-1} year^{-1})	−0.7	−0.3	−1.6	−0.9	0.0	−1.4	−0.4	−0.8	0.6	−0.0	0.7	1.0

Ando (2014)

*Annual SC (soil C stock) change was calculated by subtracting soil C stock in the previous year from that in the respective year. For example, SC change in 1C was the difference between soil C stock in long fallow and that in 1C

Table 12.7 Annual N budget during short cropping in Bur L

	2010/2011				2011/2012			
	1C	2C	3C	4C	2C	3C	4C	5C
Output (kg N ha^{-1} year^{-1})								
Harvested grain	62	24	19	11	31	23	29	19
Maize stover and weed	34	29	26	23	26	23	21	18
Leachate	45	8	8	5	21	12	6	5
Sum	142	60	52	39	79	59	56	42
Input (kg N ha^{-1} year^{-1})								
Ash or plant residue	9	34	29	26	34	26	23	21
Rainfall	12	12	12	12	10	10	10	10
Sum	21	46	41	38	44	37	34	31
Budget (input-output) (kg N ha^{-1} year^{-1})	−121	−14	−12	−2	−34	−22	−22	−11
SN change[*] (kg N ha^{-1} year^{-1})	−55	2	9	−8	−11	20	31	−23

Ando (2014)

*Annual SN (soil N stock) change was calculated by subtracting soil N stock in the previous year from that in the respective year. For example, SN change in 1C was the difference between soil N stock in long fallow and that in 1C

Table 12.8 Annual N budget during short cropping at Unburned

	2010/2011				2011/2012			
	1C	2C	3C	4C	2C	3C	4C	5C
Output (kg N ha^{-1} year^{-1})								
Harvested grain	12	10	12	13	14	13	11	14
Maize stover and weed	43	36	29	27	40	31	29	27
Leachate	7	3	9	11	12	6	6	7
Sum	62	48	50	50	65	50	46	48
Input (kg N ha^{-1} year^{-1})								
Plant residue	81	40	32	29	43	36	29	27
Rainfall	12	12	12	12	10	10	10	10
Sum	93	52	44	41	54	46	39	37
Budget (input-output) (kg N ha^{-1} year^{-1})	30	4	−6	−9	−12	−4	−7	−11
SN change (kg N ha^{-1} year^{-1})	20	−29	3	39	−39	−4	−31	6

Ando (2014)

to inorganic N during dry season. Then, the even small input N via rainfall could largely affect the N balance. Therefore, soil N stock did not decrease during short cropping at spots unburned with and burned with emergent and bush tree piles.

After return to fallow at Unburned, soil N stock did not change because of small loss via leaching and woody increment (Table 12.9). The budget tended to be negative with increase in fallow period because large amount of absorbed N by woody plants was stored in stem than in litter (Fujii et al. 2013). With longer fallow period, N budget could be positive with increase in litter fall (Hartemink et al. 2000). Those small decreases in soil N stock might be compensated by free-living N2 fixation about 3–30 kg N ha^{-1} reported in tropical savanna (Reed et al. 2011).

Table 12.9 Annual N budget during short fallow after different cropping periods at Unburned

	2010/2011						2011/2012					
	1C1F	1C2F	1C3F	2C1F	2C2F	3C1F	1C2F	1C3F	1C4F	2C2F	2C3F	3C2F
Output (kg N ha^{-1} year^{-1})												
Plant increment	17	32	51	19	24	29	42	44	80	26	38	26
Leacjate	4	3	4	1	1	2	1	1	3	1	1	2
Sum	21	35	55	20	25	32	44	45	84	27	39	28
Input (kg N ha^{-1} year^{-1})												
Litter	15	26	32	17	25	18	26	32	43	25	28	36
Rainfall	12	12	12	12	12	12	10	10	10	10	10	10
Sum	27	38	44	29	37	30	36	42	54	35	38	46
Budget (input-output) (kg N ha^{-1} year^{-1})	6	3	−11	9	12	−2	−7	−3	−30	8	−1	18
SN change (kg N ha^{-1} year^{-1})	−40	−3	−46	−13	−22	−31	−47	5	−36	−54	4	51

Ando (2014)

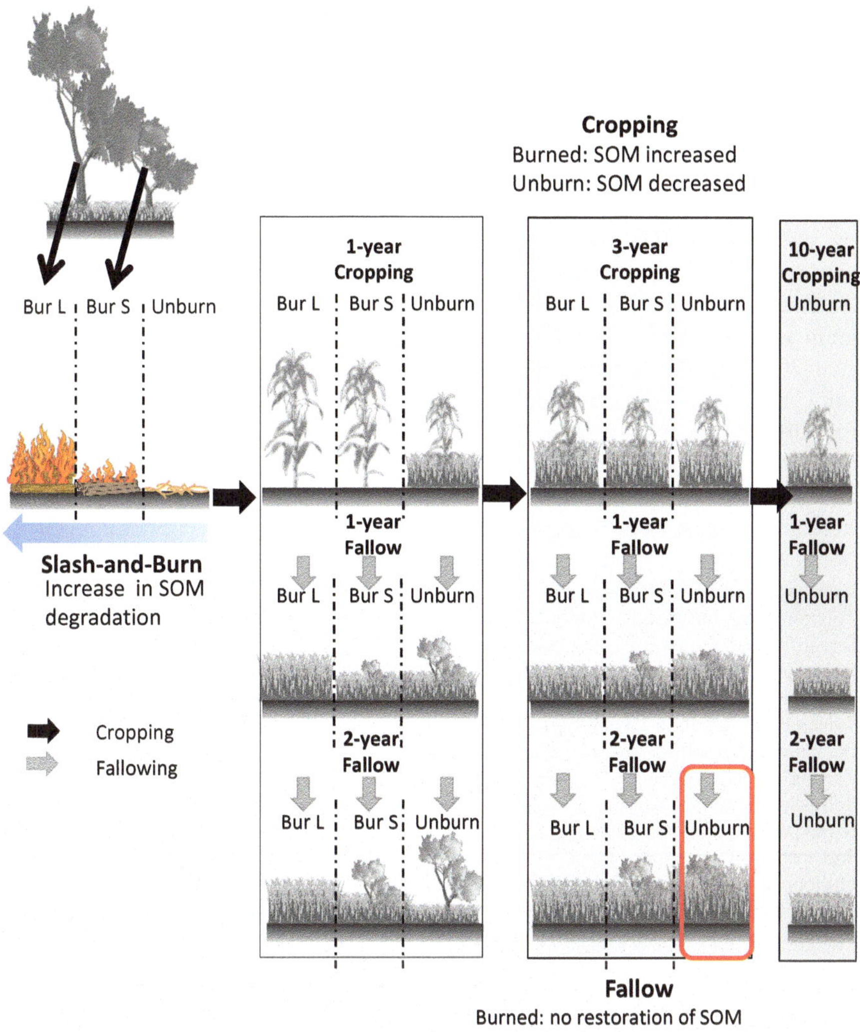

Fig. 12.10 Changes of soil organic matter through vegetation during cropping and fallow duration

12.4 Conclusion: Evaluation of Cropping and Short Fallow Rotation Under Slash-and-Burn Agriculture in a Semiarid Woodland

The results of cropping and short fallow rotation are schematically illustrated in Fig. 12.10. More than 10-year cropping exhausted SOM and woody biomass, SOM and woody biomass were not restored during short fallow. At spots burned with emergent and bush trees, SOM decreased only by burning and could be restored during short cropping, although N contents of SOM and woody biomass were not restored even during fallow. Without burning, 2-year fallow after 3-year cropping is relatively useful management in terms of maintenance of POM, SOM, and grain yield, although woody biomass decreased compared to 2-year fallow after 1-year cropping. The C and N stock could not be fully recovered, especially burned spots.

Under rainfed agriculture without fertilizer in semiarid region, this type of slash and burn was relatively suitable to maintain grain yield and SOM contents although the yield and the SOM were low level compared to other regions where slash and burn is practiced. In other regions such as Central Africa, Southeast Asia, and South America, soil C and N stock decreased rapidly even during short cropping, while grain yield was higher than this semiarid region (Kendawang et al. 2004; Tanaka et al. 2004; Funakawa et al. 2006). However, in the semiarid region, the decrease in soil C and N stock during cropping was small with return of plant residue because long dry season constrains loss of SOM through leaching and decomposition and N_2 fixation by free-living N_2 fixation. Therefore, SOM will not decrease drastically under short cropping and short fallow rotation because recent decrease in emergent trees and increase in bush trees brought small loss of SOM during burning.

Acknowledgment This chapter is partly derived from Ando (2014) and Ando et al. (2014a, b). This work was financially supported by the Kyoto University Foundation, a Grant-in-Aid for Scientific Research (B) from the Japan Society for the Promotion of Science (No. 23310027), and the projects on Vulnerability and Resilience of Social-Ecological Systems and Desertification and Livelihoods in Semi-Arid Afro-Eurasia, at the Research Institute for Humanity and Nature. Tables 12.1, 12.2 and 12.3 and Figs. 12.3, 12.4 and 12.5 are reprinted from "Short term effects of fire intensity on soil organic matter and nutrient release after slash-and-burn in Eastern Province, Zambia", Ando, K., Shinjo, H., Noro, Y., Takenaka, S., Miura, R., Sokotela, S.B., Funakawa, S., Soil Science and Plant Nutrition, copyright © Japanese Society of Soil Science and Plant Nutrition, reprinted by permission of Taylor & Francis Ltd, www.tandfonline.com on behalf of Japanese Society of Soil Science and Plant Nutrition. Figures 12.6, 12.7 and 12.8 are reprinted from "Effects of cropping and short-natural fallow rotation on soil organic carbon in the Eastern Province of Zambia", Ando K., Shinjo H., Kuramitsu H., Miura R., Sokotela S., Funakawa S., Agriculture, Ecosystem and Environment, 196, 36–41, copyright (2014), with permission from Elsevier.

References

Ando K (2014) Dynamics of soil organic matter under slash-and-burn agriculture in a semiarid woodland of Zambia. PhD thseis, Graduate School of Agriculture, Kyoto University

Ando K, Shinjo H, Kuramitsu H, Miura R, Sokotela S, Funakawa S (2014a) Effects of cropping and short-natural fallow rotation on soil organic carbon in the eastern province of Zambia. Agric Ecosyst Environ 196:34–41

Ando K, Shinjo H, Noro Y, Takenaka S, Miura R, Sokotela S, Funakawa S (2014b) Short-term effects of fire intensity on soil organic matter and nutrient release after slash-and-burn in Eastern Province, Zambia. Soil Sci Plant Nutr 60:173–182

Aweto AO (1981) Secondary succession and soil fertility restoration in southwestern Nigeria: II. Soil fertility restoration. J Ecol 69:609–614

Bray RH, Kurtz LT (1945) Determination of total, organic and available forms of phosphorus in soils. Soil Sci 59:39–45

Chidumayo EN (1997) Miombo ecology and management an introduction. Intermediate Technology Publications, London

Chirwa TS, Mafongoya PL, Mbewe DNM, Chishala BH (2004) Changes in soil properties and their effects on maize productivity following *Sesbania sesban* and *Cajanus cajan* improved fallow systems in eastern Zambia. Biol Fertil Soils 40:20–27

DeBano LF, Neary DG, Folliott PF (1998) Fire's effects on ecosystems, Wiley, New York, pp 71–80

Fujii K, Funakawa S, Hayakawa C, Sukartiningsih KT (2013) Fluxes of dissolved organic carbon and nitrogen in cropland and adjacent forest in a clay-rich Ultisol of Thailand and a sandy Ultisol of Indonesia. Soil Tillage Res 126:267–275

Funakawa S, Hayashi Y, Tazaki I, Sawada K, Kosaki T (2006) The main functions of the fallow phase in shifting cultivation by the Karen people in northern Thailand: a quantitative analysis of soil organic matter dynamics. Tropics 15:1–27

Giardina CP, Sanford RL, Døckersmith IC, Jaramillo VJ (2000) The effects of slash burning on ecosystem nutrients during the land preparation phase of shifting cultivation. Plant Soil 220:247–260

González-Pérez JA, González-Vila FJ, Almendros G, Knicker H (2004) The effect of fire on soil organic matter—a review. Environ Int 30:855–870

Hartemink AE, Buresh RJ, Van Bodegom PM, Braun AR, Jama B, Janssen BH (2000) Inorganic nitrogen dynamics in fallows and maize on an Oxisol and Alfisol in the highlands of Kenya. Geoderma 98:11–33

Hatten JA, Zabowski D (2009) Changes in soil organic matter pools and carbon mineralization as influenced by fire intensity. Soil Sci Soc Am J 73:262–273

Hauser S, Nolte C, Carsky RJ (2006) What role can planted fallows play in the humid and sub-humid zone of west and Central Africa? Nutr Cycl Agroecosyst 76:297–318

Hertel D, Moser G, Culmsee H, Erasmi S, Horna V, Schuldt B, Leuschnner C (2009) Below- and above-ground biomass and net primary production in a paleotropical natural forest (Sulawesi, Indonesia) as compared to neotropical forests. For Ecol Manag 258:1904–1912

Jobbagy EG, Jackson RB (2000) The vertical distribution of soil organic carbon and its relation to climate and vegetation. Ecol Appl 10:423–436

Kendawang JJ, Tanaka S, Ishihara J, Shibata K, Sabang J, Ninomiya I, Ishizuka S, Sakurai K (2004) Effects of shifting cultivation on soil ecosystems in Sarawak, Malaysia. I. Slash and burning at Balai Ringin and Sabal experimental sites and effect on soil organic matter. Soil Sci Plant Nutr 50:677–687

Luoga EJ, Witkowski ETF, Balkwill K (2004) Regeneration by coppicing (resprouting) of miombo (African savanna) trees in relation to land use. For Ecol Manag 189:23–35

Martin RE, Miller RL, Cushwa CT (1975) Germination response of legume seeds subjected to moist and dry heat. Ecology 56:1441–1445

Mobbs DC, Cannell MG (2006) Optimal tree fallow rotations: some principles revealed by modeling. Agrofor Syst 29:113–132

Nandwa SM (2001) Soil organic carbon management for sustainable productivity of cropping and agro-forestry systems in eastern and Southern Africa. Nutr Cycl Agroecosyst 61:143–158

Norton JB, Mukhwana EJ, Norton U (2012) Loss and recovery of soil organic carbon and nitrogen in a semiarid agroecosystems. Soil Sci Soc Am J 76:505–514

Ouattara B, Ouattara K, Serpantié G, Mando A, Sédogo MP, Bationo A (2006) Intensity cultivation induced effects on soil organic carbon dynamic in the western cotton area of Burkina Faso. Nutr Cycl Agroecosyst 76:331–339

Rasul G, Thapa GB (2003) Shifting cultivation in the mountains of south and Southeast Asia: regional patterns and factors influencing the change. Land Degrad Dev 14:495–508

Reed SC, Cleveland CC, Townsend AR (2011) Functional ecology of free-living nitrogen fixation: a contemporary perspective. Annu Rev Ecol Evol Syst 42:489–512

Six J, Conant RT, Paul EA, Paustian K (2002) Stabilization mechanisms of soil organic matter: implications for C-saturation of soils. Plant Soil 241:155–176

Soil Survey Staff (2006) Keys to soil taxonomy, 10th edn. USDA-ARS, Washington
Strømgaard P (1984) The immediate effect of burning and ash fertilization. Plant Soil 80:307–320
Tanaka S, Kendawang JJ, Ishihara J, Shibata K, Kou A, Jee A, Ninomiya I, Sakurai K (2004) Effects of shifting cultivation on soil ecosystems in Sarawak, Malaysia. II. Changes in soil chemical properties and runoff water at Balai Ringin and Sabal experimental sites. Soil Sci Plant Nutr 50:689–699

Chapter 13
Comparison of Nutrient Utilization Strategies of Traditional Shifting Agriculture Under Different Climatic and Soil Conditions in Zambia, Thailand, Indonesia, and Cameroon: Examples of Temporal Redistribution of Ecosystem Resources

Shinya Funakawa

Abstract Based on the results of field studies in different regions of the tropics, land management strategies using shifting cultivation under different bioclimatic and soil conditions are comparatively discussed. Four regions are included in this study: eastern Zambia, northern Thailand, East Kalimantan of Indonesia, and eastern Cameroon. An increase in N mineralization and resulting increase in N flux, after the original vegetation was removed for cropping, were the highest for the eastern Cameroon forest, followed by northern Thailand, eastern Cameroon savanna, East Kalimantan of Indonesia, and eastern Zambia. This trend, however, was considered to coincide with mineral nutrient loss due to leaching. This loss would be more detrimental for the strong-weathered soils, such as Oxisols in Cameroon. The function of the fallow phase was the most clearly observed in shifting cultivation in northern Thailand; the soil fertility was reinstated in the later stage of the young fallow, around 8 years. A similar trend was also found for the soils in eastern Cameroon, in which the soil fertility was the highest in the young fallow forest. Such improvement in soil fertility during the fallow stage was not observed for the plots in East Kalimantan, Indonesia, and eastern Zambia, presumably due to the specific climatic conditions. Generally, the farmers' practices and land uses during shifting cultivation in Southeast Asia were well adapted to the respective soil and climatic conditions. However, the present pattern of land use in eastern Cameroon has not been controlled solely by natural conditions; human factors, such as activities of neighboring agropastoralists, have a greater impact on these dynamics. In the case of eastern Zambia, limited and fluctuating precipitation may be the most important factor in agricultural production, and because primary

S. Funakawa (✉)
Graduate School of Global Environmental Studies, Kyoto University, Yoshida-Honmachi, Sakyo-ku, Kyoto 606-8501, Japan
e-mail: funakawa@kais.kyoto-u.ac.jp

© Springer Japan KK 2017

S. Funakawa (ed.), *Soils, Ecosystem Processes, and Agricultural Development*, DOI 10.1007/978-4-431-56484-3_13

production and soil organic matter decomposition rate are regulated by this, soil fertility status would not change considerably under the present land use practices. Therefore, the classical interpretation of slash-and-burn agriculture is not applicable to the agricultural practices in the latter two African cases.

Keywords Cameroon • Indonesia • Shifting agriculture • Thailand • Zambia

13.1 Introduction

Shifting cultivation, often referred to as slash-and-burn or swidden agriculture, is a rather extensive farming system that incorporates both cropping and fallow phases in its land rotation (Kunstadter and Chapman 1978; Nye and Greenland 1960). Due to the low fertility of Ultisols or Oxisols in the tropics (Driessen et al. 2001; Kyuma 1982), it was believed that, in traditional shifting cultivation, sustainable crop production requires added nutrients via slash-and-burn after certain periods of fallow. As reported in previous studies, burning is considered to add phosphorus and basic compounds to the soil (Alegre et al. 1988; Ewel et al. 1981; Nye and Greenland 1960; Raison et al. 1985; Tanaka et al. 1997). Additionally, N supply will be increased by enhanced mineralization after burning (Knoepp and Swank 1993; Marion et al. 1991; Nye and Greenland 1960; Sakamoto et al. 1991; Tanaka et al. 2001). On the other hand, the fallow phase functions in accumulating enough nutritional elements into the forest vegetation (Richter et al. 2000) and SOM via litter input (Ruark 1993) for the next cropping phase to be successful. Overall, shifting agriculture is regarded as low-input agricultural land management with in situ nutrient replenishing process by fallow vegetation, in other words, a temporal redistribution of ecosystem resources, fitting the natural climatic and soil conditions.

The climatic conditions, as well as soil fertility, may vary in different regions of the tropics; therefore, land management practices in respective regions would also vary. In this chapter, based on the results of field studies in different regions of the tropics as presented in the preceding chapters, land management strategies in shifting cultivation under different bioclimatic and soil conditions are comparatively discussed in terms of climatic conditions, soil properties, impact of slash-and-burn practices, and fallow practices.

13.2 Description of Study Sites

Four regions were included in this study: eastern Zambia, northern Thailand, East Kalimantan of Indonesia, and eastern Cameroon. The natural conditions and basic soil properties at the study sites are presented in Tables 13.1 and 13.2, respectively.

Table 13.1 Site description of survey sites in Zambia, Thailand, Cameroon, and Indonesia

	Zambia	Thailand		Indonesia	Cameroon	
	Eastern Province	DP, Mae Hong Son Province	RP, Chiang Rai Province	East Kalimantan	ADf, East Province	ADs, East Province
Coordinates	S14°08′, E31°43′	N18°24′, E98°05′	N19°50′, E100°20′	S00°53′,E115°57′	N04°32′, E13°15′	N04°35′, E13°16′
Mean annual air temperature (°C)	26	20	25	27	23	23
Mean annual precipitation (mm)	860	1200	2100	2200	1500	1500
Elevation (m)	890	1200	700	50	680	700
Soil type[a]	Typic Plinthustalfs	Ustic Haplohumults	Typic Haplustults	Typic Paleudults	Typic Kandiudox	Typic Kandiudox
Parent materials	Granite	Sedimentary rocks and granite	Sedimentary rocks	Sedimentary rocks	Metamorphic rocks	Metamorphic rocks
Vegetation	*Brachystegia manga, Julbernardia globiflora*	*Lithocarpus* sp.	*Lithocarpus* sp., *Eugenia* sp.	*Shorea laevis, Dipterocarpus cornutus*	*Albizia zygia, Myrianthus arboreus*	*Chromolaena odorata, Imperata cylindrica*

[a]Soils were classified according to soil taxonomy (Soil Survey Staff 2014)

Table 13.2 Physicochemical properties of representative soil profiles at the study sites in Zambia, Thailand, Indonesia, and Cameroon

	Horizon	Depth (cm)	pH (H$_2$O)	(KCl)	Total C (g kg^{-1})	Total N	Exchangeable Bases (cmol$_c$ kg^{-1})	Al	CEC	Particle size distribution Sand (%)	Silt	Clay
Zambia	A	0–5	6.8	5.9	14.0	0.9	6.8	n.d.	7.4	75	14	11
	BA	5–17	6.8	5.5	5.8	0.4	3.3	n.d.	4.6	62	16	22
	Btl	17–30	6.6	5.1	4.4	0.4	3.1	n.d.	5.9	60	19	20
	Bv2	30–60	6.5	5.1	4.5	0.4	3.1	n.d.	7.5	57	30	13
Thailand, DP site	A	0–10	6.1	4.6	43.0	2.7	6.4	0.6	15.5	32	27	41
	BA	10–20	5.6	4.2	17.0	1.1	1.7	2.2	11.2	26	29	45
	Btl	20–30	5.4	4.1	12.0	0.8	0.9	2.8	10.5	26	26	48
	Bt2	30–40	5.5	4.1	9.3	0.7	0.8	2.8	10.7	23	27	50
	Bt3	40–50	5.5	4.2	9.5	0.7	0.9	2.8	12.5	23	25	52
Thailand, RP site	A	0–7	5.0	4.1	62.6	3.8	5.8	1.9	27.6	5	25	70
	BA	7–20	4.9	3.9	19.8	1.5	2.0	3.5	19.9	4	23	73
	Bt	20–45 +	4.6	4.0	8.9	1.0	1.4	3.2	20.1	6	19	75
Indonesia	A	0–5	4.5	3.8	30.8	2.6	3.7	4.1	16.1	34	30	36
	BA	5–10	4.3	3.8	13.7	1.5	1.0	9.9	12.0	29	38	33
	Bl	10–20	4.6	3.9	8.4	1.1	1.0	6.6	12.1	30	32	38
	Btl	20–40	4.6	3.9	7.2	1.0	0.6	7.7	14.1	25	30	44
	Bt2	40–60	4.7	3.9	5.1	0.9	0.5	8.9	17.9	18	41	42

Cameroon, forest plot (ADf)	A	0–10	4.2	3.9	18.8	1.6	2.3	3.0	8.0	40	5	55
	BA	10–20	4.2	4.0	13.6	1.1	1.6	3.2	6.1	39	6	54
	Btl	20–35	5.0	4.3	9.5	0.9	1.8	1.6	7.0	29	4	68
	Bt2	35–50	4.3	4.3	6.9	0.6	1.1	1.8	6.2	25	4	70
	Bt3	50–80 +	4.9	4.7	5.7	0.6	1.7	0.7	6.4	23	3	73
Cameroon, savanna plot (ADs)	A	0–8	5.5	4.4	29.1	1.8	4.5	1.0	10.3	39	8	53
	BA	8–19	4.7	4.1	18.0	1.2	2.5	2.4	7.6	37	6	57
	Btl	19–38	4.6	4.1	9.4	0.8	1.8	2.1	5.6	27	4	69
	Bt2	38–65	4.7	4.4	6.9	0.6	1.8	1.3	6.8	23	3	74
	Bt3	65–90 +	4.9	4.7	5.2	0.5	2.0	0.6	6.5	23	3	74

13.2.1   Study Site in Zambia

The study site in Zambia was located in the Eastern Province; the climate has a unimodal distribution of annual rainfall, with a rainy season from November to April and a dry season during the remaining months. The mean annual precipitation is 860 mm, which varies throughout the year, and the annual temperature is 26 °C. The woodland vegetation type is classified as eastern dry miombo (Chidumayo 1997), dominated by *Brachystegia manga* Dwild., *Julbernardia globiflora* Benth., and *Diplorhynchus condylocarpon* (Mull. Arg.) Pichon. The soil was classified as Typic Plinthustalfs (Soil Survey Staff 2014) (Table 13.2). The soil at the long fallow plot had a pH of 6.5–6.8, a texture of sandy loam (A horizon) to sandy clay loam (Bt horizon), and total content of C (14 g kg^{-1}) and N (0.9 g kg^{-1}) at the surface layers. The woodland biomass of the study site was 38.3 Mg ha^{-1}, which is much lower than that reported by Chidumayo (1997); this suggests that emergent trees have already been lost due to higher demands in land use within the region.

In the Eastern Province of Zambia, maize is planted after a slash-and-burn treatment. Trees are usually cut only within the field being cleared, and the cut trees are collected and piled only in part of the field because of low levels of available biomass. Therefore, only the spots with the tree piles can be burned; the remaining cleared areas without piles are not burned. Traditionally, most of the piles were composed of emergent trees because more emergent trees were present in the past. Recently, because most emergent tree biomass has disappeared as a result of shorter fallow periods, the trees piles are increasingly composed of bush trees. Therefore, two kinds of tree piles, emergent and bush, could be present simultaneously within the cleared field. These recent changes may increase the spatial variability of soil fertility and crop production due to the fluctuations in burned and unburned areas and type of biomass being burned (Fig. 13.1).

13.2.2   Study Sites in Thailand

Two study sites were examined in northern Thailand. One was Du La Poe (DP) village in Mae Hong Son Province, introduced in detail in Chap. 11, and the other was Rakphaendin (RP) village in Chiang Mai Province, which was selected for examining the direct effect of burning on nutrient dynamics. The basic information on the two villages is given in Tables 13.1 and 13.2. Both sites have a Köppen's Aw climate type, with a distinct dry season, although the former village is closer to the border with a Cw climate type due to its high elevation. The annual precipitation and mean annual temperature are 1200 mm and 20 °C at DP and 2100 mm and 25 °C at RP, respectively. The dominant vegetation was *Lithocarpus* sp. in both the villages. The parent materials of the soils were mainly sedimentary rocks in both villages and granite in particular at DP. Most of the soils studied were

Eastern Zambia; Typic Plinthustalfs, MAT: 26ºC,
AP: 860 mm, Biomass: 39 Mg C ha^{-1}

Northern Thailand;
Ustic Haplohumults,
MAT: 20ºC,
AP: 1,200 mm,
Biomass: 169 Mg C ha^{-1}

East Kalimantan,
Indonesia;
Typic Paleudults,
MAT: 26ºC,
AP: 2,500 mm,
Biomass: 293 Mg C ha^{-1}

Fig. 13.1 Landscape of study plots

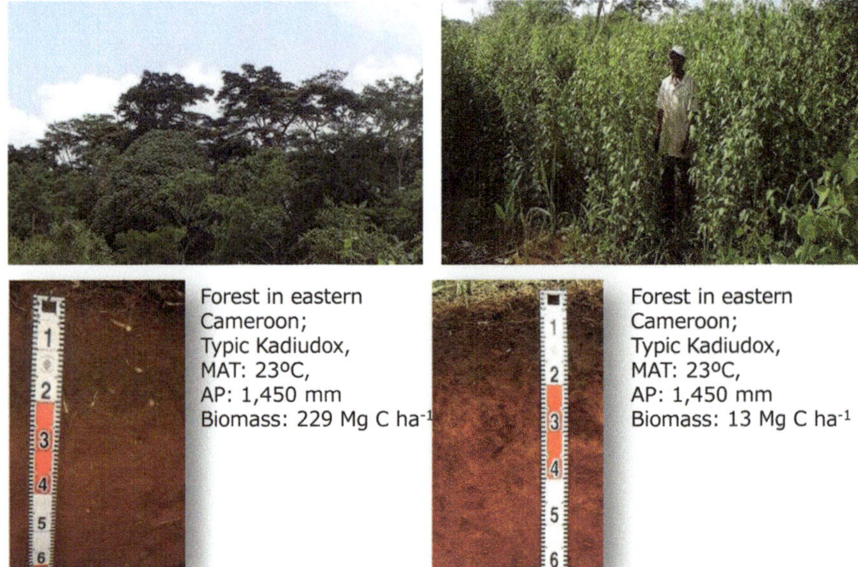

Fig. 13.1 (continued)

classified into Ustic Haplohumults, Lithic/Typic Haplustults, or Typic Dystrustepts in the USDA classification system (Soil Survey Staff 2014).

The shifting cultivation system for the DP village was similar to a traditional land rotation system in this region, in which cropping for upland rice is limited to only 1 year, followed by at least 7 years of fallow. According to our field observations, this fallow period seems to be the longest one in this area. In fact, in many of the nearby villages, the fallow stage has recently been shortened to around 4 years. During the slash and burn, farmers only burn branches, and the remaining stems are typically used for firewood. Therefore, the amount of materials actually burned was less than 10 Mg ha^{-1}, implying that the ash input and/or soil burning effect was not expected much in this system. Only a limited area is used for the cultivation of cash crops, such as cabbage, in this village.

13.2.3 Study Site in Indonesia

In the East Kalimantan Province, Indonesia, the study was carried out in three villages and a nearby experimental station, which was located along a small tributary of R. Mahakam with an elevation of less than 100 m. The area has a Köppen's Af climate type, with no distinct dry season and an udic soil moisture regime. Mean annual temperature is above 27 °C and annual precipitation exceeds 2000 mm. Original vegetation in this region is a typical lowland dipterocarp forest, dominated by *Shorea laevis, Dipterocarpus cornutus*, and others. Parent materials are mainly tertiary sandstone and/or mudstone, and soils distributed were mostly Typic Paleudults. According to the World Reference Base for Soil Resources (WRB) (IUSS Working Group WRB 2014), most of the soils studied are classified as Alisols, typically characterized by high CEC values, with CEC/clay exceeding 24 $cmol_c\ kg^{-1}$, with strong acidity (Funakawa et al. 2008).

In these regions, after clearing and burning the forest cover, farmers usually plant upland rice for 1 year only, and then leave the land for more than 20 years as fallow. On relatively fertile soils near the river, however, shorter period of fallow (around 8 years) is implemented, but such soils were excluded in the present study. According to the personal interviews, however, there was no clear evidence that farmers preferred riverside soils. A detailed data analysis on the physicochemical properties, as well as the mineralization process, of organic C and N in the soils from East Kalimantan was given in our previous reports (Funakawa et al. 2007, 2009).

13.2.4   Study Sites in Cameroon

The study was carried out in the Andom village, East Province, Cameroon, where a semi-deciduous forest comprises the main vegetation (Zapfack et al. 2002). It was located in a transitional area between the "Northwestern Congolian lowland forests" and "Northern Congolian forest-savanna mosaic," according to the "ecoregions" defined by the World Wildlife Fund for Nature (WWF), on the north of the South Cameroon Plateau—a dominant geographical feature of Cameroon. The elevation is approximately 650–700 m above sea level. The mean annual air temperature and annual precipitation are 23 °C and 1500 mm, respectively. Experimental sites were installed in both the forest and savanna, which were located closely within the borders between the tropical monsoon (Am) and tropical savanna (Aw) climates in the Köppen system. The dominant vegetation in forest was *Albizia zygia*, while that in the savanna was *Chromolaena odorata*. The soils distributed here are usually reddish brown (2.5YR 4/6) below the surface and are classified as Typic Kandiudox (Soil Survey Staff 2014), which were predominantly distributed on Neoproterozoic (Pan-African) granitoids. The detailed analysis of the soil properties in the forest and savanna for this region was given in Chap. 9 and Sugihara et al. (2014).

In this "Northern Congolian forest-savanna mosaic" region, cassava is planted as a dominant tuber crop for the cropland of the savanna territory, whereas various crops such as banana (dominantly Plantain), cassava, maize, and cacao were planted in the cropland prepared after forest reclamation. According to our field observations and interviews of farmers, savanna-like vegetation seemed to be maintained by frequent fires, often caused by the activity of neighboring agropastoralists from the north. The deterioration of forest vegetation, due to agricultural activities by local farmers, could not be determined with considerable certainty.

13.3   Effects of Reclamation of Original Vegetation on Nutrient Dynamics

Figure 13.2 comparatively illustrates the stocks and annual fluxes of C and N in soil ecosystems with original vegetation and reclaimed cropland.

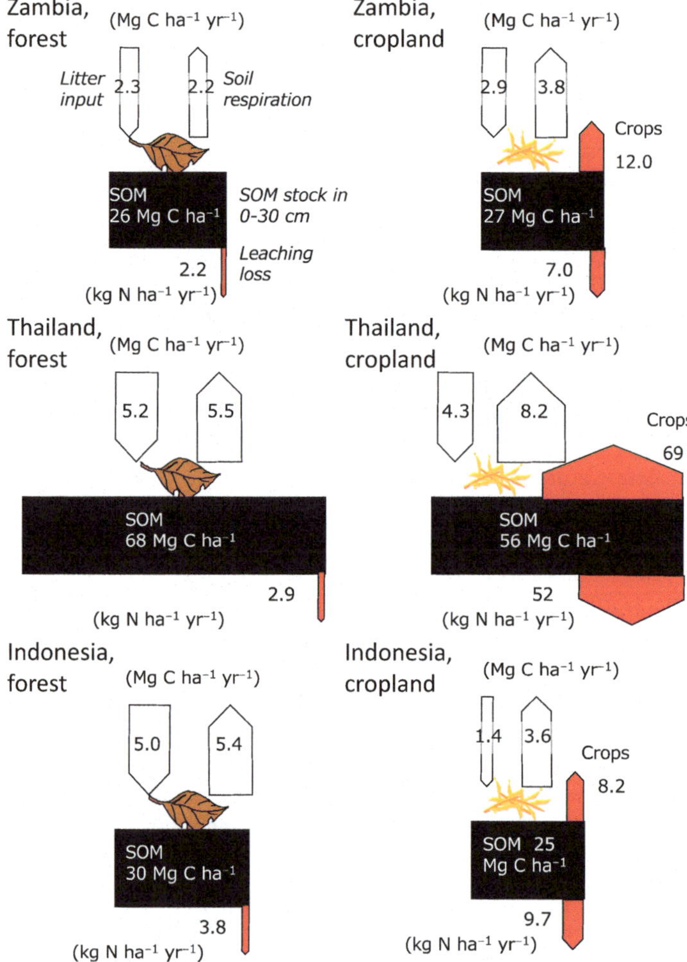

Fig. 13.2 Alteration of C and N dynamics after reclamation of forest at each of study sites

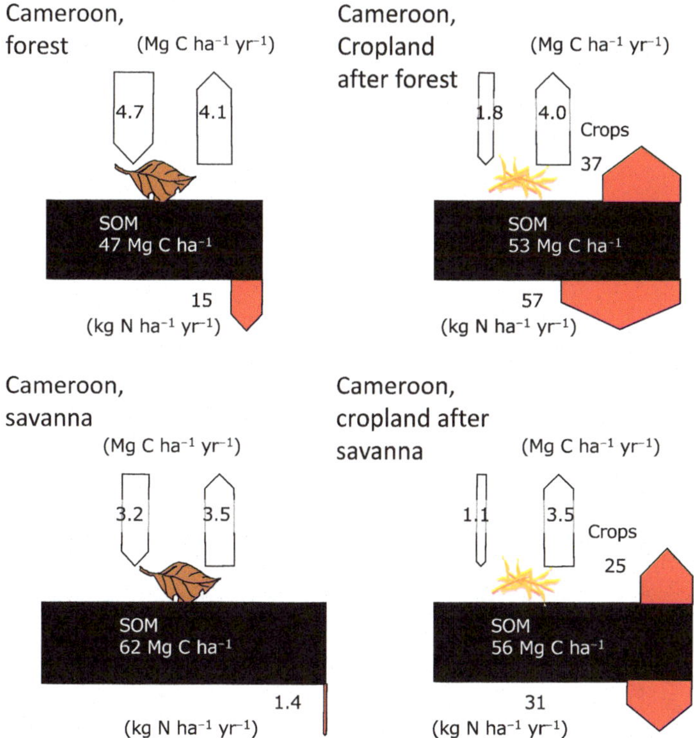

Fig. 13.2 (continued)

In the left figures of Fig. 13.2, C and N fluxes under natural vegetation are described, together with the total quantity of soil organic C. It was often observed that for natural vegetation the amount of litter produced and the microbial decomposition of SOM were almost balanced. Among the study sites, the C fluxes, which measure the litter input and organic matter decomposition/soil respiration, were as low as 2.3 Mg ha^{-1} year^{-1} and 2.2 Mg C ha^{-1} year^{-1}, respectively, for the miombo forest in Zambia. This was presumably due to a low precipitation and consequently lower net primary production. In all the study sites, except for the forest plot of Cameroon, N leaching deeper than 30 cm in the forest-soil ecosystems was noticeably low, as less than 4 kg N ha^{-1} year^{-1}. Only in the Cameroonian forest, N leaching totaled 15 kg N ha^{-1} year^{-1}, presumably due to the N-rich environment (CN ratio of around 11) through the dominant activities of leguminous trees (*Albizia zygia*, *Myrianthus arboreus*, etc.).

The reclamation of original forests, often accompanied by slash-and-burn practices, resulted in drastic changes in nutrient dynamics. The right figures of Fig. 13.2 indicate a decreasing input of plant litter to soil (except for Zambia); an accelerated decomposition of SOM would rapidly decrease SOM. During the course of SOM decomposition in the cropped field, mineralized N was released into the soil solution and then partially absorbed by crops, but comparable amounts in the remaining portion were lost through leaching. It should be noted that leached N was mostly in the form of nitrate (NO_3^-), and equivalent amounts of cations such as K, Mg, and Ca were lost together with NO_3^-. Such concomitant loss of essential mineral nutrients would be especially serious in strongly weathered soils, such as Oxisols in Cameroon, because the nutrient replenishment from these soils is slow and limited.

13.4 Recovery of Soil Fertility Status During the Fallow Phases

Figure 13.3 compares the fluctuations of organic C content and pH on the surface soils in the respective study sites in eastern Zambia, northern Thailand, East Kalimantan of Indonesia, and eastern Cameroon, with respect to the fallow and consecutive cropping periods, if any. These parameters were chosen as representative indicators of soil fertility status.

It is widely believed that fallow practices may contribute toward renewing soil fertility, which was once exhausted during the cropping phase. Such a trend was

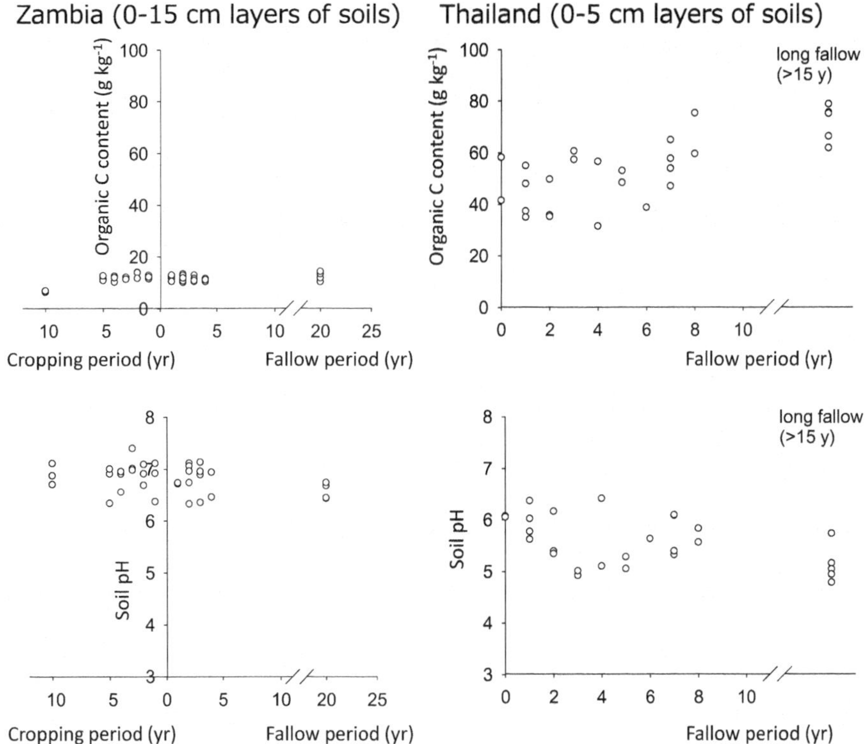

Fig. 13.3 Fluctuation of organic C content and pH during different stages of land use in each of study sites

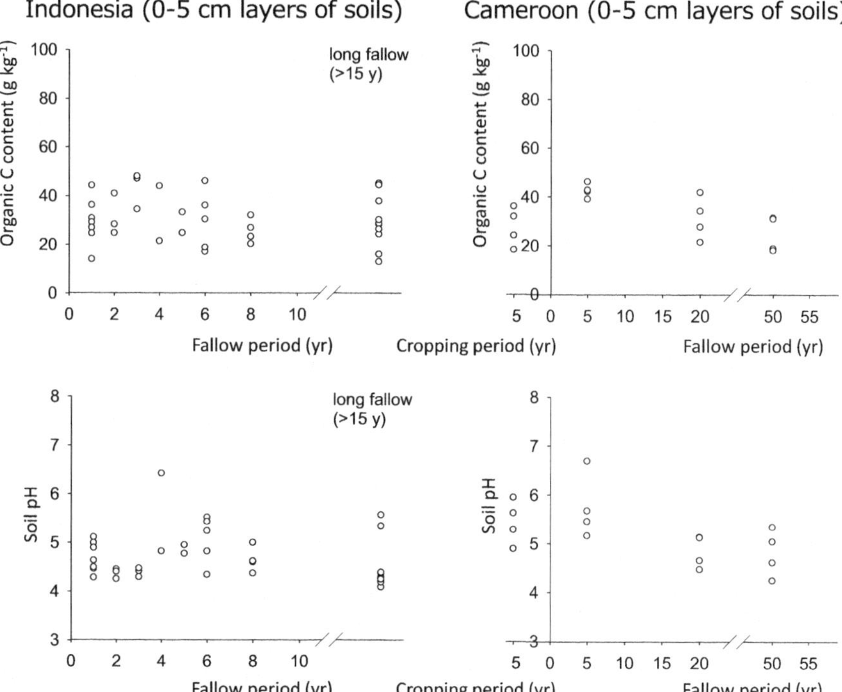

Fig. 13.3 (continued)

observed most clearly for ustic Ultisols (Ustults) in northern Thailand. In the study site of Zambia, where Alfisols dominate, neither the depletion of fertility during the cropping phase nor fertility improvement during the fallow phase was observed. This could be due to low decomposition rates of SOM and litter input under prolonged dry climates with low precipitation. The SOM level was consistently low, and soil pH was high, regardless of the phase of shifted agriculture for the dry climatic conditions in Zambia. On the contrary, the Udults of tropical rain forests in East Kalimantan, typically characterized by high temperatures and humidity, were considered to be quite favorable for soil microbial activities. Under such conditions, SOM did not seem to accumulate, even after a long period of fallow. Soil pH was consistently low, except for some soils affected by parent materials of limestone and/or topography at the riverside.

The Cameroonian soils exhibited unique characteristics concerning soil fertility fluctuations during different stages of land use, i.e., both the levels of SOM and soil pH were highest at the young fallow stage (around 5 years), and they decreased during the period of prolonged fallow. Figure 13.4 also indicates several parameters relating soil fertility status on an acreage basis (within the first 30 cm of soil) for Cameroonian soils. The amount of soil organic matter, level of exchangeable bases, and soil acidity were often remarkably high in the soils of the young fallow forest compared with that of the other land use stages, presumably due to an active supply of forest litter to the surface soil as well as the pump-up effect of deep-soil nutrients

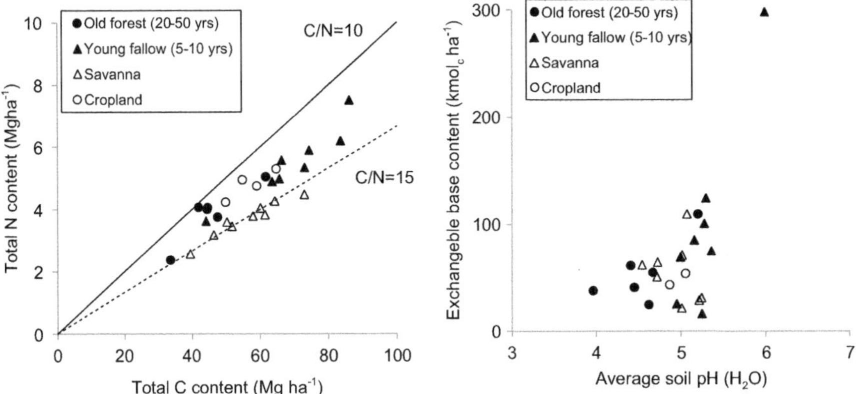

Fig. 13.4 Amounts of soil organic C and N, CN ratio, soil pH, and the content of exchangeable bases in the first 30 cm of soil at the Cameroonian study site

by root systems of forest vegetation, which was also observed for the soils in northern Thailand, as was discussed in detail in Chap. 11.

Since the Oxisol soils in eastern Cameroon (composed of kaolin minerals and sesquioxides) and Ultisols in northern Thailand (composed of mica and kaolin minerals) were not rich in surface negative charges of soil colloids that could retain exchangeable bases or Al and hence were poor in their acid-buffering capacity, their acidity-related properties such as soil pH could easily be changed through the process of soil acidification (net addition of protons) or soil alkalization (net addition of bases). The improving trend in terms of soil pH during young fallow stages (up to 10 years) may therefore be detectable only in these soils. The values for SOM and soil pH may subsequently change during the establishment of a mature forest based on the climatic conditions of respective regions; that is, soil pH would reach from 4.5 to 5.5 and 5.0 to 6.0 in eastern Cameroon and northern Thailand, respectively, and SOM levels at the top 5 cm of soils increase from 20 to 30 g kg^{-1} and 60 to 80 g kg^{-1} in eastern Cameroon and northern Thailand, respectively. In this sense, the climax values of soil pH and the SOM level at East Kalimantan were considered to be, with exception of limestone-derived soils and riverside soils, 4 to 4.5 and 20 to 40 g kg^{-1}, respectively.

13.5 Farmer's Strategies for Replenishing Nutrient Resources in Slash-and-Burn Agriculture Under Different Climatic and Soil Conditions

The positive function of the fallow phase was most clearly observed for shifting cultivation in northern Thailand, i.e., soil fertility status was reinstated during the later stage of the young fallow phase, around 8 years. Shifting farmers usually started to open the fallow forest again at this stage. According to the analysis in

Chap. 11, soil fertility in terms of the "P and NC ratio" did not exhibit any clear trend throughout the land use stages, including long fallow. The slash-and-burn practice, therefore, was considered to be indispensable to maintain the soil properties for the next crop phase of the shifting cultivation strategy practiced by the Karen people. The acceleration of N flux after forest reclamation and slash-and-burn practice was most drastically observed in northern Thailand, also indicating the effectiveness of this particular land use. According to our field observations, the farmers have strictly maintained only 1 year of cropping for their land rotation system. Since the cropping field in the study area was generally situated on a steep slope, soil erosion control may be a primary concern (Lal 1986). Short cropping may promise a good recovery of the next fallow vegetation and seems to decrease the soil erosion risk. The shifting cultivation method implemented by Karen farmers in northern Thailand, in which only 1 year's rice cropping is practiced on sloped lands with a relatively short fallow period, is a reasonable one for utilizing relatively fertile soils developed under monsoon subtropics, while mitigating the soil erosion risk.

In contrast with the agricultural practices of northern Thailand, the upland rice cropping in East Kalimantan is planted after reclaiming a relatively old fallow forest, usually more than 20 years old after the previous cultivation. The plant biomass supplied for burning amounted to 100–300 Mg ha^{-1} (Funakawa et al. unpublished data). Soils in this region were typically strongly acidic, and soil fertility status, including the N-supplying potential of the soils, was considered to be low compared to the soils of northern Thailand (Funakawa et al. 2008, 2009). The agricultural practices in this region, which utilize higher amounts of plant biomass for burning after a longer period of fallow, were therefore adapted to this particular soil condition (Udults with strongly acidic nature) and tropical rain forest climate.

The climatic pattern of eastern Cameroon (MAT: 23 °C; AP: 1450 mm) is somewhat intermediate between that of East Kalimantan (MAT: 26 °C; AP: 2500 mm) and northern Thailand (MAT: 20 °C; AP: 1200 mm) in terms of severity of dry season and annual temperatures. The fluctuation of SOM levels and soil pH during different stages of land use in this region was similar to the cases in northern Thailand, i.e., the soil fertility status was relatively high at the stage of young fallow forest (5–10 years). However, the major crops and dominant soils, as well as human activities including neighboring agropastoralists, are very much different from the Asian countries. The activity of agropastoralists often included burning every year, which would contribute toward maintaining savanna vegetation for long period of time regardless of the high precipitation in this region. The land use patterns of the neighboring farmers in this study seem to be strongly influenced by the presence of this human-induced savanna vegetation; that is, cassava is planted as a dominant tuber crop in cropland of the savanna territory, whereas various crops such as banana (dominantly Plantain), cassava, maize, and cacao were planted in cropland prepared after forest reclamation. Thus, the present land use systems in this region have not been controlled by natural conditions only, but more by human factors. Even so, the agricultural production should be supported by reasonable resource-replenishing processes, such as chemical fertilizers and/or manures, or through utilizing natural ecosystem processes, which was observed for the shifting cultivation in Southeast Asia. In the latter case, further positive utilization of the function of young fallow forest could be incorporated into the land rotation. Especially in this region, it is possible that a savanna-based cassava planting system could be converted into a forest-based multi-cropping system if the nutrient-replenishing function of young fallow vegetation could be incorporated into the present savanna-based system.

In the case of eastern Zambia, limited and fluctuating precipitation may be an important factor and a serious threat at the same time, for the agricultural production, unlike the other regions in this comparative study. As far as primary production and the SOM decomposition rates are regulated by this specific climatic condition, it could be said that soil fertility status would not change considerably by the present land use practices. In turn, the SOM levels and soil acidity, including base status, have been maintained under these conditions. Therefore, the classical interpretation of slash-and-burn agriculture is not applicable for the agricultural practices of this region. The hasty application of modernized agriculture, such as chemical fertilizers, irrigation, the introduction of improved varieties, etc., may bring other problems such as the rapid decline of SOM, soil acidification, soil fertility decline, and more. We should be cautious of these disadvantages, even as modern agricultural practices continue to be introduced.

13.6 Conclusion

Based on the results of field studies in different regions of tropics, as presented in the preceding chapters, shifting cultivation strategies under different bioclimatic and soil conditions were compared. Four regions were included in this study: eastern Zambia, northern Thailand, East Kalimantan of Indonesia, and eastern Cameroon. The natural conditions and basic soil properties at the study sites are different from each other. Then the effects of reclamation of the original vegetation on nutrient dynamics and the recovery of soil fertility during the fallow phases were comparatively analyzed. Finally, farmer's strategies for replenishing nutrient resources in slash-and-burn agriculture under different climatic and soil conditions were discussed.

The stimulation of N mineralization, and resulting N flux after reclamation of original vegetation, from the greatest to least was clearly observed in the order of forest in eastern Cameroon and northern Thailand and savanna in eastern Cameroon, East Indonesian Kalimantan, and eastern Zambia. This trend, however, coincided with the risk of mineral nutrient loss due to leaching. Such risk would be more detrimental in the strongly weathered soils, such as the Oxisols in Cameroon.

Additionally, the benefits of the fallow phase were the most clearly observed for the shifting cultivation in northern Thailand, i.e., soil fertility status was reinstated in the later stage of the young fallow, around 8 years. A similar trend was also traced for the soils in eastern Cameroon, in which the soil fertility status was the highest during the young fallow forest. Such improvement in soil fertility status was, however, not clearly observed for the East Kalimantan, Indonesia, and eastern Zambia, presumably due to the prolonged humid and hot climate in the former and seriously dry climate in the latter.

Next, the farmers' practices in the respective regions were analyzed. The shifting cultivation practice by Karen farmers in northern Thailand, in which only 1 year's rice cropping is planted on sloped lands with a relatively short fallow period, is a reasonable one for utilizing relatively fertile soils developed under monsoon subtropics, while mitigating soil erosion. However, the upland rice cropping in East Kalimantan is practiced after reclaiming a relatively old fallow forest, usually more than 20 years old after previous cultivation. Since the soils were typically strongly acidic, and soil fertility status was low, the agricultural practice in this region, in which higher amounts of plant biomass were supplied for burning after longer period of fallow, was considered to be adapted toward this particular soil and tropical rain forest climate condition.

Compared to these examples in Southeast Asia, the present pattern of land use in eastern Cameroon has not been controlled by natural conditions only, but more by human factors, in that the yearly burning practice of neighboring agropastoralists has contributed toward maintaining savanna vegetation for a long period, and the land use patterns of farmers seem to be strongly influenced by the presence of this human-induced savanna vegetation. Cassava is planted as a dominant tuber crop in cropland of the savanna territory, whereas various crops such as banana (dominantly Plantain), cassava, maize, and cacao were planted in cropland prepared after forest reclamation. In the case of eastern Zambia, limited and fluctuating precipitation may be an important factor for agricultural production. As far as the primary production and the SOM decomposition rate are regulated by this specific climatic condition, soil fertility would not change considerably under the present land use practices. In turn, the SOM level and soil acidity, including base status, could be maintained under this condition. Therefore, the classical interpretation of slash-and-burn agriculture is not applicable for the agricultural practice in eastern Zambia.

References

Alegre JC, Cassel DK, Bandy DE (1988) Effect of land clearing method on chemical properties of an Ultisol in the Amazon. Soil Sci Soc Am J 52:1283–1288

Chidumayo EN (1997) Miombo Ecology and Management an Introduction. Intermediate Technology Publications, London

Driessen PM, Deckers JA, Spaargaren OC, Nachtergaele FO (eds) (2001) Lecture notes on the major soils of the world. World Soil Resources Reports 94, Food and Agriculture Organization of the United Nations, Rome

Ewel J, Berish C, Brown B, Price N, Raich J (1981) Slash and burn impacts on a Costa Rican wet forest site. Ecology 62:816–829

Funakawa S, Tachikawa S, Kadono A, Pulunggono HB, Kosaki K (2007) Factors controlling soil organic matter decomposition in small home gardens in different regions of Indonesia. Tropics 17:59–72

Funakawa S, Watanabe T, Kosaki T (2008) Regional trends in the chemical and mineralogical properties of upland soils in humid Asia: With special reference to the WRB classification scheme. J Soil Sci Plant Nutr 54:751–760

Funakawa S, Makhrawie M, Puluggono HB (2009) Soil fertility status under shifting cultivation in East Kalimantan with special reference to mineralization patterns of labile organic matter. Plant Soil 319:57–66

IUSS Working Group WRB (2014) World Reference Base for Soil Resources 2014. International soil classification system for naming soils and creating legends for soil maps. World Soil Resources Reports No. 106. FAO, Rome

Knoepp JD, Swank WT (1993) Site preparation burning to improve southern Appalachian pine-hardwood stands: nitrogen responses in soil, water, and streams. Can J For Res 23:2263–2270

Kunstadter P, Chapman EC (1978) Problem of shifting cultivation and economic development in northern Thailand. In: Kunstadter P, Chapman EC, Sabhasri S (eds) Farmers in the forest. The University Press of Hawaii, Honolulu, pp 3–23

Kyuma K (1982) Capability considerations for tropical soils. Tropical Agriculture Research Series No. 15. Tropical Agriculture Research Center, Ministry of Agriculture, Forestry and Fisheries, Tsukuba, Japan, pp 105–117

Lal R (1986) Soil surface management in tropics for intensive land use and high and sustained production. In: Stewart RA (ed) Advances in soil science, vol 5. Springer New York, pp 1–109

Marion GM, Moreno JM, Oechel WC (1991) Fire severity, ash deposition, and clipping effects on soil nutrients in chaparral. Soil Sci Soc Am J 55:235–240

Nye PH, Greenland DJ (1960) The Soil under Shifting Cultivation. CBS Tech. Commun. No. 51, Harpenden, UK, pp 1–12

Raison RJ, Khanna PK, Woods PV (1985) Mechanisms of element transfer to the atmosphere during vegetation fires. Can J For Res 15:132–140

Richter DD, Markewitz D, Heine PR, Jin V, Raikes J, Tian K, Wells CG (2000) Legacies of agriculture and forest regrowth in the nitrogen of old-field soils. For Ecol Manag 138:233–248

Ruark GA (1993) Modeling soil temperature effects on in situ decomposition rates for fine roots of loblolly pine. For Sci 39:118–129

Sakamoto K, Yoshida T, Satoh M (1991) Comparison of carbon and nitrogen mineralization between fumigation and heating treatments. Soil Sci Plant Nutr 38:133–140

Soil Survey Staff (2014) Keys to soil taxonomy, Twelfth edn. United States Department of Agriculture, Natural Resources Conservation Service, Washington, DC

Sugihara S, Shibata M, Mvondo Ze A, Araki S, Funakawa S (2014) Effect of vegetation on soil C, N, P and other minerals in Oxisols at the forest-savanna transition zone of central Africa. J Soil Sci Plant Nutr 60:1–15

Tanaka S, Funakawa S, Kaewkhongkha T, Hattori T, Yonebayashi K (1997) Soil ecological study on dynamics of K, Mg, and Ca, and soil acidity in shifting cultivation in northern Thailand. Soil Sci Plant Nutr 43:695–708

Tanaka S, Ando T, Funakawa S, Sukhrun C, Kaewkhongka T, Sakurai K (2001) Effect of burning on soil organic matter content and N mineralization under shifting cultivation system of Karen people in northern Thailand. Soil Sci Plant Nutr 47:547–558

Zapfack L, Engwald S, Sonke B, Achoundong G, Madong BA (2002) The impact of land conversion on plant biodiversity in the forest zone of Cameroon. Biodivers Conserv 11:2047–2061

Chapter 14
Interactions Between Agricultural and Pastoral Activities in the Sahel with Emphasis on Management of Livestock Excreta: A Case Study in Southwestern Niger

Hitoshi Shinjo

Abstract An investigation conducted in southwestern Niger, Africa, revealed livestock management practices based on the relationship between farmers and sedentary pastoralists to promote efficient use of livestock excreta through corralling in the Sahel region. Transhumance was found to be essential for households residing in an area with a high rate of cropping because they had to remove their herds from cropped fields to a drier region in the rainy season. Corralling, parking herds in harvested fields at night, by the sedentary pastoralists was carried out mainly in fields rented from farmers. Another type of corralling was performed on a contract basis in farmers' fields, on the condition that the farmers provide food for the pastoralists staying for corralling. About half of the sedentary pastoralists were not engaged in the practice of contract corralling or contracted livestock, suggesting the existence of a loose relationship between the farmers and sedentary pastoralists in terms of corralling and livestock grazing. In half of the households with contract corralling, the practice changed in two consecutive seasons. This implies that contract corralling is a flexible practice to compensate for the shortage of millet production and forage resources. Considering the amount of livestock excreta applied and the percentage of the area corralled, two-thirds of the households had the potential to promote contract corralling for more efficient use of livestock excreta.

Keywords Fulani • Land use • Mixed farming system • Transhumance

H. Shinjo (✉)
Graduate School of Global Environmental Studies, Kyoto University, Yoshida-Honmachi, Sakyo, Kyoto 606-8501, Japan
e-mail: shinhit@kais.kyoto-u.ac.jp

© Springer Japan KK 2017
S. Funakawa (ed.), *Soils, Ecosystem Processes, and Agricultural Development*,
DOI 10.1007/978-4-431-56484-3_14

14.1 Introduction

The Sahel region, in the southern periphery of the Sahara Desert, is marginal for cropping due to the limited and erratic amount of rainfall, ranging from 150 to 500 mm per annum. Thus, in this region, livestock husbandry has long been the major subsistence activity besides cropping. Traditionally, pastoralists in this region have been nomads, moving from place to place where forage resources are available. They may stay in crop farmers' fields in the dry season to enable livestock to graze stubbles and stalks of crop residues in harvested fields and also travel to fallow land within a day's distance (hereafter, crop farmer is referred to simply as "farmer"). During their stay, livestock excreta, i.e., feces and urine, are spread on the fields. This practice is referred to as corralling, which is beneficial for both pastoralists and farmers. The pastoralists can utilize crop residues as forage and the farmers can apply the livestock excreta as organic fertilizer. Since the use of chemical fertilizers in this region is still limited (Mokwunye and Hammond 1992), the application of livestock excreta through corralling is one of the major soil fertility maintenance/improvement practices and was found to be effective by Schlecht et al. (2004). Their trials of livestock excreta application to pearl millet ranging from 200 to 1400 g m^{-2} revealed the beneficial effects on grain and stalk yields which increased proportionally to the rate of livestock excreta applied in the first year. The residual effect of livestock excreta was found to be significant and showed a linear relation with the amount applied for up to 4 years after the application in the first year.

However, these specialized forms of crop and livestock production by different ethnic groups of pastoralists and farmers are undergoing a transition (Powell and Williams 1995). Increase of human population combined with long-term weather changes is transforming the specialized systems to more intensively managed systems. As livestock husbandry assumes a more settled pattern than the nomadic one, it increasingly incorporates crop production into the management system. This incorporation intensifies livestock husbandry by supplying forage crops and crop residues as feed. In this process, many traditional exchange relationships between pastoralists and farmers are disappearing.

This transformation seems to be taking place in the Fakara region, southwestern Niger, a part of the Sahel region (Fig. 14.1). During the last several decades, it was observed that Fulani pastoralists became sedentary in the area where farmers had already settled (hereafter this group of pastoralists who became sedentary is referred to as the "sedentary Fulani"). The sedentary Fulani started cropping on land rented from the farmers in addition to their original activity of livestock husbandry. Because they became sedentary for the purpose of growing their own crops in the rainy season (Fig. 14.2a), their seasonal grazing patterns have changed. Only one shepherd was engaged in the transhumance practice consisting of taking the herd to a drier region during the rainy season (Fig. 14.2b), while the remaining household members stayed in the villages to manage cropping and take care of the milking cows and calves that were not taken for transhumance. Hence, the

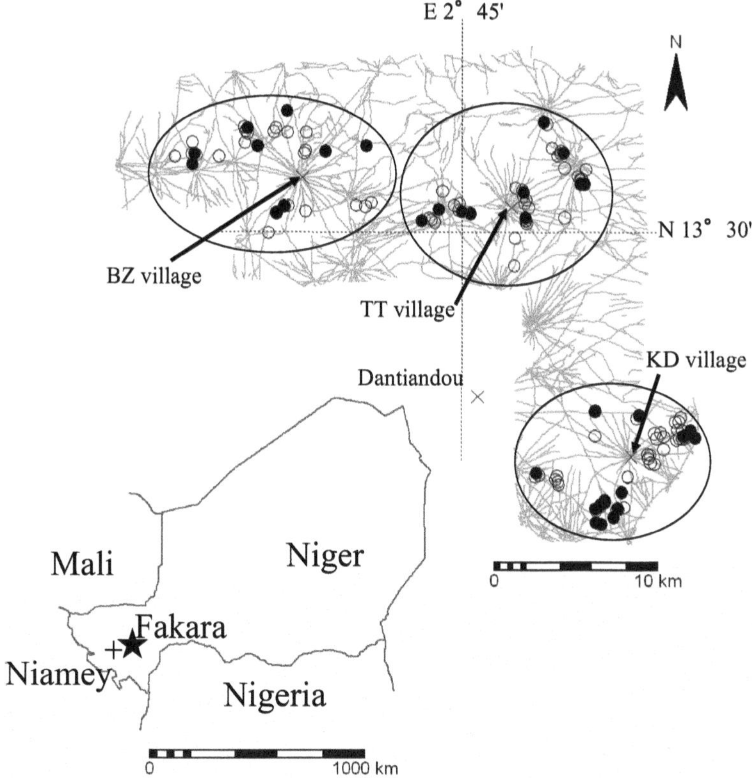

Fig. 14.1 Location of the study area (Fakara) in Niger, West Africa (*bottom*), and households of sedentary Fulani in Fakara (*top*). Open (O) and closed (●) *circles* indicate the location of the households not studied and studied, respectively. Curved lines in the study area show roads and lanes (Shinjo et al. 2008)

sedentary pattern of the pastoralists near farmers may change their relationship in terms of land use, corralling, and livestock management. The review paper by Powell and Williams (1995) characterized this change as one of the steps in the evolution of mixed farming systems from specialized to integrated systems. Nevertheless, because each household has different settings, not all the households may transform their production systems following this evolutionary course. Thus, when realistic options for sustainable resource management are proposed, they should be based on the actual relationship between farmers and pastoralists and their controlling settings. This paper focuses on the livestock management practices of the sedentary Fulani, especially corralling, to promote efficient use of livestock excreta through corralling.

Fig. 14.2 Some practices outlined in the present study (Shinjo et al. 2008). (**a**) Crop production. Pearl millet (*Pennisetum glaucum*) is the major crop cultivated in the rainy season. (**b**) Transhumance. Herds of livestock are taken from the study area to a drier region at the onset of the rainy season. (**c**) Corralling. Livestock excreta are spread on the surface of cropland during the dry season. Cows are tied to standing wooden sticks at night

14.2 Materials and Methods

14.2.1 Study Area

The study area, Fakara, was located about 60 km east from Niamey, the capital of Niger (Fig. 14.1), and about 200 km south from the border between Niger and Mali. As reported by Matsunaga et al. (2006), who conducted studies in the same area, the climate is typical of southern Sahel with 416 mm average annual rainfall from 2001 to 2004. A rainy season normally starts in June and ends in October. The landscape consists of plateaus, sandy skirts, and valley bottoms (D'Herbès and Valentin 1997). Most soils in the sandy skirt and valley bottoms are very sandy (more than 90 % sand). The plateau is rarely arable and is mainly used for grazing due to the presence of an ironstone layer on the surface. Sandy skirts and valley bottoms are used for rainfed farming. Major crops consist of pearl millet (*Pennisetum glaucum*) and cowpea (*Vigna unguiculata*) (Fig. 14.2a). Some fields are left fallow to allow the regrowth of native plants. The cropping season corresponds to the rainy season with sowing in May/June and harvest in October. Draft animals are rarely used for cultivation. Instead, practices such as sowing, weeding, and harvest are carried out manually by most households. Major livestock found here include humped cattle (zebu), sheep, goats, chicken, and guinea fowls. Cattle are raised mainly for milking, while sheep and goats for milking and meat.

In the study area, the farmers, who belong to the Zarma ethnic group, settle in village centers, while the sedentary Fulani inhabit areas away from the village centers. This is partly because they cannot rear a large number of livestock in the villages. Due to the lack of apparent village boundaries, I deliberately assigned the sedentary Fulani to the three main villages, namely, BZ, TT, and KD villages, based on the distance between their settlements and the village centers. This practice followed the method applied in a previous study (Hiernaux and Ayantunde 2004) (Fig. 14.1). These three villages were selected for varying cropping rates. By interpreting the aerial photography, Hiernaux and Ayantunde (2004) reported that the rate of cropping area to the total area increased in the order of 24 % (BZ), 37 % (TT), and 66 % (KD). The cropping rate differed among the three villages partly due to the difference in the area of non-arable plateau. BZ is occupied by the plateau to the largest extent among the three villages. KD is located immediately west of the Dallal Bosso region, a fossil alluvial plain with shallow water table. In the Dallal Bosso region, most arable lands had already been cultivated continuously and livestock could take water and forage in some ponds even in the middle of the dry season.

14.2.2 *Interview with Sedentary Pastoralists and Estimation of Their Land Acreage According to Land Use*

From 129 households of the total sedentary Fulani in the three villages in 2003, 36 households (28 % of the total households) were selected by clustering analysis, in terms of millet production and number of livestock. The number of households selected from each village was proportional to that of all the households in each village: 9 for BZ, 16 for TT, and 11 for KD. Prior to the main survey in 2006, 23 of the 36 households were preliminarily surveyed in April 2005, at the end of the dry season. In this survey, the areas cropped, fallowed, and corralled were determined by tracing their boundary with a GPS (Global Positioning System) receiver (Garmin Co.), whose positional error was at least 5 m. The area corralled was identified by the presence of animal feces excreted in that dry season (Fig. 14.2c). The feces in the fields were mainly excreted at nighttime when livestock slept. To keep livestock in the fields at nighttime, livestock were tied to wooden sticks (Fig. 14.2c), or calves, lambs, or kids were kept in the enclosures so that the rest of the herd would gather around the enclosures. At daytime, livestock grazed crop residues in other fields or native plant leaves in fallow lands. The households were also interviewed on years of settlement, number of livestock (cows, sheep, and goats), transhumance, corralling, and land renting. The number of livestock was expressed in tropical livestock unit (TLU) of 250 kg live weight (ILRI 1995). One cow, one sheep, and one goat with an average body weight corresponded to 1.2, 0.24, and 0.16 TLU, respectively. Numbers of chickens and guinea fowls were omitted because these livestock were kept around houses and were not corralled in the fields. In May 2006, all the 36 sample households were investigated on the abovementioned aspects. The households not surveyed in 2005 were interviewed in 2006 on transhumance, corralling, and livestock management for 2005, so that these practices could be traced over 2 years for all the sample households. Instead of direct measurement, the total amount of fecal excreta by livestock in a field in a dry season was estimated by multiplying the number of cows, sheep, and goats, the days of their stay in the field, and the average amount of fecal excreta by one cow, one sheep, and one goat. The last figures were 1.19 kg dry matter per day per cow, 0.18 kg dry matter per day per sheep, and 0.10 kg dry matter per day per goat, half of the daily amount of excreta per each animal with an average body weight (Fernández-Rivera et al. 1995). Because livestock stayed in the field only at nighttime, I assumed the half of the daily excreta would be applied on the field. All the interviews were carried out in the Fulani language with a Fulani-French interpreter.

14.2.3 Statistical Analysis

All the variables were first tested for distribution normality using the chi-square test for goodness of fit. For the variables normally distributed, statistical difference was determined by two-sided Student's t-test. For the variables not normally distributed, the statistical difference was determined by two-sided Mann-Whitney's U test.

14.3 Results and Discussion

14.3.1 Characteristics of Livestock Management by the Sedentary Fulani

The households were categorized by their livestock management practices (Table 14.1). Regarding transhumance, a clear contrast among the villages was found. No households in BZ practiced transhumance, unlike all the households in KD and TT, mainly due to the difference in the area occupied by crops, as described in the section "Study Area." During the rainy season, when the livestock were not permitted to enter the cropped fields, the households in BZ allowed the livestock to graze a relatively larger area of fallow or non-arable land surrounding the cropped fields. In contrast, since most lands in KD and TT were occupied by crops, herds were removed from this area. The second criterion involved contract corralling, namely, corralling in the fields of the Zarma farmers on a contract basis. When corralling in farmers' fields was applied, the sedentary Fulani were paid for corralling by the farmers with food and did not pay a fee for corralling. Only 10 households in the three villages practiced contract corralling. The third criterion involved contracted livestock. In this scheme, sedentary Fulani managed some of the client farmers' livestock in their herds, most of whom were Zarma farmers in this area. The contract for managing the livestock of the Zarma farmers was usually kept through a year. However, the client farmers had to pay a fee to the herder only in the rainy season when serious control of the grazing area was necessary, so that livestock would not enter the cropped fields. The amount of the fee was determined according to the number of the clients' livestock. On the other hand, the clients were not requested to pay a fee in the dry season when all the parcels of the land could be grazed freely. Besides the fee, the sedentary Fulani were able to acquire dairy products from the contracted livestock through a year. Half of the 36 households practiced contracted livestock (Table 14.1). As also shown in Tables 14.1 and 14.3, households in BZ, 8 in KD, and 5 in TT, totaling 16 households, were not engaged in contract corralling or contracted livestock. This fact suggested that the sedentary Fulani did not always have a close relationship with the Zarma farmers in terms of corralling or livestock grazing.

14.3.2 Characteristics of Contract Corralling

To analyze the features of contract corralling in more detail, some characteristics of the corralling practices were statistically compared for the two types of households grouped based on the contract practice (Table 14.2). No difference among the villages was found. The households with contract corralling conducted corralling in their fields during a 5.0-month period, which was shorter than that of the households without contract corralling, namely, 7.3 months. This was because households with contract corralling took their herd to the fields of client farmers for corralling, resulting in the reduction of the duration of the corralling period in their own fields. However, the total input of excreta for households with contract corralling, namely, 4.0 Mg, was not statistically different from that for households without contract corralling, namely, 5.3 Mg. These statistical results suggest that households with contract corralling did not conduct corralling in the clients' fields long enough to reduce the excreta input in their own fields. In other words, client farmers could expect only a limited excreta input under the current practice of contract corralling. Furthermore, since the average number of livestock owned by the Zarma farmers in this area was reported to be as small as 2.3 TLU (Tropical Livestock Unit) of cows, 0.5 TLU of sheep, and 0.3 TLU of goats (Abdoulaye et al. unpublished), the application of excreta by corralling was very limited in the fields of the Zarma farmers.

For the households with contract corralling, the number of cows contracted, namely, 9 TLU for cows, and the ratio of contracted cows to total cows, namely, 45 %, were larger than those without contract corralling (Table 14.2). These

Table 14.1 Number of respondent households of sedentary Fulani on livestock management practices

		Village designation			
		BZ	KD	TT	Total
Households studied		9	16	11	36
Transhumance[a]		0	16	11	27
Contract corralling[b]	Contracted livestock[c]				
Yes	Yes	2	6	0	8
Yes	No	0	0	2	2
No	Yes	4	2	4	10
No	No	3	8	5	16

Shinjo et al. (2008)

[a]Transhumance: transfer of livestock to drier regions in northern Niger during the cropping season
[b]Some Fulani practice corralling not only in their fields but also in Zarma fields for food
[c]Some Zarma farmers make a contract with Fulani to raise their livestock

differences corresponded to the result that eight of the ten households with contract corralling raised contracted livestock (Table 14.1). Of these eight households, three households practiced contract corralling in the fields of the client farmers for contracted livestock. However, it should be noted that this finding did not directly indicate the existence of a close relationship between contract corralling and contracted livestock, since the contract for corralling and raising livestock was made separately. In fact, one household refused to conduct corralling in the field of a cow owner due to the lack of millet to be provided by the cow owner as the payment for corralling.

Yearly changes in contract corralling implied that contract corralling seemed to be a more flexible practice than transhumance and contracted livestock (Table 14.3). Five households with contract corralling during the 2004/2005 dry season abandoned this practice in the following dry season. Three of these households explained that the reason was to concentrate livestock excreta on their own fields for maintaining millet yield. On the other hand, six households without contract corralling in the preceding dry season practiced it in the 2005/2006 dry season, two of whom pointed out that the reason was to compensate for the low millet production in their own fields. Contract corralling could provide them with food from client farmers and saved their own millet stock. Some households practiced contract corralling when asked by a farmer near land where they happened to find adequate forage resources. As indicated in these reasons, contract corralling could be applied by the herders and may depend on other purposes such as grazing. On the other hand, the households did not change the transhumance and contracted livestock practice during the 2 years, because transhumance was essential due to the shortage of available forage resources in the rainy season, and the contracted livestock practice was obligatory for the Zarma farmers who could not manage livestock during the cropping season.

Table 14.2 Characteristics of corralling practices

	Whole households $n = 36$		Households with contract corralling $n = 10$		Households without contract corralling $n = 26$		Difference[g]
	Mean	STD[f]	Mean	STD[f]	Mean	STD[f]	
Total area managed (ha)	6.6	4.8	5.5	2.9	7.0	5.4	–
Area cropped (ha)	6.4	4.6	5.5	2.9	6.8	5.2	–
Area corralled (ha)	2.4	1.9	2.4	1.9	2.4	1.9	–
Period of corralling (month)[a]	6.7	2.3	5.0	2.5	7.3	1.9	**
Cows managed (TLU)[b]	19.1	9.6	23.4	8.4	17.5	9.6	–
Cows owned (TLU)	12.5	9.2	9.9	10.0	13.5	8.9	–
Cows contracted (TLU)[c]	4.4	6.7	10.1	7.8	2.2	4.8	**
Cows entrusted (TLU)[d]	2.3	3.2	3.4	3.8	1.8	2.9	–
Small ruminants managed (TLU)	5.9	7.2	3.9	4.5	6.7	8.0	–
Sheep owned (TLU)	1.9	2.7	1.8	2.9	2.0	2.6	–
Sheep contracted (TLU)[c]	0.0	0.2	0.0	0.0	0.0	0.2	–
Sheep entrusted (TLU)[d]	0.4	0.6	0.5	0.8	0.3	0.5	–
Goats owned (TLU)	3.1	5.4	1.2	1.1	3.9	6.2	–
Goats contracted (TLU)[c]	0.0	0.0	0.0	0.0	0.0	0.0	–
Goats entrusted (TLU)[d]	0.4	0.5	0.4	0.3	0.5	0.5	–
Total input of livestock excreta (Mg)	5.0	2.9	4.0	2.5	5.3	3.0	–
Input rate of livestock excreta ($\times 10^2$ g m^{-2})[e]	2.5	1.8	2.1	1.9	2.7	1.7	–
Ratio of corralled area to cropped area (%)	49	36	49	40	49	35	–
Ratio of contracted cows to total cows (%)	21	31	45	37	12	23	**
Ratio of cows managed to area cropped (TLU/ha)	3.9	2.9	5.5	3.9	3.3	2.1	–

Shinjo et al. (2008)

[a]Corralling period in fields of households

[b]Tropical livestock unit

[c]Cows, sheep, or goats of Zarma managed on contract basis

[d]Cows, sheep, or goats of other Fulani households

[e]Total manure input divided by area corralled

[f]Standard deviation

[g]** indicates statistically significant difference between the households with and without contract corralling

14.3.3 Pattern of Application of Livestock Excreta

To promote efficient use of livestock excreta through corralling, the pattern of application of livestock excreta was analyzed. As shown in Fig. 14.3, the households with contract corralling did not differ from those without contract corralling

Table 14.3 Yearly change in the number of households in terms of transhumance, contract corralling, and contracted livestock

		Transhumance	Contract corralling	Contracted livestock
04–05	05–06			
Yes	Yes	26	4	18
Yes	No	1	5	1
No	Yes	1	6	0
No	No	8	21	17

Shinjo et al. (2008)

in this aspect. Schlecht et al. (2004) suggested that the application of livestock excreta should be limited to 300 g m^{-2}, with effects expected over four cropping seasons, to obtain a maximal rate of return in terms of yield and to minimize the leaching of nutrients. Following this suggestion, a household with an input rate of 300 g m^{-2} in 25 % of the area cropped utilized livestock excreta most efficiently, so that it could end the application of livestock excreta for the whole field within 4 years. Thus, 7500 g m^{-2} %, computed by multiplying the application rate, namely, 300 g m^{-2}, by the percentage of the area corralled, namely, 25 %, could be used as a criterion for effective management of corralling. As shown in Fig. 14.3, in 23 households, corresponding to almost two-thirds of the whole households, the values were larger than 7500 with a potential to conduct corralling in the fields of other households.

Figure 14.3 also depicts the two groups of households, based on the ratio of the area corralled to that cropped. In the first group of 12 households, corralling was conducted at a rate of more than 70 % of the area cropped, with a relatively small input rate of livestock excreta, namely, 300 g m^{-2} at most. They applied the livestock excreta during both the 2004/2005 dry season and the 2005/2006 dry season at a rate of almost 100 % of their lands. On the other hand, in the second group consisting of the other households, corralling was conducted at a rate of less than 50 % of the area cropped, although the input rate varied from almost zero to more than 600 g m^{-2}. In this group of households with an input rate of more than 400 g m^{-2}, the areas for application of livestock excreta within a field changed from the 2004/2005 dry season to the 2005/2006 dry season, as they recognized the residual effect of application of livestock excreta. The reasons for the differences in this application pattern remain to be clarified.

14.3.4 Increasing Efficiency of Application of Livestock Excreta Under the Actual Relationship Between the Farmers and the Sedentary Fulani

The study revealed that a small amount of livestock excreta was applied in the farmers' fields by the sedentary Fulani, as scheme of contract corralling

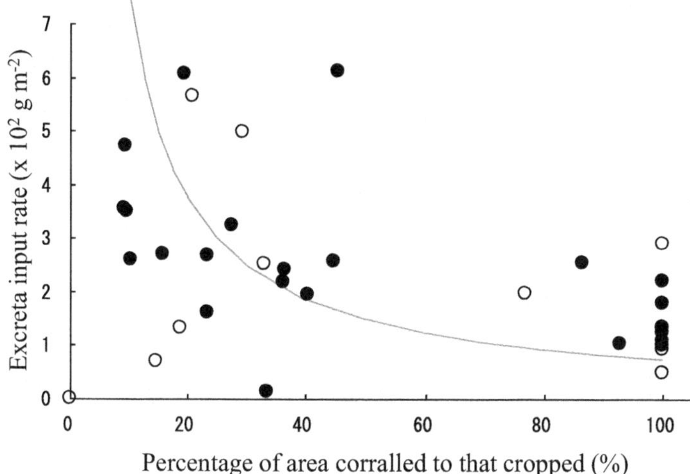

Fig. 14.3 Relationship between the percentage of the area corralled to that cropped and the excreta input rate in the fields managed by the sedentary Fulani in the 2005/2006 dry season (Shinjo et al. 2008). Open (○) and closed (●) *circles* indicate households with and without contract corralling, respectively. Solid line depicts the constant product of both values. The constant product used is 7500, computed by multiplying 300 g m^{-2} by 25 %, as the criterion for the effective management of corralling, following the suggestion of Schlecht et al. (2004)

(Table 14.2), and that for the majority of the Zarma farmers, corralling was not beneficial for crop production. The same finding was reported for the Hausa farmers by Oyama (2015). He described that only the farmers rich enough to pay for contract corralling by cash or millet can ask pastoralists for it. He depicted even the worst case that one of the poor farmers seriously injured the Fulani who had contract corralling in the field of the rich farmer. He suggested that the poor farmer might be frustrated by the contact corralling. Because livestock can graze freely in the dry season, it grazed crop residues in the field of the poor farmers and it delivered its excreta in the field of the rich farmers with the contract corralling. The poor farmer may think that contract corralling removed the resource as crop residues from his field and enrich the field of the rich farmers.

However, almost two-thirds of the sedentary Fulani were found to have the potential to practice corralling in the fields of other farmers (Fig. 14.3). Thus, if these sedentary Fulani could promote the contract corralling practice, a larger number of Zarma farmers could benefit from contract corralling. For this promotion, the proposal by Thiombiano (personal communication) to develop a mutual assistance system between the farmers and the sedentary pastoralists in Burkina Faso, West Africa, should be considered. The proposal, which is based on the current situation of land tenure for the sedentary Fulani, is compatible with this study. Most sedentary Fulani in this area cultivate fields with the permission of the Zarma land title holders with rent in kind of about one-tenth of millet grain production. In the present study, only two households purchased the land. Under the increasing pressure of land use, the land title holders seem more likely to ask sedentary Fulani to return the land. Thus, Thiombiano suggested that the sedentary Fulani should conduct corralling in the fields of the Zarma land title holders as the cost of assurance of land renting, leading to benefits for both. Nevertheless, the Zarma land title holders, who are chief or seniors in villages in many cases, have usually large herds of livestock and can apply livestock excreta from their own herds. If this is the case, the contract corralling in the field of the land title holders might lead again to nutrient enrichment in the field of rich farmers, which in turn may enhance the discrimination between the rich and poor farmers. Hence, this proposal needs further refinement to enable the sedentary Fulani to conduct corralling in the fields of the Zarma farmers without livestock, who might be different from the land title holders but account for the majority of the farmers and seriously need the application of livestock excreta. The other nutrient management practices other than corralling should also be sought for them.

14.4 Farmer's Strategies for Replenishing Nutrient Resources in Slash-and-Burn Agriculture Under Different Climatic and Soil Conditions

Due to the sedentarization of previously nomadic people, benefits of corralling for crop production were found to be mostly limited to the fields managed by the sedentary Fulani. However, since some sedentary Fulani had the potential to conduct corralling in the fields of other farmers to promote efficient use of livestock excreta, a refined relationship between the farmers and the sedentary Fulani that could enhance the contract corralling practice should be realized.

Acknowledgment This chapter is based on Shinjo et al. (2008). This study was conducted as part of a collaborative research project between JIRCAS and ICRISAT on the "Improvement of Fertility of Sandy Soils in the Semi-Arid Zone of West Africa through Organic Matter Management." I thank the staff members of ICRISAT, especially Mr. Sodja Amadou, for their assistance during the investigation. I also express my gratitude to Dr. Lamourdia Thiombiano, FAO Regional Office for Africa, for the fruitful discussion.

References

D'Herbès JM, Valentin C (1997) Land surface conditions of the Niamey region: ecological and hydrological implications. J Hydrol 188–189:18–42
Fernández-Rivera S, TO Williams, Hiernaux P, Powell JN (1995) Faecal excretion by ruminants and manure availability for crop production in semi-arid West Africa. In Powell JM, Fernández-Rivera S, Williams TO, Renard C (eds) Livestock and sustainable nutrient cycling in mixed farming systems of sub-Saharan Africa. Volume II: Technical Papers. Proceedings of

an international conference held in Addis Ababa, Ethiopia, 22–26 November 1993 (ILCA (Addis Ababa) 149–169

Hiernaux, P, Ayantunde A (2004) The Fakara: a semi-arid agro-ecosystem under stress. International Livestock Research Institute (Niamey), p 95

ILRI (International Livestock Research Institute) (1995). Livestock Policy Analysis. ILRI Training Manual 2. ILRI (Nairobi), p 264

Matsunaga R, Singh BB, Moutari A, Tobita S, Hayashi K, Kamidohzono A (2006) Cowpea cultivation in the Sahelian region of West Africa: farmers' preferences and production constraints. Jpn J Trop Agric 50:208–214

Mokwunye AV, Hammond LL (1992) Myths and science of fertilizer use in the tropics. Lal R, Sanchez PA (eds) SSSA Special Publication No 29. (Madison, Wisconsin, USA), pp 121–135

Oyama S (2015) Combatting desertification in the Sahel of West Africa – utilization of waste for greening, hunger alleviation and conflict prevention. 315pp. Showado, Kyoto, Japan (in Japanese)

Powell JM, TO Williams (1995) An overview of mixed farming systems in sub-Saharan Africa. In: Powell JM, Fernández-Rivera S, Williams TO, Renard C (eds) Livestock and sustainable nutrient cycling in mixed farming systems of sub-Saharan Africa. Volume II: Technical Papers. Proceedings of an international conference held in Addis Ababa, Ethiopia, 22–26 November 1993 (ILCA (Addis Ababa), pp 21–36

Schlecht E, Hiernaux P, Achard F, Turner MD (2004) Livestock related nutrient budgets within village territories in western Niger. Nutr Cycl Agroecosyst 68:199–211

Shinjo H, Hayashi K, Abdoulaye T, Kosaki T (2008) Management of livestock excreta through corralling practice by sedentary pastoralists in the Sahelian region of West Africa – a case study in southwestern Niger. Trop Agric Dev 52:97–103

Possible Strategies for Controlling Nutrient Dynamics in Future Agricultural Activities in the Tropics

Chapter 15
Control of Wind Erosion, Loss of Soils, and Organic Matter Using the "Fallow Band System" in Semiarid Sandy Soils of the Sahel

Kenta Ikazaki

Abstract In the Sahel region of West Africa, wind erosion has a significant impact on soil nutrient dynamics. However, they had not been accurately evaluated because of the disadvantages of the conventional sediment samplers. Therefore, the Aeolian Materials Sampler (AMS) was developed to compensate for the disadvantages, and field scale nutrient flow by wind erosion was measured using the AMSs. It was revealed that (1) the annual loss of soil nitrogen from a cultivated field by wind erosion was two to three times greater than the annual absorption by the staple crop; (2) however, when there was a herbaceous fallow land (more than 5-m wide in the east-west direction) at the west side of the field, most of the blown soil nutrients were captured by the fallow land. On the basis of this high trapping capacity of the fallow land, "Fallow Band System" was designed to control nutrient dynamics as well as improve the crop production. A field experiment showed that (a) a single fallow band with a width of 5 m trapped 74 % of annual incoming soil particles and 58 % of annual incoming coarse organic matter (COM), suggesting that the "Fallow Band System" can control the nutrient flow by wind erosion and (b) owing to the enormous deposition of the trapped soil materials by the fallow band, soil nutrient status and hydraulic conductivity of the surface soil were greatly improved in the former fallow band, which resulted in the improved crop yield compared with the field without the system.

Keywords Desertification • Feasibility of technique • Nutrient and carbon cycling • Sub-Saharan Africa

K. Ikazaki (✉)
Japan International Research Center for Agricultural Sciences, 1-1 Ohwashi Tsukuba, Ibaraki 305-8686, Japan
e-mail: ikazaki@affrc.go.jp

© Springer Japan KK 2017
S. Funakawa (ed.), *Soils, Ecosystem Processes, and Agricultural Development*,
DOI 10.1007/978-4-431-56484-3_15

15.1 Development of a New Sediment Sampler to Measure the Flux of Soil Nutrients and Carbon by Wind Erosion

The Sahel region is located at the south fringe of the Sahara Desert with a mean annual rainfall of between 200 and 600 mm and includes parts of Senegal, Mauritania, Mali, Burkina Faso, Niger, and Chad, the least developed countries of the world. The rainfall in this region is concentrated from July to September, and almost no rainfall is observed from November to March. Pearl millet (*Pennisetum glaucum* (L.) R. Br.), as a staple food, is grown in the short rainy season with almost no chemical fertilizer and animal or mechanical traction. The parent materials of the Sahelian sandy soils are predominantly Aeolian deposits carried by the wind from the present Sahara Desert around the last glacial maximum (Kadomura 1992; approximately between 20,000 and 12,000 years ago). The soils, therefore, are mainly classified as Arenosols according to the World Reference Base for Soil Resources 2014 (International Union of Soil Sciences Working Group WRB 2015) and have low-soil nutrients and carbon (Manu et al. 1991). Also, they are prone to wind and water erosion because of the unstable structure.

The Sahel has been considered one of the most vulnerable areas to desertification (United Nations Environment Programme 1997). Desertification, in this region, is mainly caused by wind erosion (Lal 1993; Fig. 15.1) which reduces the soil nutrients (Daniel and Langham 1936; Zobeck and Fryrear 1986; Sterk et al. 1996; Larney et al. 1998; Bielders et al. 2002) and soil productivity (Lyles 1975; Larney et al. 1998) by the removal of fertile topsoil. Previous studies (Sterk et al. 1996; Bielders et al. 2002; Visser et al. 2005) estimated the soil nutrient or carbon flux associated with the movement of soil particles (SP) by using the Modified Wilson and Cooke (MWAC) sampler (Wilson and Cooke 1980) or the Big Spring Number Eight (BSNE) sampler (Fryrear 1986). However, the flux calculated using only the MWAC sampler or the BSNE sampler would not be reliable because of the following reasons: (1) Although large sediment flux occurs near the soil surface (Stout and Zobeck 1996; Rasmussen and Mikkelsen 1998), these samplers cannot measure the flux at the height of 0–0.05 m above the ground (their lowest trap is located at 0.05 m above the ground), and thus, the flux below 0.05 m has to be calculated by extrapolating the regression curve derived from the relationship between the flux and trap height above 0.05 m. (2) Previous studies did not take coarse organic matter (COM; Fig. 15.2) into account. This would be a problem because the COM could contain large proportion of the soil nutrients and carbon in the topsoil (preferentially blown part by the wind) of the Sahelian sandy soil (see 15.1.1 for more information).

In this context, the Aeolian Materials Sampler (AMS) (Ikazaki et al. 2009a; Fig. 15.3), which is a nonrotating sediment sampler that can measure the surface flux of SP and COM below 0.05 m, was designed with reference to the previous designs by Bagnold (1941), Greeley et al. (1982), Fryrear (1986), Stout and Fryrear (1989), Nickling and McKenna Neuman (1997), Bauer and Namikas (1998), and

Fig. 15.1 Erosive storm in the Sahel

Fig. 15.2 Coarse organic matter

Rasmussen and Mikkelsen (1998), and its performances were evaluated with wind tunnel experiments and field experiments. The COM flux above 0.05 m can be measured with the BSNE sampler owing to its near-isokinetic characteristic (Fryrear 1986; Shao et al. 1993; Goossens et al. 2000) and enough large inlet (the inlet of the MWAC sampler might be too small to measure the COM flux).

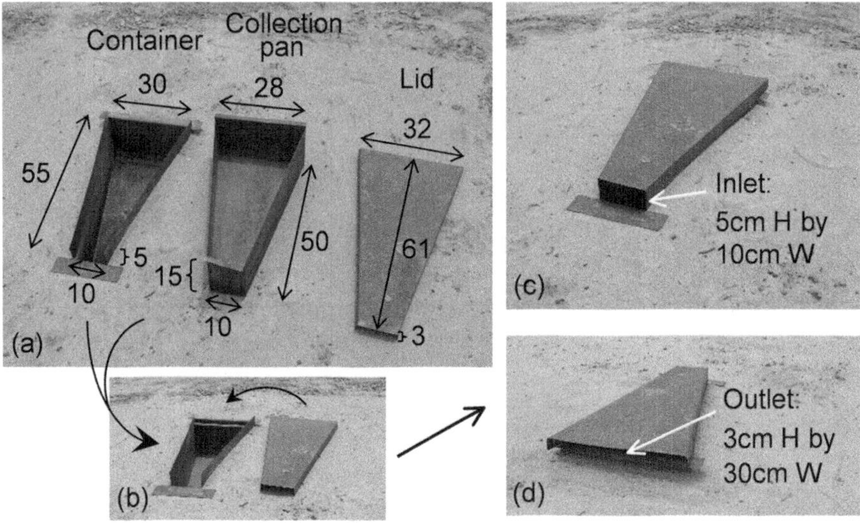

Fig. 15.3 Aeolian Materials Sampler, unit in (**a**) is centimeter

Wind tunnel experiments under the field conditions of the Sahel showed that the trapping efficiency (TE) of the AMS for the mass flux of SP (TEsp_mass) and COM (TEcom_mass) is

$$TE_{\text{sp\_mass}} = -2.40u_{2\text{m}} + [96.1 + 12.7\exp(-0.0523\theta)] \quad (0 \le \theta \le 45)$$
$$(R^2 = 0.84, P < 0.01) \tag{15.1}$$

$$TE_{com\_\text{mass}} = 61.0 + 18.0\exp(-0.0614\theta) \quad (0 \le \theta \le 45)$$
$$(R^2 = 0.94, P < 0.01) \tag{15.2}$$

where TE_{mass} is the trapping efficiency for the mass flux (%), $u_{2\text{m}}$ is the wind velocity at 2-m height (m s^{-1}), and θ is the incident wind angle to the AMS (°) (Ikazaki et al. 2009a). The $TE_{\text{sp\_mass}}$ and $TE_{\text{com\_mass}}$ were determined separately since the sensitivities of SP and COM to airflow are different. Although both $TE_{\text{sp\_mass}}$ and $TE_{\text{com\_mass}}$ are not 100 %, Eqs. 15.1 and 15.2 suggested that they can be easily and accurately calibrated using wind data to reliably estimate the mass flux at the height of 0–0.05 m in the Sahel.

Another wind tunnel experiment conducted by Ikazaki et al. (2011a) showed that (a) the TEs of the AMS for the total nitrogen (TN) and total carbon (TC) contents of SP ($TE_{\text{sp\_tn}}$ and $TE_{\text{sp\_tc}}$, respectively) are constant (72 % and 65 %, respectively, on average), irrespective of the $TE_{\text{sp\_mass}}$, wind velocity, and wind direction, and (b) those of COM ($TE_{\text{com\_tn}}$ and $TE_{\text{com\_tc}}$) are also constant (99 % and 98 %, respectively, on average), irrespective of the $TE_{\text{com\_mass}}$, wind velocity, and wind direction. In the study by Ikazaki et al. (2011a), soil nitrogen was assessed because it is a severe limiting factor in crop production in the Sahel (Bacci et al. 1999; Bationo and Ntare 2000; Pandey et al. 2001).

On the basis of the abovementioned result (a) and Eq. 15.1, it was found that the surface flux of TN and TC associated with the movement of SP at the height of 0–0.05 m can be estimated by using the following formulae:

$$F_{\text{sp\_tn}} = \left[M_{\text{sp}}/\left(TE_{\text{sp\_mass}}/100\right)\right] \times \left[TN_{\text{sp}}/(72/100)\right] \tag{15.3}$$

$$F_{\text{sp\_tc}} = \left[M_{\text{sp}}/\left(TE_{\text{sp\_mass}}/100\right)\right] \times \left[TC_{\text{sp}}/(65/100)\right] \tag{15.4}$$

where $F_{\text{sp\_tn}}$ and $F_{\text{sp\_tc}}$ are the flux of TN and TC by the movement of SP (g m^{-2} s^{-1}), M_{sp} is the SP mass flux determined using the AMS (g m^{-2} s^{-1}), and TN_{sp} and TC_{sp} are the TN and TC contents of the trapped SP by the AMS (g g^{-1}). Similarly, the surface flux of TN and TC associated with the movement of COM below 0.05 m can be estimated by using the following formulae:

$$F_{\text{com\_tn}} = \left[M_{\text{com}}/(TE_{\text{com\_mass}}/100)\right] \times \left[TN_{\text{com}}/(99/100)\right] \tag{15.5}$$

$$F_{\text{com\_tc}} = \left[M_{\text{com}}/(TE_{\text{com\_mass}}/100)\right] \times \left[TC_{\text{com}}/(98/100)\right] \tag{15.6}$$

where $F_{\text{com\_tn}}$ and $F_{\text{com\_tc}}$ are the flux of TN and TC by the movement of COM (g m^{-2} s^{-1}), M_{com} is the COM mass flux determined using the AMS (g m^{-2} s^{-1}), and TN_{com} and TC_{com} are the TN and TC contents of the trapped COM by the AMS (g g^{-1}).

Field experiments at the International Crops Research Institute for the Semi-Arid Tropics (ICRISAT) West and Central Africa in Niger (IWCA, 13° 14′ N, 2°17′ E, 235 m above sea level) were also conducted by Ikazaki et al. (2010, 2011a) for evaluating the performance of the AMS under the actual field conditions of the Sahel. Ikazaki et al. (2011a) found that the mass transport rate of SP, SP loss from a cultivated field, and SP gain in a fallow land during erosion events cannot be accurately estimated without measuring the mass flux at the height of 0–0.05 m using the AMS. Ikazaki et al. (2010) also reported that the COM loss from the cultivated field calculated by employing only the BSNE sampler was not consistent with the actual value, but that calculated by using the combination of AMS and BSNE samplers agreed with the actual value.

15.1.1 Coarse Organic Matter (COM)

In Sect. 15.1, COM is defined as free organic debris with particle diameter greater than 200 μm and mainly consisting of plant residues and SP as the remaining soil components with reference to Ikazaki et al. (2009a). COM is conceptually similar to the free light fraction (free LF) in Six et al. (1998) that is particulate organic matter existing between aggregates. The difference between COM and free LF is that COM includes free organic debris >2.0 mm in diameter, but free LF does not. In the Sahel, the potential soil loss from a cultivated field by wind erosion was estimated to be 5.0 mm year^{-1} (Bielders et al. 2000), and in the 5.0-mm-thick topsoil, COM accounted for 35–52 % of the total carbon and 22–34 % of the total nitrogen (Ikazaki et al. 2009a).

15.2 Field Estimation of Movements of Soil Nitrogen and Carbon During Wind Erosion Events

Ikazaki et al. (2011b, 2012) estimated the movements of SP and COM during wind erosion events by using the combination of AMS and BSNE samplers. They carried out 3-year field experiments at the ICRISAT West and Central Africa and revealed the following points:

1. The annual loss of SP from the cultivated field was 60–80 Mg ha^{-1} year^{-1} (assuming that the bulk density is 1.6 Mg m^{-3}, it corresponds to 4–5 mm year^{-1}). It is similar to the potential soil loss (79 Mg ha^{-1}) estimated by Bielders et al. (2000). However, it is larger than the annual SP loss reported by Sterk and Stein (1997) and Bielders et al. (2002) and the proposed values (2.5–12.4 Mg ha^{-1} year^{-1}) by Renard et al. (1997) above which an economic impact on crop production is highly expected.
2. The annual loss of COM from the cultivated field was 0.3–1.3 Mg ha^{-1} year^{-1}, which was 50–75 % of the initial amount of COM that existed in the loose surface soil before the first wind erosion event in each year. These high proportions can be explained by the fact that COM is generally not mixed into the soil by the Sahelian farmers who grow pearl millet with no tillage and spot sowing.
3. The annual loss of soil nitrogen from the cultivated field associated with the movements of SP and COM was roughly 25–40 kg ha^{-1} and 5–20 kg ha^{-1}, respectively. The total annual nitrogen loss was 40–50 kg ha^{-1}, which is approximately 2–3 times as much as the annual nitrogen uptake by pearl millet, thereby implying that wind erosion greatly affects the soil nutrient status in the Sahel.
4. The loss of soil carbon from the cultivated field associated with the movements of SP and COM was roughly 300–400 kg ha^{-1} and 100–400 kg ha^{-1}, respectively. The total annual carbon loss was 500–700 kg ha^{-1}.

From the results (3) and (4), it was shown that the loss of soil nitrogen and carbon associated with the movement of COM is not negligible as expected and should be taken into account in the Sahel, especially in a cultivated field where a large amount of COM exists in the loose surface soil (e.g., fields with manure application).

As described above, a huge amount of SP, COM, and associated soil nitrogen and carbon was lost from the cultivated fields. Now, the question is where these windblown materials go. Ikazaki et al. (2011b, 2012) measured the flux of SP, COM, and associated soil nitrogen and carbon both in a cultivated field (source) and fallow land (sink) and found that most of them was trapped by the 5-m-wide fallow land (i.e., most of them settled into the fallow land within the first 5 m from the boundary between the field and fallow). The TEs of the 5-m-wide fallow land for the incoming SP from the cultivated field in 2005, 2006, and 2007 were estimated to be 93 %, 79 %, and 77 % and those for the incoming COM were 97 %, 87 %, and 81 %, respectively. These efficiencies for the incoming SP were much higher than that reported by Bielders et al. (2002), though the results of both studies cannot be directly compared because (1) Bielders et al. (2002) did not mention the biomass and vegetation coverage rate of the fallow land, and (2) methods of estimating the SP flux in the fallow land in both studies are not the same. The results in Ikazaki et al. (2011b, 2012) suggested that the SP, COM, and associated soil nitrogen and carbon are not lost but localized when viewed on a larger scale (e.g., a village scale).

15.3   Development of a New Practice for Controlling the Loss of Soil Nutrients and Carbon by Wind Erosion

Wind erosion in the Sahel can be controlled by reducing wind velocity at the soil surface with ridging or windbreaks, or by creating sufficient resistances to wind forces on the soil surface with mulching with post-harvest crop residue. Bielders et al. (2000) described that the 0.2-m-high ridges perpendicular to the prevailing erosive wind direction (east) reduced soil loss in a cultivated field by 57 % at maximum compared with the control plots (no ridged). However, most Sahelian farmers do not possess sufficient labor (Matlon 1987; Nagy et al. 1988; Ruthven and David 1995) and animal traction (Nagy et al. 1988; Baidu-Forson and Renard 1996; Hiernaux and Ayantunde 2004) to employ this method. The effect of windbreaks on wind erosion control in the Sahel has been also studied. Michels et al. (1998) reported that a windbreak of *Andropogon gayanus* Kunth and *Bauhinia rufescens* Lam. reduced annual soil flux by 6–55 % and 47–77 %, respectively. However, windbreaks, as a countermeasure against wind erosion, have not been adopted by Sahelian farmers (Bielders et al. 2001). According to Bielders et al. (2001), the reasons for this non-adoption could be lack of plant material, lack of training and development project support, reduction in cropping area, land tenure problems (ownership of trees), and the need for protection against grazing livestock. The effect of mulching with post-harvest millet residue on wind erosion control in the Sahel has been reported by many authors (i.e., Geiger et al. 1992; Michels et al. 1995; Lamers et al. 1995; Sterk and Spaan 1997; Buerkert and Lamers 1999; Bielders et al. 2000). To put it briefly, mulching with 1500–2000 kg ha^{-1} of pearl millet residue reduced the soil flux by 46–64 % compared with the control plots. In contrast with that, mulching with 500–1000 kg ha^{-1} of post-harvest millet was not always effective in reducing the soil flux. However, mulching with 1500–2000 kg ha^{-1} of pearl millet residue is not applicable for the Sahelian farmers. This is because (1) the productivity of the Sahelian sandy soils is too low, and the biomass of millet stem and leaf is sometimes less than 1500–2000 kg ha^{-1}. (2) Pearl millet residue is generally used extensively as animal feed, fuel, and constructing material (Lamers and Feil 1993; Michels et al. 1995; Schlecht and Buerkert 2004). As described above, effective countermeasures against wind erosion have been proposed, but none of them have been adopted by Sahelian farmers (Rinaudo 1996) since they would not be practical for the Sahelian farmers who are tight in terms of cash and labor (Matlon 1987; Nagy et al. 1988; Ruthven and David 1995). Therefore, Ikazaki et al. (2011c) designed a new practice for controlling the loss of soil nutrients and carbon by wind erosion on the basis of the results in Sect. 15.2 and termed it "Fallow Band System." In designing the "Fallow Band System," the practicality for Sahelian farmers was made the first priority.

The outline of the "Fallow Band System," a shifting herbaceous windbreak, is illustrated in Fig. 15.4 which shows a temporal sequence of land use in this system. (A) Because erosive storms, harmattan winds in the dry season, and convective storms in the early rainy season (Sterk 2003) are mainly east wind, 5-m-wide herbaceous fallow bands are established in the first year in a north-south direction (a right angle to the erosive wind direction) in a cultivated field during the rainy season. It is notable that fallow bands can be easily created by skipping the usual seeding and weeding (by doing nothing). Therefore, this system does not impose any additional expenses and labor requirements on Sahelian farmers who are economically challenged and have limited manpower (Matlon 1987; Nagy et al. 1988; Ruthven and David 1995), and thus, the practicality of this system for Sahelian farmers will be high. The remainder of the field is cultivated using a conventional method and called "cultivated band." Fallow bands are maintained during the subsequent dry season and early rainy season, so that they will capture windblown SP, COM, and associated soil nutrients and carbon. (B) In the next rainy season, new fallow bands are arranged windward side of the former fallow bands, on which in turn crops are cultivated as well as in the cultivated bands. (C) Step (B) is repeated every year.

According to Ikazaki et al. (2011c), the "Fallow Band System" will not only control wind erosion but also improve soil chemical and physical properties, and thus crop production. Soil chemical properties will be improved owing to the settlement of windblown materials (SP and COM), which contain considerable amount of soil nutrients and carbon, in the former fallow bands. Buerkert and Lamers (1999) reported that the windblown soil materials trapped by polyethylene plastic tubes placed in a cultivated field improved soil chemical properties and promoted early-season crop growth.

Soil physical properties can be improved by covering the crust layer(s) of the topsoil with the trapped SP and COM. Figure 15.5 shows a topsoil of the Sahelian sandy soils. As described by Rajot et al. (2003) and Ikazaki (2015), a loose sand layer (a few centimeters thick in a well-managed field) with high hydraulic conductivity and high-nutrient contents lies on top of a crust layer(s) with low

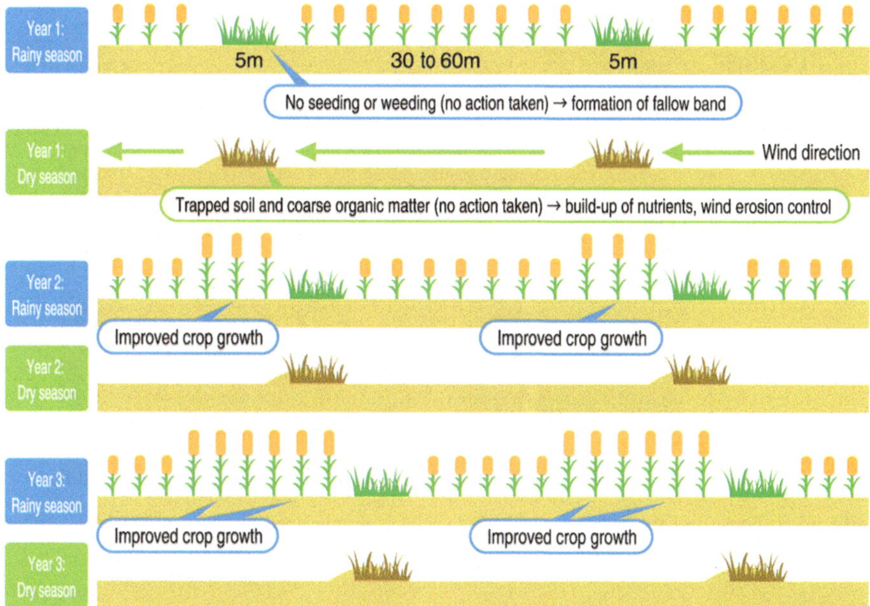

Fig. 15.4 Outline of the "Fallow Band System" (partly modified from Global Environmental Forum 2013)

hydraulic conductivity and low soil nutrient contents. As wind erosion progresses, the loose sand layer will be removed, and consequently, the crust layer(s) will be exposed as in Fig. 15.5, resulting in the sharp decrease in soil permeability. In the "Fallow Band System," it is highly possible that covering the exposed crust layer(s) with the trapped SP and COM increases the hydraulic conductivity of the surface soil and improves the rainfall infiltration into the soil. This hypothesis is supported by the results in Buerkert and Lamers (1999) and Ribolzi et al. (2006) that soil water content or soil permeability was increased in the surface soil where windblown soil materials had settled.

The "Fallow Band System" has a potential of decrease in crop yield because of the reduction in cropping area (if 5-m-wide fallow bands are created at an interval of 45 m, then the reduction in cropping area is 10 %). However, increased crop yield that will be resulted from the improvement of soil chemical and physical properties in the former fallow bands can be more than offset of this potential decrease in crop yield of the system.

Ikazaki et al. (2011c) evaluated the effectiveness of the "Fallow Band System" and showed that (1) a single fallow band trapped 74 % of annual incoming SP and 58 % of annual incoming COM. The TE for the incoming SP (74 %) is similar to that of a windbreak of 0.6-m-high and 5-m-wide fallow vegetation (70 %) reported by Banzhaf et al. (1992) and higher than that of the surface mulching (46–64 %). Interestingly, the TEs for the annual incoming SP and COM (74 % and 58 %,

Fig. 15.5 Topsoil of the Sahelian sandy soil, crust is emphasized by removing skeleton grains (coarse particles) between dense plasmic (fine particle) layers

respectively) in Ikazaki et al. (2011c) were relatively lower than those in Ikazaki et al. (2012) (93 %, 79 %, and 77 % for SP and 97 %, 87 %, and 81 % for COM, as mentioned above). This might be because the biomass of the fallow band in Ikazaki et al. (2011c) was lower than that in Ikazaki et al. (2012). (2) The total mass of the SP trapped by a single fallow band was 1830 Mg ha^{-1} (expressed in a hectare of a fallow band), containing 910 kg ha^{-1} of TN and 8.7 Mg ha^{-1} of TC (C:N ratio was 9.5). That of the COM was 4.0 Mg ha^{-1}, containing 60 kg ha^{-1} of TN and 1.3 Mg ha^{-1} of TC (C:N ratio was 21.5). Owing to the enormous deposition of the SP and COM on the fallow band, total soil respiration (root respiration was excluded) in the former fallow band was much higher than that in the cultivated band (mainly due to the additional decomposition of the trapped COM), suggesting that soil nutrient status in the former fallow band was greatly improved. (3) The hydraulic conductivity of surface soil of the former fallow band was increased mainly by covering the crust layer(s) with the trapped SP and COM of which hydraulic conductivity is high. This resulted in higher water content in the surface soil of the former fallow band than that of the cultivated band, and therefore, the amount of soil water available for crops was increased in the former fallow band. From these results (1–3), it was concluded in Ikazaki et al. (2011c) that the "Fallow Band System" can be useful for controlling wind erosion and improving soil fertility in the sandy soil of the Sahel.

The effect of the "Fallow Band System" on the crop yield was reported in Ikazaki et al. (2009b). It was found that (a) the crop yield was maximized when the interval of the fallow bands was 30–36 m in most cases. (b) Although the decrease in crop yield in the first year is inevitable, the crop yield from the entire field with the "Fallow Band System" can be increased by 36 to 81 % (ten-year average) compared with the field without the system when the fallow bands are created at an interval of 30–36 m. Therefore, it was concluded that the "Fallow Band System" can be useful for improving crop yield, which can be a strong incentive for the Sahelian farmers for practicing the system (the "Fallow Band System" was disseminated by a project of the Japan International Corporation Association (JICA) and adopted by about 500 farmers from 89 villages, 23 districts, 5 regions in Niger).

According to Ikazaki (2015), the "Fallow Band System" could be improved by introducing *Chamaecrista mimosoides* (L.) E. Greene into the fallow bands. *Chamaecrista mimosoides*, one of the leguminous plants fixing atmospheric nitrogen (Tobita et al. 2011), is not easily edible for livestock and has a hard stem that resists to physical damage caused by windblown SP. Therefore, it is possible that *C. mimosoides* can be used to strengthen both effects of wind erosion control and yield improvement in the system.

References

Bacci L, Cantini C, Pierini F et al (1999) Effects of sowing date and nitrogen fertilization on growth, development and yield of a short day cultivar of millet (*Pennisetum glaucum* L.) in Mali. Eur J Agron 10:9–21

Bagnold RA (1941) The physics of blown sand and desert dunes. Mathuen, London

Baidu-Forson J, Renard C (1996) Comparing productivity of millet-based cropping systems in unstable environments of the Sahel: possibilities and challenges. Agric Syst 51:85–95

Banzhaf J, Leihner DE, Buerkert A et al (1992) Soil tillage and windbreak effects on millet and cowpea: I. Wind speed, evaporation, and wind erosion. Agron J 84:1056–1060

Bationo A, Ntare BR (2000) Rotation and nitrogen fertilizer effects on pearl millet, cowpea and groundnut yield and soil chemical properties in a sandy soil in the semi-arid tropics, West Africa. J Agric Sci 134:277–284

Bauer BO, Namikas S (1998) Design and field test of a continuously weighing, tipping-bucket assembly for aeolian sand traps. Earth Surf Process Landf 23:1177–1183

Bielders CL, Michels K, Rajot JL (2000) On-farm evaluation of ridging and residue management practices to reduce wind erosion in Niger. Soil Sci Soc Am J 64:1776–1785

Bielders CL, Alvey S, Cronyn N (2001) Wind erosion: the perspective of grass-roots communities in the Sahel. Land Degrad Dev 12:57–70

Bielders CL, Rajot JL, Amadou M (2002) Transport of soil and nutrients by wind in bush fallow land and traditionally managed cultivated fields in the Sahel. Geoderma 109:19–39

Buerkert A, Lamers JPA (1999) Soil erosion and deposition effects on surface characteristics and pearl millet growth in the West African Sahel. Plant Soil 215:239–253

Daniel HA, Langham WH (1936) The effect of wind erosion and cultivation on the total nitrogen and organic matter content of soils in the Southern High Plains. J Am Soc Agron 28:587–596

Fryrear DW (1986) A field dust sampler. J Soil Water Conserv 41:117–120

Geiger SC, Manu A, Bationo A (1992) Changes in a sandy Sahelian soil following crop residue and fertilizer additions. Soil Sci Soc Am J 56:172–177

Global Environmental Forum (2013) Lifestyles and measures against desertification. Ministry of the Environment of Japan, Tokyo https://www.env.go.jp/en/nature/desert/download/against%20desertification_eng.pdf. Accessed 29 Nov 2016

Goossens D, Offer Z, London G (2000) Wind tunnel and field calibration of five aeolian sand traps. Geomorphology 35:233–252

Greeley R, Leach RN, Williams SH et al (1982) Rate of wind abrasion on Mars. J Geophys Res 87 (B12):10009–10024

Hiernaux P, Ayantunde A (2004) The Fakara: a semi-arid agro-ecosystem under stress. International Livestock Research Institute, Nairobi. http://www.jircas.affrc.go.jp/project/africa_dojo/Metadata/grad_research/15.pdf. Accessed 29 Nov 2016

Ikazaki K (2015) Desertification and a new countermeasure in the Sahel, West Africa. Soil Sci Plant Nutr 61:372–383

Ikazaki K, Shinjo H, Tanaka U et al (2009a) Sediment catcher to trap coarse organic matter and soil particles transported by wind. Trans ASABE 52:487–492

Ikazaki K, Shinjo H, Tanaka U et al (2009b) The "Fallow Band" system: a proposed new low-input technology option contributing both to the prevention of desertification and the increase of pearl millet yield in the Sahel, West Africa. JIRCAS Res. Highlights 16. http://www.jircas.affrc.go.jp/english/publication/highlights/2008/2008_10.html. Accessed 29 Nov 2016

Ikazaki K, Shinjo H, Tanaka U et al (2010) Performance of aeolian materials sampler for the determination of amount of coarse organic matter transported during wind erosion events in Sahel, West Africa. Pedologist 53:126–134

Ikazaki K, Shinjo H, Tanaka U et al (2011a) Aeolian materials sampler for measuring surface flux of soil nitrogen and carbon during wind erosion events in the Sahel, West Africa. Trans ASABE 54:983–990

Ikazaki K, Shinjo H, Tanaka U et al (2011b) Field-scale soil particle movement during wind erosion in the Sahel, West Africa. Soil Sci Soc Am J 75:1885–1897

Ikazaki K, Shinjo H, Tanaka U et al (2011c) "Fallow band system," a land management practice for controlling desertification and improving crop production in the Sahel, West Africa: 1. Effectiveness in desertification control and soil fertility improvement. Soil Sci Plant Nutr 57:573–586

Ikazaki K, Shinjo H, Tanaka U et al (2012) Soil and nutrient loss from a cultivated field during wind erosion events in the Sahel, West Africa. Pedologist 55:355–363

IUSS Working Group WRB (2015) World reference base for soil resources 2014, update 2015. World Soil Resources Reports, No. 106. FAO, Rome

Kadomura H (1992) The Sahara. In: Kadomura H, Katsumata M (eds) On the shores of the Sahara: the nature and people of the Sahel. TOTO Publishers, Tokyo, pp 36–43 (Japanese translation and reorganization of Rognon P (1985) Désert et désertification. Total Inform 100:4–10)

Lal R (1993) Soil erosion and conservation in West Africa. In: Pimentel D (ed) World soil erosion and conservation. Cambridge University Press, Cambridge, pp 7–25

Lamers JPA, Feil RP (1993) The many uses of millet residues. ILEIA Newsletter 9:15 http://www.agriculturesnetwork.org/magazines/global/after-harvest/the-many-uses-of-millet-residues/at_download/article_pdf. Accessed 29 Nov 2016

Lamers JPA, Michels K, Feil PR (1995) Wind erosion control using windbreaks and crop residues: local knowledge and experimental results. Der Tropenlandwirt 96:87–96

Larney FJ, Bullock MS, Janzen HH et al (1998) Wind erosion effects on nutrient redistribution and soil productivity. J Soil Water Conserv 53:133–140

Lyles L (1975) Possible effects of wind erosion on soil productivity. J Soil Water Conserv 30:279–283

Manu A, Bationo A, Geiger SC (1991) Fertility status of selected millet producing soils of West Africa. Soil Sci 152:315–320

Matlon P (1987) The West African semiarid tropics. In: Mellor JW, Delgado CL, Blackie MJ (eds) Accelerating food production in Sub-Saharan Africa. Johns Hopkins University Press, Baltimore, pp 59–77

Michels K, Lamers JPA, Buerkert A (1998) Effects of windbreak species and mulching on wind erosion and millet yield in the Sahel. Exp Agric 34:449–464

Michels K, Sivakumar MVK, Allison BE (1995) Wind erosion control using crop residue: I. Effects on soil flux and soil properties. Field Crop Res 40:101–110

Nagy JG, Sanders JH, Ohm HW (1988) Cereal technology interventions for the West African semi-arid tropics. Agric Econ 2:197–208

Nickling WG, McKenna Neuman C (1997) Wind tunnel evaluation of a wedge-shaped aeolian sediment trap. Geomorphology 18:333–345

Pandey RK, Maranville JW, Bako Y (2001) Nitrogen fertilizer response and use efficiency for three cereal crops in Niger. Commun Soil Sci Plant Anal 32:1465–1482

Rajot JL, Alfaro SC, Gomes L et al (2003) Soil crusting on sandy soils and its influence on wind erosion. Catena 53:1–16

Rasmussen KR, Mikkelsen HE (1998) On the efficiency of vertical array aeolian field traps. Sedimentology 45:789–800

Renard KG, Foster GR, Weesies GA et al (1997) Soil-loss tolerance. In: Predicting soil erosion by water: a guide to conservation planning with the revised universal soil loss equation (RUSLE). USDA Agric. Handb. 703. US Government Printing Office, Washington DC, pp 12–13

Ribolzi O, Hermida M, Karambiri H et al (2006) Effects of aeolian processes on water infiltration in sandy Sahelian rangeland in Burkina Faso. Catena 67:145–154

Rinaudo T (1996) Tailoring wind erosion control methods to farmers' specific needs. In: Buerkert B, Allison BE, Von Oppen M (eds) Wind erosion in West Africa: the problem and its control. Margraf Verlag, Weikersheim, pp 161–171

Ruthven O, David D (1995) Benefits and burdens: researching the consequences of migration in the Sahel. IDS Bull 26:47–53

Schlecht E, Buerkert A (2004) Organic inputs and farmers' management strategies in millet fields of western Niger. Geoderma 121:271–289

Shao Y, McTainsh GH, Leys JF et al (1993) Efficiency of sediment samplers for wind erosion measurement. Aust J Soil Res 31:519–532

Six J, Elliott ET, Paustian K et al (1998) Aggregation and soil organic matter accumulation in cultivated and native grassland soils. Soil Sci Soc Am J 62:1367–1377

Sterk G (2003) Causes, consequences and control of wind erosion in Sahelian Africa: a review. Land Degrad Dev 14:95–108

Sterk G, Spaan WP (1997) Wind erosion control with crop residues in the Sahel. Soil Sci Soc Am J 61:911–917

Sterk G, Stein A (1997) Mapping wind-blown mass transport by modeling variability in space and time. Soil Sci Soc Am J 61:232–239

Sterk G, Herrmann L, Bationo A (1996) Wind-blown nutrient transport and soil productivity changes in southwest Niger. Land Degrad Dev 7:325–335

Stout JE, Fryrear DW (1989) Performance of a windblown-particle sampler. Trans ASAE 32:2041–2045

Stout JE, Zobeck TM (1996) The Wolfforth field experiment: a wind erosion study. Soil Sci 161:616–632

Tobita S, Shinjo H, Hayashi K et al (2011) Identification of plant genetic resources with high potential contribution to the soil fertility enhancement in the Sahel—with special interest in fallow vegetation. In: Bationo A, Waswa B, Okeyo JM et al (eds) Innovations as key to the green revolution in Africa: exploring the scientific facts. Springer, Berlin, pp 701–706

United Nations Environment Programme (UNEP) (1997) Middleton N, Thomas D (eds) World atlas of desertification, 2nd edn. Arnold, London

Visser SM, Stroosnijder L, Chardon WJ (2005) Nutrient losses by wind and water, measurements and modeling. Catena 63:1–22

Wilson SJ, Cooke RU (1980) Wind erosion. In: Kirkby MJ, Morgan RPC (eds) Soil erosion. Wiley, Chichester, pp 217–251

Zobeck TM, Fryrear DW (1986) Chemical and physical characteristics of windblown sediment: II. Chemical characteristics and total soil and nutrient discharge. Trans ASAE 29:1037–1041

Chapter 16
Process of Runoff Generation at Different Cultivated Sloping Sites in North and Northeast Thailand

Shinya Funakawa

Abstract The conditions for surface runoff generation, which is usually strongly related to the process of soil erosion, were analyzed in three plots at different cultivated sloping sites in North and Northeast Thailand using a runoff gauge connected to a data logger. In most of the cases, the rainfall intensity and the surface soil moisture contributed significantly to surface runoff generation. The rainfall intensity in the Khon Kaen plot on sandy soils was higher than that in the other two plots with fine-textured soils in the northern region, and surface runoff occasionally occurred throughout the rainy season with no clear seasonal trend, unlike in the other two plots, where surface runoff occurred more often during the latter half of the rainy season due to the higher rainfall intensity and/or capillary saturation of surface soils. The proportion of the surface runoff generated in relation to the amount of rainfall increased with the increase of the slope gradient of the plots. The proportion of the amount of soil erosion in relation to the amount of surface runoff was, however, the largest in the sandy plot of Khon Kaen with the lowest slope gradient, indicating that the sandy soils were more easily eroded than the clayey soils presumably due to the weakly organized structure of the soil aggregates. Therefore, the conditions that enhance the risk of surface runoff and soil erosion were found to vary and should be taken into account for agricultural management in the respective regions.

Keywords Runoff gauge • Rainfall intensity • Soil moisture • Surface runoff • Soil texture

S. Funakawa (✉)
Graduate School of Global Environmental Studies, Kyoto University, Yoshida-Honmachi, Sakyo-ku, Kyoto 606-8501, Japan
e-mail: funakawa@kais.kyoto-u.ac.jp

© Springer Japan KK 2017
S. Funakawa (ed.), *Soils, Ecosystem Processes, and Agricultural Development*,
DOI 10.1007/978-4-431-56484-3_16

16.1   Introduction

The recent economic development in Thailand has resulted in a drastic change of agricultural production systems, especially in upland cropping areas. Land management strategies that can provide long-term sustainability of agricultural production should be adopted. In the present study, we investigated the process of surface runoff generation, which is a driving force of soil erosion, under different conditions in North and Northeast Thailand in order to examine the possibility of decreasing the soil erosion risk.

In the mountainous region of North Thailand, shifting cultivation, which has often been referred to as "slash-and-burn" or "swidden agriculture," has long been one of the main production systems of subsistence agriculture (Kunstadter and Chapman 1978, Chap. 11 in this volume). However, due to the improvement of the transport infrastructure and the increase in population, traditional shifting cultivation with an adequate long fallow period has been replaced by more intensive cropping with a short fallow period or by almost continuous cultivation (Kyuma 2001). Since the agricultural land in this area is mostly opened on steep hill slopes, erosion control would be one of the most serious problems to be solved during the intensification of agriculture.

On the other hand, the agroecological conditions of Northeast Thailand are characterized by a wide distribution of sandy soils, irregular occurrence of rainfall events during the rainy season, and undulating topography with relatively gentle slopes. These conditions might traditionally have been unfavorable for agricultural development in the area. Presently, due to the development of the transport infrastructure and promotion of domestic/international market economy, market-oriented agriculture is being widely adopted, and plantations of sugarcane and/or cassava, as commercial crops on the middle to upper slopes, are widely observed in Northeast Thailand.

The intensification of land use, which is commonly observed in North and Northeast Thailand, may lead to serious soil degradation problems, especially on sloping land, if appropriate technology is not applied. One of the most serious problems after forest clearing for agricultural production is soil erosion, whereby the fertility of surface soils decreased within a short period of time. Several factors can affect the intensity of soil erosion, i.e., rainfall intensity, length of slopes, slope gradient, and surface coverage (Wischmeier and Smith 1978; Sonneveld and Nearing 2003). In the present study, we investigated the conditions conductive to surface runoff generation, which can be closely related to soil erosion, in three plots at different cultivated sloping sites in North and Northeast Thailand, with special reference to meteorological factors such as rainfall intensity and/or soil conditions.

16.2   Description of Study Sites and Research Methodology

Three experimental plots were set up. The first consisted of a sugarcane field on sandy soils in Ban Sam Jan (SJ), Khon Kaen Province, Northeast Thailand. The soil was classified into Typic Ustipsamments, according to soil taxonomy (Soil Survey Staff 2014). Sugarcane (*Saccharum officinarum* L.) was planted during the year of the experiment (2002). Detailed information on soils and landscape were previously reported (Funakawa et al. 2006a). The slope gradient was about 5 %. The second plot was set up in Ban Nam Rin (NR), upper northern part of Mae Hong Son Province, in which marketed vegetables were cultivated at a moderately high elevation (800 m). The soils in this area were derived from limestone and, hence, not strongly acidic and suitable for the cultivation of annual crops. They were classified into Udic Haplustalfs, according to US soil taxonomy (Soil Survey Staff 2014). The slope gradient of the NR plot was approximately 35 % (20 °). The soil surface was kept bare during the experiment (2002). The last plot was set up in Ban Du La Poe (DP), Mae Hong Son Province, where Karen people practice traditional shifting cultivation in this mountainous region (1200 m above sea level). The soils were derived from fine-textured sedimentary rocks and classified into Ustic Haplohumults. Detailed information on the soils and agriculture was given in our previous report (Funakawa et al. 2006b, Chapter 11 in this volume). The slope gradient of the DP plot was about 60 % (30 °). Upland rice (*Oryza sativa* L.) was planted after reclamation of fallow forest (7 years) during the study period (2001).

One 2.5 m × 2.5 m plot at the SJ site and duplicate 1 m × 1 m plots at the NR and DP sites were set up and surrounded by stainless plates on the upper border and both sides to prevent runoff water from penetrating into the plots from outside (Fig. 16.1a). Then the water budget in the small square area of the plots was calculated at 10-min intervals, based on the measured values of rainfall (using a tipping bucket rain gauge), volumetric soil moisture content in 0–15 and 15–45 cm layers of soil (using TDR probes), and the amount of surface runoff, which were recorded on a data logger (CR10X, Campbell Co. Ltd.). The amount of surface runoff was measured using a handmade runoff gauge, which was designed to count a switch closure for each bucket tip in the same way as a tipping bucket rain gauge (Fig. 16.1b).

Soil physical and physicochemical properties were determined using undisturbed core samples and air-dried soil samples (< 2 mm), respectively. The eroded soils trapped in the bucket just below the runoff gauge were collected after the end of the experiment, and the total amount of soil erosion throughout the rainy season was determined in each plot.

A statistical software SYSTAT 8.0 was used for the statistical analysis (SPSS Inc. 1998).

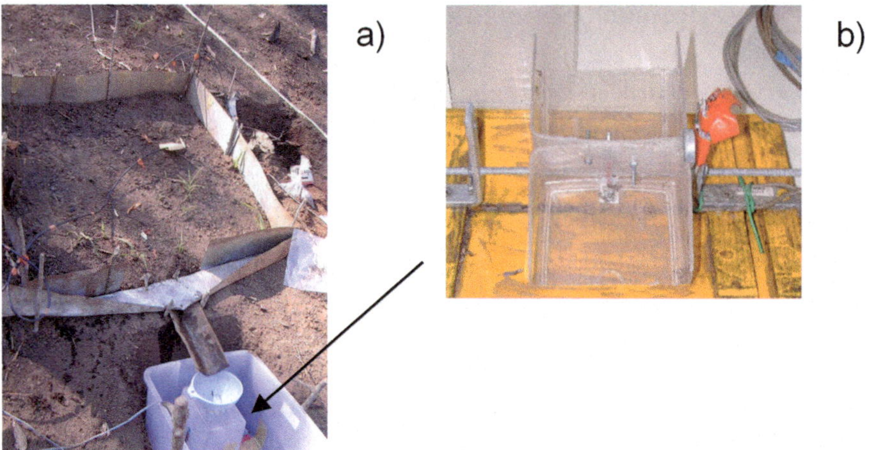

Fig. 16.1 Experimental plot, DP (**a**) and handmade runoff gauge (**b**)

16.3 General Soil Properties and Rainfall Characteristics of the Study Plots

Table 16.1 lists selected physical and physicochemical properties of the soils studied. The soils in the SJ plot were characterized by a sandy texture, low organic matter content, high bulk density, and lower hydraulic conductivity with a low total porosity below 0.5 L L^{-1}, compared to those in the NR and DP plots. It should be noted that the moisture content at field capacity (-6 kPa or pF 1.8) was approximately 0.4 L L^{-1} in the clayey soils (NR and DP plots), whereas 0.25 L L^{-1} in the sandy soils of the SJ plot.

Figure 16.2 shows the daily rainfall and fluctuations of soil moisture contents in the study plots during the experiment. During a certain period of time, the data were missing due to malfunction of the data logger. Even in the rainy season, the SJ plot experienced a clear drought from mid-June to July with little rainfall. Surface soil moisture fell to a level equivalent to that observed in the dry season, i.e., around 0.05 L L^{-1}. After mid-August, as regular rainfall events occurred, the moisture content in the subsoils was continuously maintained at field capacity or capillary saturation ($\theta = 0.25$). In the NR plot, capillary saturation occurred earlier, that is, in mid-June, the moisture conditions of the subsoils reached the level of field capacity ($\theta = 0.4$), and though data were missing during a certain period of time, such a situation was considered to be maintained until the end of the rainy season. After September, the moisture conditions of the surface soils also reached the level of the field capacity. In contrast, the moisture conditions of both the surface and subsoils in the DP plot reached capillary saturation immediately after the start of the rainy season (i.e., mid-May), and such a situation persisted until the end of the rainy season.

Table 16.1 Physical and physicochemical properties of the soils in the three experimental plots at the sites in North and Northeast Thailand

Site	Depth (cm)	Bulk density (g cm^{-3})	Solid phase (L L^{-1})	Total porosity (L L^{-1})	Moisture content at field capacity (−6 k Pa or pF 1.8) (L L^{-1})	Saturated hydraulic conductivity (m s^{-1})	Particle size distribution			Total C (g kg^{-1})
							Sand (%)	Silt (%)	Clay (%)	
SJ	0–7	1.41	0.52	0.48	0.25	1.2×10^{-5}	92	5	3	2.71
	7–15	1.56	0.57	0.43	0.26	6.4×10^{-6}	89	6	5	1.22
	15–30	1.72	0.64	0.36	0.25	7.5×10^{-6}	88	6	7	0.24
	30–45	1.80	0.68	0.32	0.24	1.1×10^{-6}	83	5	12	0.53
NR	0–15	0.98	0.34	0.66	0.43	6.1×10^{-5}	23	20	57	33.4
	30–40	1.09	0.37	0.63	0.45	1.9×10^{-5}	23	11	66	14.1
DP	0–15	0.83	0.29	0.71	0.44	5.8×10^{-5}	40	13	47	43.5
	30–40	1.23	0.45	0.55	0.39	2.5×10^{-5}	36	13	51	19.7

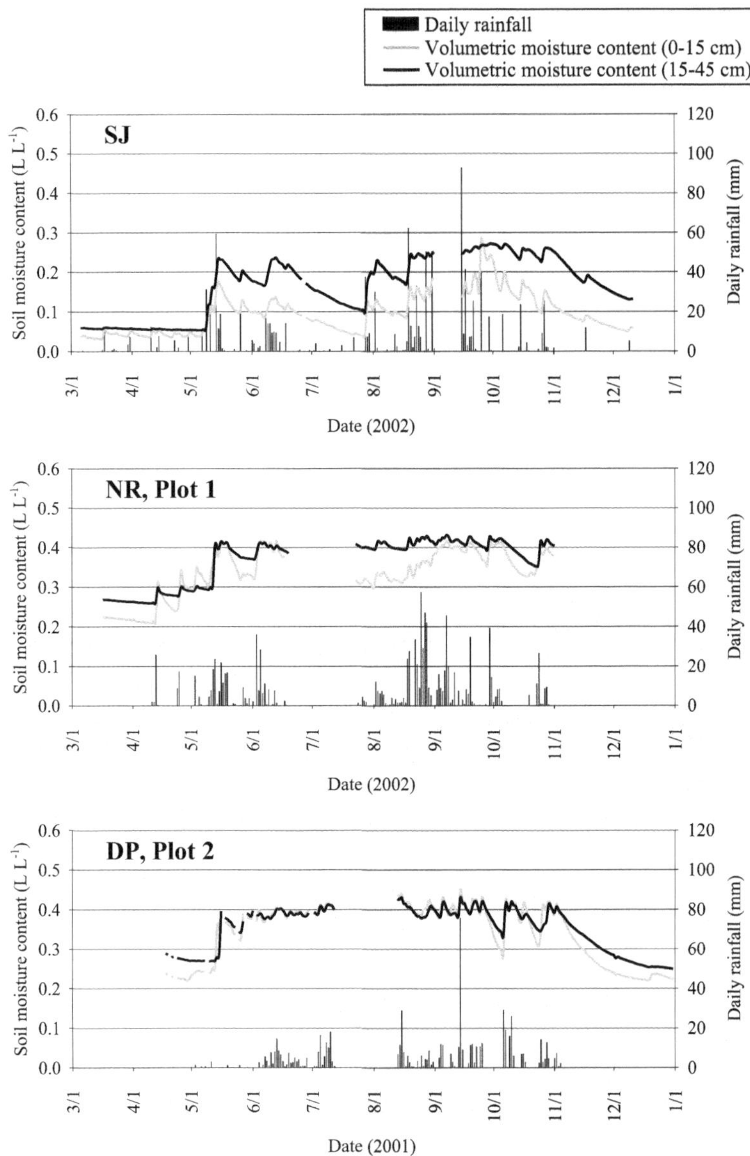

Fig. 16.2 Fluctuations of soil moisture content with daily precipitation in the SJ, NR, and DP plots

Figure 16.3 depicts the proportion of rainfall distribution observed for rainfall events with different 10-min rainfall intensities, i.e., 0–1, 1–2, 2–5, 5–10, and >10 mm per 10 min, respectively, monitored in each experimental plot over two consecutive years. The SJ plots were characterized by a higher occurrence of more intensive rainfall than the other plots, e.g., rainfall with an intensity of >10 mm per

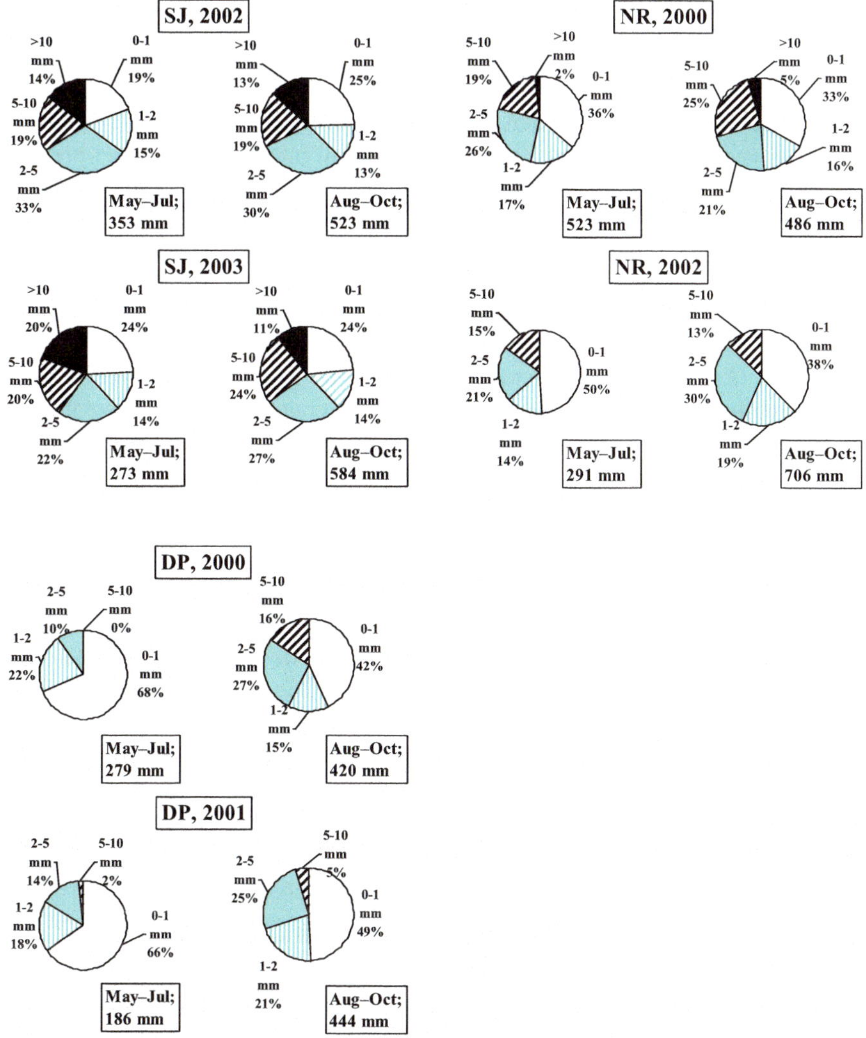

Fig. 16.3 Proportion of the amounts of rainfall with different 10-min rainfall intensities in different seasons in the SJ, NR, and DP plots

10 min, accounting for more than 10 % of the total rainfall. In contrast, in the DP plot situated in the highlands, low-intensity rainfall events of <1 mm per 10 min were dominant and almost no rainfall events of >10 mm were observed. A conspicuous feature of the rainfall distribution in the DP plot was that intensive rainfall events were distributed more often in the latter half of the rainy season (Aug–Oct). The conditions in the NR plot were intermediate between those in the SJ and DP plots in terms of rainfall distribution.

16.4 Water Movement During Rainfall Events Depending on the Soil Conditions and Rainfall Intensity

In order to analyze the water movement and runoff generation from each rainfall event under different soil conditions and/or rainfall intensities, Fig. 16.4 presents three representative cases of water movement during each rainfall event as follows: rainfall with different intensities on the dry surface of the sandy soil in the SJ plot (a and b) and that on the wet surface of the clayey soil in the NR plot (c). In the upper figures, the amount of rainfall and fluctuations in the soil moisture contents at 0–15 and 15–45 cm depths are shown together with the surface runoff recorded. The processed data are presented in the lower figures, that is, cumulative amounts of rainfall (RF_c) and surface runoff (RO_c) and total increase in the soil moisture content at 0–45 cm depth (θ_c), since the start of the rainfall event. The cumulative water loss (WL_c) was calculated as:

$$WL_c = RF_c - \theta_c - RO_c \tag{16.1}$$

Using these data, the fate of water associated with a rainfall event could be traced. For example, the rainfall event (a) started 23:30 on March 6 in the late dry season in the SJ plot. Both the surface and subsoils were extremely dry as the θ values were 0.04 and 0.06 L L^{-1}, respectively. Due to low rainfall intensity (1.1 mm per 10 min at the highest rate), the soil moisture content increased only in the surface layers. The moisture content of the subsoil did not change appreciably. The amount of surface runoff was very small throughout the rainfall event.

The rainfall event (b), which was characterized by rather heavy rainfall, started at 18:10 on May 9. Both the surface and subsoils were dry as the θ values were 0.04 and 0.05 L L^{-1}, respectively. Even under such conditions, when an intensive rainfall event (11.9 mm per 10 min) was observed at 19:40, water appeared to remain on the soil surface, based on the fact that WL_c increased sharply and subsequently decreased, as shown in the lower figure, and surface runoff occurred, though the amount was rather small, namely, 0.2 L m^{-2} (only 1.7 % of the amount of rainfall water) (in the upper figure). At this stage, water percolated into the soils rather easily, as shown in the upper figure, that is, the moisture content of the surface soil increased rapidly and then decreased slowly. On the other hand, the subsoil moisture content increased slowly as water percolated from the overlying layers. Since WL_c, which initially increased, subsequently decreased to a value close to zero, as puddle water on the surface percolated into the soils, downward movement of water further down to the soil layers at a 45 cm depth, or internal drainage, was limited.

In the rainfall event (c), both the surface and subsoils were already wet ($\theta = 0.40$ and 0.42 L L^{-1}, respectively) at the start of rainfall at 17:40 on September 6. During the following 2 days, the total rainfall amounted to 77 mm with moderate intensities (5.6 mm per 10 min at the highest rate). Even so, a very large amount of surface

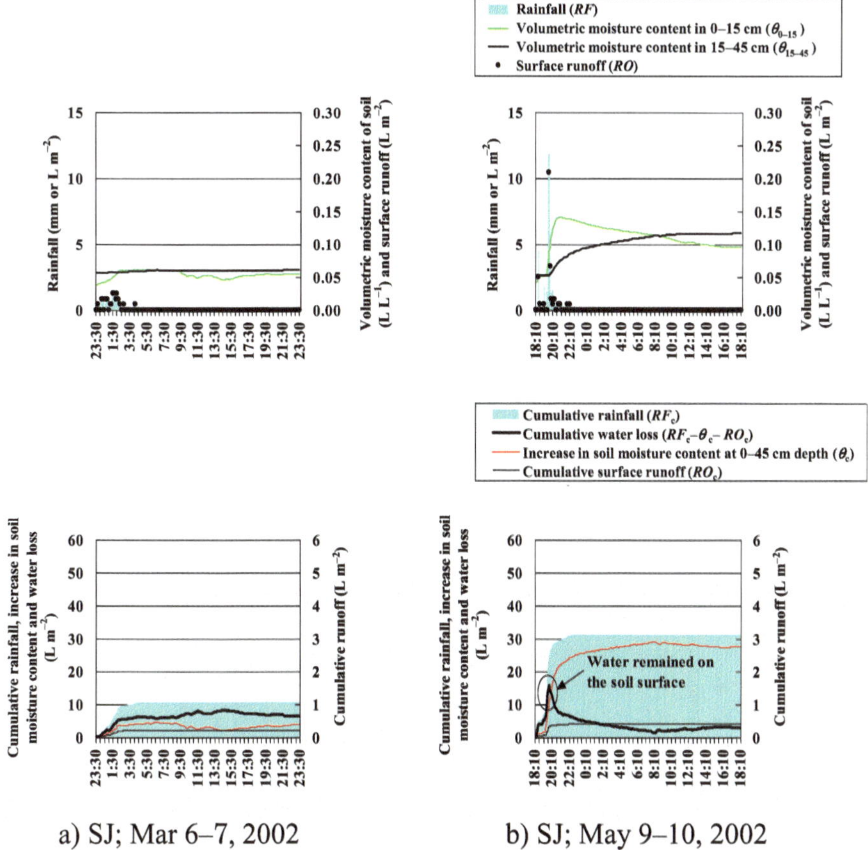

a) SJ; Mar 6–7, 2002   b) SJ; May 9–10, 2002

Fig. 16.4 (**a**) Examples of water dynamics for rainfall events under different conditions. (**b**) Rainfall on dry surface of sandy soil in the SJ plot. (**c**) Rainfall on wet surface of clayey soil in the NR plot

runoff was observed when the rainfall intensity increased, e.g., 1.5 L m^{-2} with 4.9 mm of rainfall at 19:00 on September 7 and 1.6 L m^{-2} with 5.6 mm of rainfall at 15:30 on September 8. In these cases, water loss through surface runoff accounted for 31 and 29 % of the total incoming rainfall, respectively. Rainfall water could not percolate immediately into the soil layers with a high moisture content. But even after such a rainfall event associated with a considerable surface runoff, a moderately rapid increase in the WL_c value was observed (in lower figure), suggesting the existence of rapid drainage from the soil layers.

16.5 Factors Affecting Surface Runoff Generation

Based on these rainfall events, both the rainfall intensity and soil moisture conditions during each rainfall event appeared to affect the generation of surface runoff. To analyze the factors controlling runoff generation in each plot in more detail,

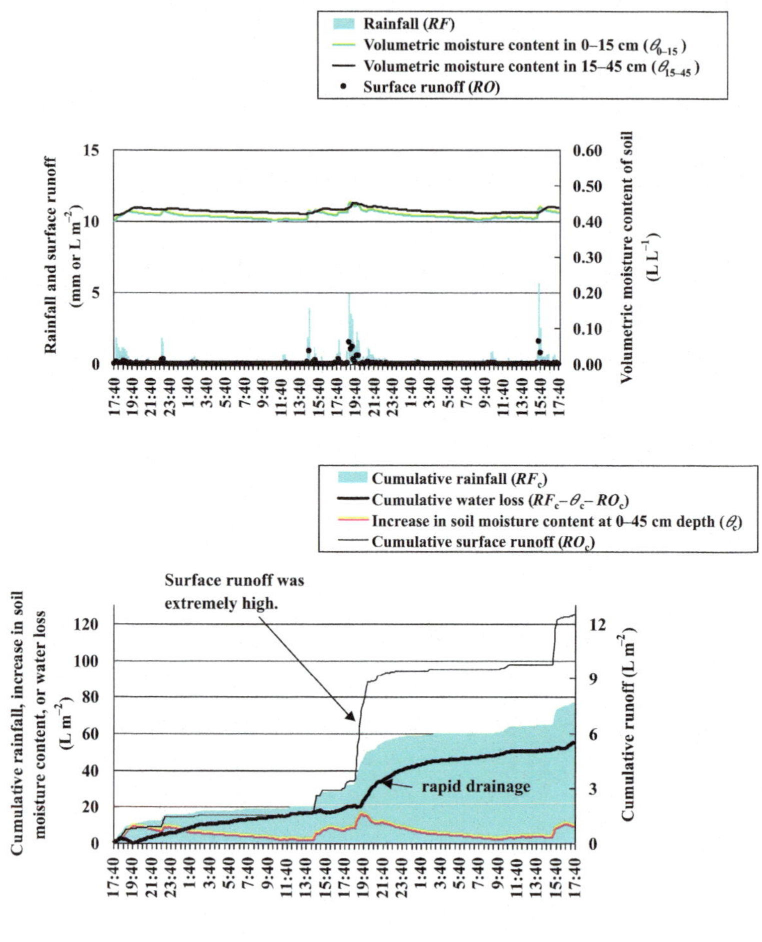

c) NR, Plot 1; Sep 6–8, 2002

Fig. 16.4 (continued)

stepwise multiple regression analysis was conducted using the dataset of the amounts of rainfall (*RF*) and surface runoff (*RO*), as well as the soil moisture content in the 0–15 and 15–45 cm layers of soil (θ_{0-15} and θ_{15-45}, respectively) and water loss (*WL*). The dataset consisted of the data collected from the onset to 24 h after the termination of the respective rainfall events, and they were sampled for rainfall events with a cumulative rainfall amount above 10 mm for each event. A total of 24, 20, and 19 rainfall events met these conditions in the SJ, NR, and DP plots, respectively, during one rainy season. In the NR and DP plots, data from the two plots were incorporated into one analysis. The model functions in the stepwise multiple regression were as follows:

$$WL = aRF + b\theta_{0-15} + c\theta_{15-45} + d \tag{16.2}$$

$$\log(RO) = aRF + b\theta_{0-15} + c\theta_{15-45} + d \tag{16.3}$$

where $WL = RF - \Delta SM - RO$ and ΔSM is the increment in the soil moisture content in the 0–45 cm layers for the time intervals. In the first equation for determining WL, datasets with 10-min intervals were converted to 4-h intervals to eliminate the influence of water puddles on the soil surface. WL was considered to be mainly composed of drainage and evapotranspiration and partially of direct evaporation from the puddles on the soil and/or plant-leaf surfaces after rainfall. On the other hand, in the second equation for RO, original dataset with 10-min intervals was used. For the logarithm transformation of the RO data, the value of zero was omitted from the dataset.

The results of the regression analysis are given in Table 16.2. In all the cases, both for WL and RO, the rainfall intensity was first selected as an independent (or explanatory) variable. Rainfall intensity is generally considered to be one of the major factors that can accelerate the runoff/erosion risks and was actually determined as a controlling factor for runoff generation in several studies (Mathys et al. 2005; Abu Hammad et al. 2006).

On the other hand, the overall significant contribution of θ_{0-15} (SJ) or θ_{15-45} (NR and DP) to WL indicated that, when the soil was dry, the rainfall water supplied was first used to fill the capillary porosity and did not drain directly through the bypass flow, which was a possible cause of WL increase (upper column in Table 16.2). It also indicated at the same time that moist conditions did not strongly interfere with internal drainage. Although the soils were thus generally well drained, the moisture content in the surface layers significantly increased the surface runoff generation ($\log(RO)$, lower column in Table 16.2), indicating that RO tended to increase when the soil was wet due to the interference of rapid percolation of rainfall water into soils. The beneficial influence of surface soil moisture on runoff generation was also reported in different environments, e.g., in Spain (Calvo-Cases et al. 2003) or in China (Chen et al. 2006).

16.6   General Discussion on the Process of Surface Runoff and Soil Erosion in Plots of Different Cultivated Sloping Sites in North and Northeast Thailand

According to the equation obtained in Table 16.2 and the data recorded at 10-min intervals that were monitored with the data logger, surface runoff generation was simulated and is presented in Fig. 16.5. Although data were missing during a certain period of time, in general, the equation obtained above simulated well the actual surface runoff measured. Based on the parameters of the regression equation and

Table 16.2 Parameters for simulating water loss (WL) from the soil profiles and surface runoff (RO) using the rainfall intensity (RF) and initial soil moisture contents (θ_{0-15} and θ_{15-45}) determined by stepwise multiple regression analysis:

$$WL = aRF + b\theta_{0-15} + c\theta_{15-45} + d$$
$$\log(RO) = aRF + b\theta_{0-15} + c\theta_{15-45} + d$$

Variables and experimental plots	a	b	c	d	r^2	Number of samplings
Water loss (WL, 4-hr interval)						
SJ	0.374***	9.87***	–	−0.409	0.70***	194
NR	0.389***	–	16.1***	−5.27***	0.56***	639
DP	0.532***	−10.8*	82.4***	−27.6***	0.57***	406
Surface runoff ($\log(RO)$, 10-min interval)						
SJ	0.145***	2.52***	−0.921**	−1.95***	0.79***	182
NR	0.184***	3.14***	–	−2.44***	0.59***	364
DP	0.194***	0.968***	–	−1.64***	0.60***	453

*,**,***: these parameters were selected as independent variables in the stepwise multiple regression at 15 %, 5 %, and 1 % levels, respectively

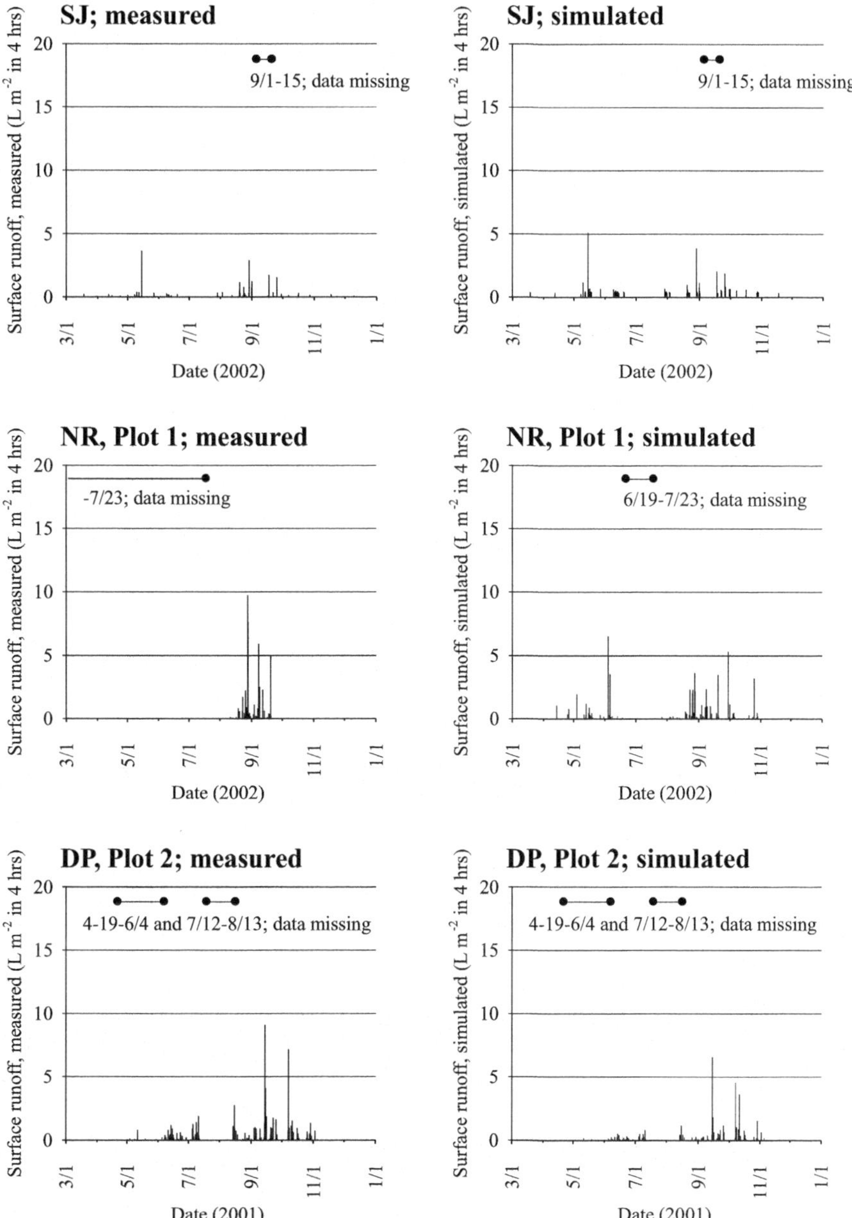

Fig. 16.5 Measured and simulated values of surface runoff in the selected SJ, NR, and DP plots

seasonal distribution pattern of the surface runoff observed, the risk of surface runoff generation in each plot as well as the possible reasons can be postulated as follows.

In the SJ plot, in spite of the much lower slope gradient (i.e., 5 %) compared to that of the other two plots (35 % in the NR plot and 60 % in the DP plot), conspicuous and frequent surface runoff occurred without any clear seasonal trend, possibly during rainfall events with a high intensity. Therefore, it may be difficult to reduce the high erosion risk using some time-course land management such as controlled seeding time of crops, etc. Spatial land management, i.e., introduction of tree vegetation into some areas, would be a suitable option to decrease the overall risk of soil loss from agricultural land through erosion. On the other hand, surface runoff occurred more frequently in the latter half of the rainy season on steeper slopes in the NR and DP plots due to the more frequent capillary saturation in the NR plot or higher distribution of heavier rainfall in the DP plot during that period, compared to the early half of the rainy season. Actually, such a condition in the DP plot is considered to have been advantageous to the traditional shifting cultivation systems, in which the soil surface appears to be susceptible to soil erosion in the early rainy season due to the thin plant cover immediately after seeding. It would, therefore, be possible to decrease the probability of soil erosion through appropriate land management, e.g., selection of crops that do not result in the land being left bare during the latter half of the rainy season.

Table 16.3 summarizes the total amounts of rainfall, surface runoff, and soil erosion using the same unit ($kg\ m^{-2}$). The proportion of surface runoff generated in relation to the amount of rainfall ($b/a \times 100$ (%) in the Table) increased in the order of SJ < NR < DP, which was consistent with the order of the slope gradient of the plots. The proportion of the amount of soil erosion in relation to the amount of surface runoff ($c/b \times 100$ (%) in the Table), however, decreased in the same order, indicating that the sandy soils in the SJ plot were more readily eroded than the clayey soils, presumably due to the weakly organized structure of the soil aggregates. The sandy soils in the SJ plot seemed to be highly susceptible to soil erosion compared to the fine-textured soils in North Thailand, i.e., in the NR and DP plots. It should be noted that the conditions that enhance the risk of surface runoff and soil erosion are thus variable and should be taken into account for agricultural management in the respective regions.

Table 16.3 Total amounts of rainfall, surface runoff, and soil erosion during the experiment

Plots	Total amount of rainfall (a)[a] ($kg\ m^{-2}$)	Amount of surface runoff generated (b) ($kg\ m^{-2}$)	Amount of soil erosion (c) ($kg\ m^{-2}$)	Proportion of surface runoff to total rainfall ($b/a \times 100$) (%)	Proportion of soil erosion to surface runoff ($c/b \times 100$) (%)	Period of measurement
SJ	963	23.8	0.76	2.47	3.21	Mar 6–Dec 31, 2002
NR	989	42.1	0.10	4.26	0.24	Mar 17–Nov 1, 2002
DP	558	49.5	0.045	8.86	0.091	Apr 18–Dec 31, 2001

[a]Amount of rainfall was expressed on an area basis on the slope

Acknowledgment This chapter is derived in part from an article published in Japanese Journal of Tropical Agriculture, 2007, copyright Japanese Society for Tropical Agriculture, available online: https://www.jstage.jst.go.jp/article/jsta1957/51/1/51_1_12/_pdf.

References

Abu Hammad AH, Børresen T, LE H (2006) Effects of rain characteristics and terracing on runoff and erosion under the Mediterranean. Soil Tillage Res 87:39–47

Calvo-Cases A, Boix-Fayos C, Imeson AC (2003) Runoff generation, sediment movement and soil water behaviour on calcareous (limestone) slopes of some Mediterranean environments in southeast Spain. Geomorphology 50:269–291

Chen H, Shao M, Wang K (2006) Effects of initial water content on hillslope rainfall infiltration and soil water redistribution. Nongye Gongcheng Xuebao/Trans Chin Soc Agric Eng 22:44–47

Funakawa S, Yanai J, Hayashi Y, Hayashi T, Noichane C, Panitkasate T, Katawatin R, Kosaki T, Nawata E (2006a) Soil organic matter dynamics on a sloped sandy cropland in Northeast Thailand with special reference to spatial distribution patterns of soil properties. Jpn J Trop Agric 50:199–207

Funakawa S, Hayashi Y, Tazaki I, Sawada K, Kosaki T (2006b) The main functions of the fallow phase in shifting cultivation by the Karen people in northern Thailand – a quantitative analysis of soil organic matter dynamics. Tropics 15:1–27

Kunstadter P, Chapman EC (1978) Problem of shifting cultivation and economic development in northern Thailand. In: Kunstadter, P., Chapman EC & Sabhasri S (eds) Farmers in the forest. The University Press of Hawaii, Honolulu, pp 3–23

Kyuma K (2001) II-4-1. Shifting cultivation in Southeast Asia. In Tropical soil science. University of Nagoya Press, pp 301–321 (in Japanese)

Mathys N, Klotz S, Esteves M, Descroix L, Lapetite JM (2005) Runoff and erosion in the Black Marls of the French Alps: observations and measurements at the plot scale. Catena 63:261–281

Soil Survey Staff (2014) Keys to soil taxonomy, Twelfth edn. U.S. Department of Agriculture and National Resources Conservation Service, Washington

Sonneveld BGJS, Nearing MA (2003) A nonparametric/parametric analysis of the universal soil loss equation. Catena 52:9–21

SPSS (1998) SYSTAT 8.0. Statistics. SPSS Inc, Chicago, 1086 p

Wischmeier WH, Smith DD (1978) Predicting rainfall erosion losses. Agricultural Handbook 537. USDA, Washington, DC

Chapter 17
Control of Water Erosion Loss of Soils Using Appropriate Surface Management in Tanzania and Cameroon

Tomohiro Nishigaki

Abstract Water erosion is one of the main concerns driving land degradation in sloping croplands in tropical countries. The objective of this study was to evaluate water erosion characteristics and the factors affecting surface runoff and soil loss and to evaluate the effect of surface mulching as a conservation management against water erosion in Tanzania and Cameroon. We installed runoff plots (width 0.8 m × slope length 2.4/2.0 m) at four sites (designated NY, TA, SO, and MA) in Tanzania and one site with three treatments in Cameroon: bare plot (CM), cassava plot (CM_C), and cassava with mulch plot (CM_{C+M}). Water budgets for rainfall, surface runoff, and soil moisture for every rainfall event and soil losses were measured over a rainy season. High rainfall amount in NY and TA characterized their high surface runoff and soil loss. High stability of soil aggregates in CM resulted in low runoff coefficient. Although sandy soils in MA had high infiltration rate and low runoff coefficient, their high susceptibility to transport by surface runoff increased its sediment concentration. Total soil loss in CM_{C+M} decreased by 49 % compared with that in CM and CM_C, despite there not being a large difference in runoff water among treatments, indicating the mulch suppressed the particle detachment by raindrops. Based on the water erosion characteristics in the five sites, surface mulching is considered to be a widely applicable management to suppress the soil loss against water erosion in tropics, where rainfall intensity is generally high.

Keywords Antecedent soil moisture condition • Climate condition • Runoff coefficient • Sediment concentration • Soil physical property

T. Nishigaki (✉)
Kyoto University, Kitashirakawaoiwake-cho, Sakyo-ku, Kyoto 606-8502, Japan
e-mail: gakkitom@gmail.com

© Springer Japan KK 2017
S. Funakawa (ed.), *Soils, Ecosystem Processes, and Agricultural Development*,
DOI 10.1007/978-4-431-56484-3_17

17.1 Introduction

Water erosion is a crucial trigger of land degradation problems all over the world due to the human activities (Lal 2001; Gabarrón-Galeote et al. 2013; Zhao et al. 2013). The loss of fertile surface soil caused by water erosion directly results in productivity losses as high as 2–3 % annually (Lal 1995; World Bank 2008). Thus, soil conservation practices for the prevention of water erosion and the achievement of sustainable agricultural systems need to be established (Tesfaye et al. 2014).

However, the environmental influential factors of water erosion, i.e., soil and rainfall properties, greatly differ with location in sub-Saharan Africa, resulting in various water erosion characteristics (Cerdà 1998; Cerdà 2002; Buendia et al. 2015). Soil physical properties, e.g., infiltration rate and water holding capacity, which control runoff generation and soil loss through water erosion are influenced by soil texture, soil structure, and clay mineralogy of soils (Greene and Hairsine 2004; Reichert et al. 2009). For example, typical sandy soils (Entisols), which are widely distributed in tropics, consist mostly of quartz with high weathering resistance and have no pedogenic horizons (Soil Survey Staff 2014). Although such sandy soils have relatively high permeability resulting in low runoff coefficient, they are generally prone to be lost by water erosion owing to its weak structure. Oxisols are also one of the common soils in tropical Africa. Clay mineralogy of Oxisols consists mainly of quartz, kaolinite, and oxides, and hence their physical properties are stable and less susceptible to seal formation (Lado and Ben-Hur 2004). Consequently, soil particle detachment is caused essentially by raindrop impact, and the particles are transported by raindrop splash rather than wash (Sutherland et al. 1996a; Greene and Hairsine 2004; Kinnell 2005). Therefore, the optimum land managements in terms of suppressing water erosion are varied with soil types with a high spatial variability in tropical areas.

Precipitation also varies with slope orientation and elevation. Moreover, it is unclear how the specific rainfall characteristics in tropical area such as short (i.e., half a day or less), high rainfall intensity, and substantial fluctuation (Moore 1979; Wainwright and Parsons 2002; Salako 2006) interact for runoff generation within rainfall events. Therefore, we measured water balance during every rainfall event, i.e., rainfall amount, soil water content, and runoff amount at 10-min intervals overcoming the technical difficulties of continuously measuring these components in the field, to reveal how they affect surface runoff generation.

It is generally said that practices to mulch the soil surface with organic materials on cultivated soils increase water availability for crops and reduce runoff and sediment concentration through the protection of surface layer against raindrops (Gholami et al. 2012; Jordán et al. 2010; Smith et al. 1992). Due to high net primary production in forest-savanna transition zone in east Cameroon, high amount of biomass of savannah vegetation is available around the fallow croplands, and therefore the use of the residue of fallow vegetation is the most reasonable for farmers as a conservation practice. However, the effect of mulching with this fallow vegetation in savanna areas on surface runoff and soil loss in croplands has not been sufficiently evaluated.

Therefore, the objective of this study was to evaluate water erosion characteristics and the factors affecting surface runoff and soil loss under different environmental conditions, and to evaluate the effect of surface mulching against water erosion using small erosion plots in Tanzania and Cameroon.

17.1.1 Materials and Methods

17.1.1.1 Description of Study Sites

Tanzania Four sites were selected to install the erosion plots in or around the Uluguru Mountains in Tanzania, designated NY, TA, SO, and MA. NY site is at the highest elevation (1600 m above sea level) on the west side of the Uluguru Mountains, while TA site is at the lowest elevation, 450 m above sea level, on the east side of the mountains. SO and MA sites are in experimental fields of the Sokoine University of Agriculture, at the foot of the Uluguru Mountains, and MA site is 7 km northwest of SO site (Fig. 17.1). These mountains lie 200 km inland from the Indian Ocean and form part of a chain of mountains in East Africa collectively called the Eastern Arc Mountains (Burgess et al. 2007). The elevation is from around 300 m at lowest point and reaching about 2600 m at highest point. The eastern slope of the mountains has higher precipitation brought by the clouds from the Indian Ocean. The site description was shown in Table 17.1. The experiment was conducted throughout a rainy season from November 2010 through May 2011 in Tanzanian sites.

Cameroon The study site was located in the village of Andom, the Diang Commune, Lom-et-Djérem Division, East Region, Cameroon (Fig. 17.1). The dominant vegetation is *Chromolaena odorata*, and *Imperata cylindrica* is also a dominant vegetation after bush fire. Soils were derived from metamorphic rocks. The site description was shown in Table 17.1. Three treatment plots were established at the same site in December 2012: bare plot (CM), cassava plot (CM_C), and cassava with mulch plot (CM_{C+M}). Residues used for mulching were collected from the adjacent savanna, *Imperata cylindrica*. The amount of applied residue was 8.0 Mg ha^{-1} in dry basis. The experimental period was throughout a rainy season from April to December 2013.

17.1.1.2 Small Erosion Plots

The sites were first cleared of natural vegetation (mainly consisting of shrub and grass) and extracted their roots using hoe and knife, and then the erosion plots were enclosed by corrugated iron sheets. The erosion plots were kept free of weeds and

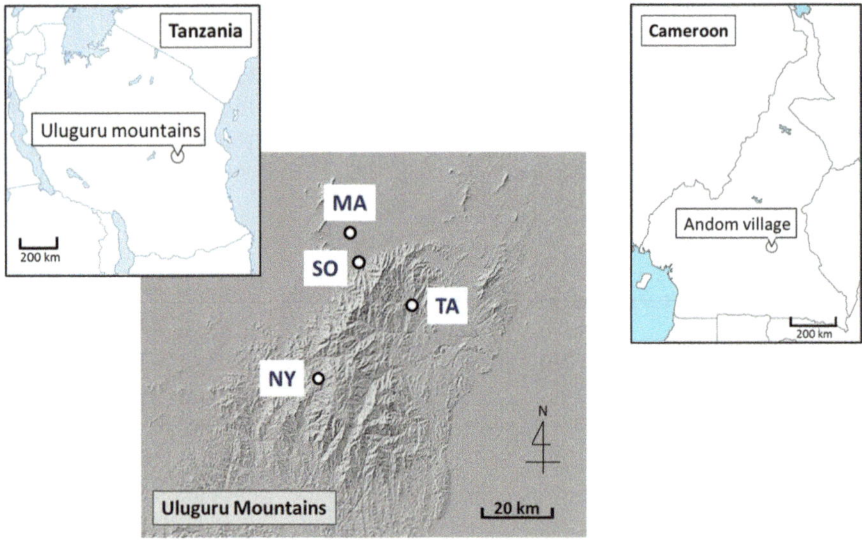

Fig. 17.1 Maps showing the location of the Uluguru Mountains and four experiment sites in/around the mountains in Tanzania and the location of Andom village in Cameroon

any other vegetation except planted cassava by hands throughout the experimental period.

Two erosion plots were established for each site and for each treatment (Fig. 17.2). The plot size was width 0.8 m × slope length 2.4 m (2.0 m for the plots at Cameroon site), which were made small enough to exclude the micro-topography on the slope influencing runoff dynamics and to clarify the water balance and the timing of runoff occurrence within the plot. The runoff gauges had the tipping bucket system and were connected to a data logger (CR1000; Campbell Scientific, Inc., Logan, UT, USA) to record cumulative runoff amount every 10 min, which is the same design as that of Funakawa et al. (2007).

17.1.1.3   Environmental Factors

During the experimental period at each site, volumetric water content (VWC) of soils at depths 0–15, 15–30, and 30–60 cm was continuously measured with time domain reflectometer probes (CS616 water content reflectometer; Campbell Scientific, Inc.) with no replication. Rainfall amount was also measured with a tipping bucket rain gauge (TE525; Campbell Scientific, Inc.) at every site. Data including surface runoff water, soil moisture, and rainfall were recorded by CR1000 data loggers at 10-min intervals.

Table 17.1 Site description, rainfall property, and chemical property of surface layer (0–15 cm) at five experimental sites

Site	Altitude (m)	Average temperature[a] (°C)	Total rainfall amount[a] (mm)	I_{10}[a,b] Average	I_{10}[a,b] Maximum	Slope gradient (degree)	Soil classification[c]	Clay (%)	Silt	Sand	pH	EC ($\mu S\ cm^{-1}$)	CEC ($cmol_c\ kg^{-1}$)	CEC/ clay ($cmol_c\ kg^{-1}$)	Total C (%)	Total N
NY	1600	18.2	980	0.72	13.7	20	Aquertic Haplustalfs	47	41	12	5.6	42.2	15.4	33	2.8	0.2
TA	450	25.4	1625	1.09	12.4	20	Typic Haplustepts	46	12	42	5.8	27.0	12.1	26	3.4	0.2
SO	550	25.4	538	0.67	12.7	5	Kanhaplic Haplustalfs	28	13	59	5.9	37.5	9.9	36	1.3	0.1
MA	500	25.0	528	0.69	12.0	5	Ustic Quartzipsamments	3	5	92	5.0	12.1	1.4	54	0.3	0.0
CM	650	23.1	1309	0.84	15.8	18	Kandiudalfic Eutrudox	49	6	45	5.0	65.6	7.8	16	1.9	0.1

[a]The value during observation period
[b]The highest volume of precipitation occurring in 10 min, in mm 10 min^{-1}
[c]Based on Soil Survey Staff (2014)

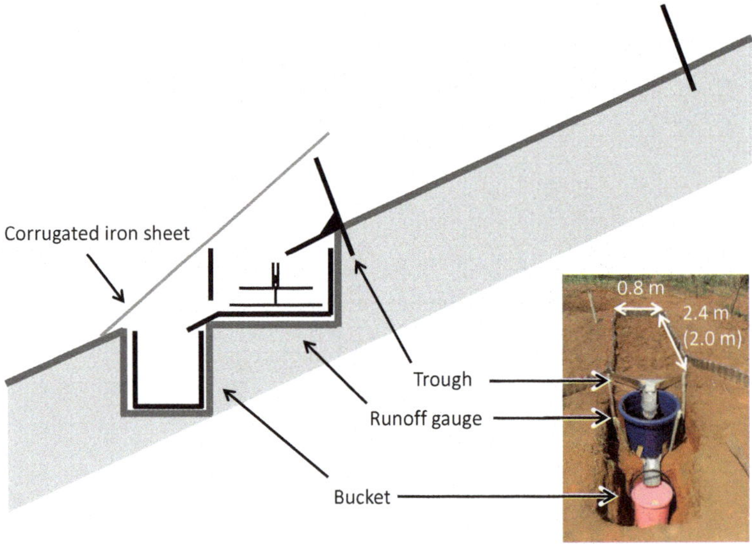

Fig. 17.2 Schematic diagram of runoff gauge and bucket at end of a runoff plot

17.1.1.4 Soil Physical Properties and Sediment Soil Analysis

Undisturbed soil core samples were collected from soil profiles at all sites at depths 0–5, 5–10, 20–25, and 40–45 cm at Tanzanian sites and 0–5, 10–15, and 25–30 cm at Cameroon site with three replications, respectively, using a 100 cm^3 (5-cm height) core sampler. All core samples were used for measuring bulk density (Grossman and Reinsch 2002) and saturated hydraulic conductivity via constant-head and falling-head methods (Klute 1965) and, except for cores from 0–5-cm depth, the soil moisture characteristic curve using a soil column (DIK-3521; Daiki Rika Kogyo Co., Ltd., Saitama, Japan) for 0 (saturated) to −3.1 kPa and pressure plate (DIK-3404, Daiki Rika Kogyo Co., Ltd.) for −9.8 to −98.0 kPa (Dane and Hopmans 2002). Sediment soils trapped in buckets were collected at intervals of about 2 weeks and then air-dried and weighed.

17.1.2 Results and Discussion

17.1.2.1 Soil and Rainfall Properties

Soil texture of surface soil at NY site consisted of 88 % of clay and silt, while that at MA site consisted of 92 % of sand (Table 17.1). Clay contents of NY, TA, and CM sites were similar, but the sand contents at TA and CM sites were higher than that of NY site. SO site had more sand content than that of TA and CM sites. Bulk density at the two mountainous sites in Tanzania, NY and TA, varied with lower value

Table 17.2 Bulk density, saturated hydraulic conductivity, and soil water constants at different depths at five experimental sites

Site	Depth (cm)	Bulk density (Mg m^{-3})		Saturated hydraulic conductivity (mm h^{-1})		Volumetric water content at saturated point (m^3 m^{-3})		Field capacity[a] (m^3 m^{-3})		Depletion of moisture content for normal growth[b] (m^3 m^{-3})		Volume of large-size pores[c] (m^3 m^{-3})
NY	0–5	1.03	(0.02)	15.1	(13.0)							
	5–10	0.98	(0.02)	88.9	(1.44)	0.56	(0.00)	0.46	(0.00)	0.40	(0.00)	0.16
	20–25	1.08	(0.01)	1.02×10^2	(46.8)	0.51	(0.01)	0.49	(0.01)	0.44	(0.01)	0.07
	40–45	0.99	(0.02)	53.5	(10.1)	0.51	(0.00)	0.47	(0.01)	0.42	(0.01)	0.09
TA	0–5	0.95	(0.05)	4.50×10^2	(1.26×10^2)							
	5–10	0.99	(0.03)	3.66×10^4	(9.72×10^3)	0.59	(0.01)	0.40	(0.01)	0.31	(0.01)	0.28
	20–25	1.07	(0.02)	3.74×10^2	(61.2)	0.55	(0.01)	0.42	(0.01)	0.33	(0.01)	0.23
	40–45	1.15	(0.03)	1.89×10^2	(57.6)	0.53	(0.01)	0.45	(0.01)	0.33	(0.01)	0.20
SO	0–5	1.40	(0.02)	99.8	(46.8)							
	5–10	1.43	(0.03)	1.56×10^4	(1.26×10^4)	0.46	(0.01)	0.34	(0.01)	0.23	(0.01)	0.23
	20–25	1.44	(0.02)	84.8	(46.8)	0.40	(0.01)	0.33	(0.00)	0.26	(0.00)	0.15
	40–45	1.46	(0.02)	55.6	(6.48)	0.41	(0.01)	0.36	(0.01)	0.26	(0.01)	0.14
MA	0–5	1.41	(0.02)	2.72×10^2	(39.6)							
	5–10	1.49	(0.02)	1.86×10^2	(1.76)	0.34	(0.00)	0.24	(0.00)	0.04	(0.00)	0.30
	20–25	1.57	(0.02)	78.0	(7.92)	0.31	(0.01)	0.22	(0.01)	0.06	(0.00)	0.25
	40–45	1.64	(0.02)	65.7	(14.4)	0.30	(0.00)	0.23	(0.01)	0.06	(0.00)	0.24
CM	0–5	1.10		7.52×10^2		0.58		0.34		0.24		0.34
	10–15	1.31		29.5		0.50		0.36		0.27		0.23
	25–30	1.34		60.8		0.49		0.37		0.30		0.19

The values given in parentheses are standard error ($N = 5$ for 0–5 cm depth and $N = 3$ for others)
[a]VWC at −3.1 kPa of matric potential
[b]VWC at −98.0 kPa of matric potential
[c]The difference of VWC between saturated point and depletion of moisture content for normal growth

(1.0–1.1 Mg m^{-3}) than that at the two foothill sites SO and MA (1.4–1.6 Mg m^{-3}), and that at CM site intermediately ranged from 1.1 to 1.3 Mg m^{-3} (Table 17.2). Saturated hydraulic conductivity (Ks) of the top layer (0–5-cm depth) was in the order CM > TA > MA > SO > NY (Table 17.2). Volume of large-size pores was large at TA, MA, and CM sites, where Ks was relatively high (Table 17.2), indicating large-size pores contributed to the water permeability.

Total rainfall during the experimental period was high in the mountainous sites in Tanzania, NY and TA, and the Cameroon site, CM (Table 17.3). Average rainfall intensity (average I_{10}) through the experimental period varied from 0.67 mm 10 min^{-1} at SO to 1.09 mm 10 min^{-1} at TA, and there was high average I_{10} in the mountainous sites in Tanzania, NY and TA, and the Cameroon site (Table 17.1). Daily rainfall and VWC are shown in Fig. 17.3. At the NY site, the surface layer was gradually saturated with water from 1 February, the beginning of the rainy season. Subsequently, the entire layer from 0- to 60-cm depth was saturated by mid-March (Fig. 17.3). At SO, VWC at 30–60-cm depth increased little over the experimental period, so little water penetrated deeper than 30 cm throughout the rainy season. VWC at MA was generally low at all depths because of the lower holding capacity of sandy soil, and it decreased immediately after rainfall events. VWC at Cameroon site was high throughout the experimental period except the short dry period in June and August in all plots. The rainfall amount was smaller in the first half (November–February in Tanzania and April–July in Cameroon) than the second half (March–May in Tanzania and August–December in Cameroon, Fig. 17.4). High portion of heavy rainfall (>5.0 mm per 10 min) was observed at TA and CM through the experimental period and at SO in the early half of the rainy season (Fig. 17.4).

17.1.2.2 Relationship Between Surface Runoff Generation and Soil and Rainfall Properties

Soil physical properties of the surface layer substantially affected runoff coefficient at each site. High Ks derived from large amount of large-size pore greatly decreased runoff coefficient at TA, MA, and particularly CM (Tables 17.2 and 17.3). On the other hand, low Ks and high VWC of the surface layer in the middle of the rainy season at NY caused low infiltration capacity, and this resulted in a large runoff coefficient. At SO, the runoff coefficient was the largest of the all sites (16.6 %). VWC below 30-cm depth did not fluctuate through the rainy season, indicating almost no water penetrated into the layer below 30-cm depth, although it had high Ks as measured for the soil core samples collected before the rainy season. This is probably because Ks was reduced by soil crust formed during the rainy season (Morin & Benyamini 1977).

Surface runoff generation was inhibited when the antecedent soil moisture condition was dry at all sites except SO, regardless of rainfall amount in the events. This is also partly indicated by the multiple regression model that showed initial VWC was one of the covariates for runoff amount at all sites except MA

Table 17.3 Total rainfall, runoff, soil loss, runoff ratio, and sediment concentration in surface runoff water during experimental period

Site	Total rainfall (mm) A	Total runoff (mm) B	Total soil loss (Mg ha^{-1}) C	Total runoff coefficient (%) 100*B/A	Total sediment concentration (g L^{-1}) 100*C/B
NY	980	157.2	36.1	16.0	23.0
TA	1625	180.5	47.3	11.1	26.2
SO	538	89.5	15.9	16.6	17.8
MA	528	62.7	22.8	11.9	36.4
CM	1309	92.8	39.2	7.1	42.3
CM$_C$	1309	104.8	38.7	8.0	36.9
CM$_{C+M}$	1309	93.8	19.1	7.2	20.4

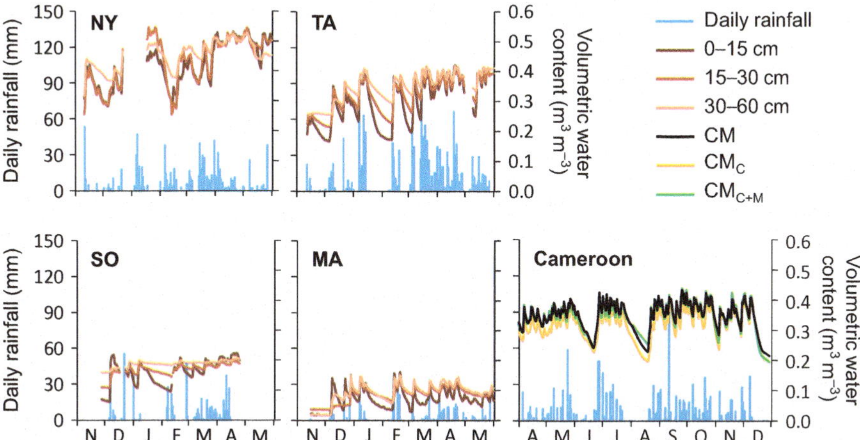

Fig. 17.3 Daily rainfall and daily average volumetric water content (VWC) of surface layers for Tanzanian sites (NY, TA, SO, and MA) and Cameroon site. Only VWC of 0–15-cm layer for CM, CM$_C$, and CM$_{C+M}$ plots is shown at Cameroon site

(Table 17.4). The trend is consistent with those reported by Le Bissonnais et al. (1995) and Ziadat and Taimeh (2013), who conducted erosion experiments using simulated rainfall in the field and the laboratory. Figure 17.5 shows the water budget during rainfall events with different antecedent surface moisture conditions at NY and CM$_C$ sites. When the antecedent soil moisture condition was dry, the stronger hydraulic gradient caused higher infiltration rate of rainfall water into the surface layer, and surface runoff generation was suppressed at both sites (Fig. 17.5a, c). Most of the infiltrated water was kept in 0–30-cm layer and did not penetrate into the deeper layer due to the large water holding capacity (i.e., unsaturated pore volume). When the antecedent soil moisture condition was wet, on the other hand, the surface layer could not hold water anymore because of little pore space, giving a large runoff ratio 57.0 % (Fig. 17.5b) at NY site. However, at

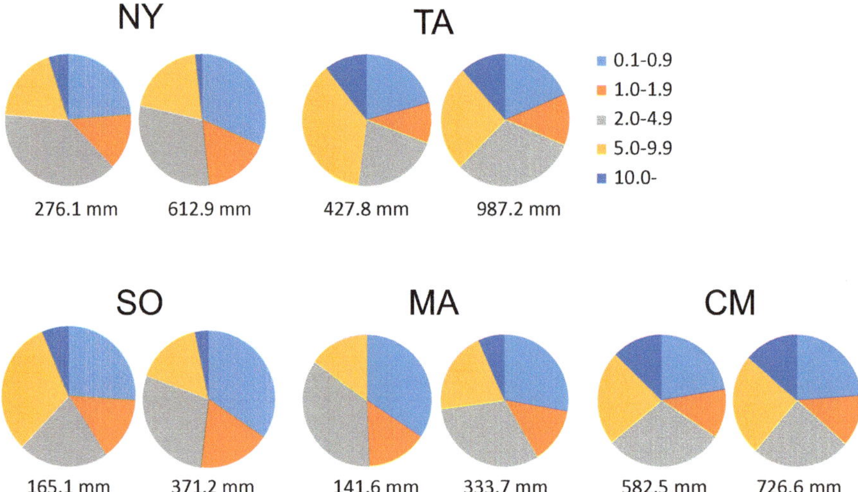

Fig. 17.4 Proportion of the amount of rainfall with different 10-min rainfall intensities in the early half (*left*) and the later half (*right*) of the rainy season. The numbers below of graphs show the total precipitation of each period

Cameroon site, rainfall water quickly penetrated into deeper layer even below 60 cm, shown as the quick decrease of total cumulative water amount of 0–60-cm layer in Fig. 17.5d. Many large-size pores and high Ks (Table 17.2), derived from the fine and strong granular structure of Oxisols in Cameroon site (Soil Survey Staff 2014), substantially contributed to the high drainage capacity and low total runoff coefficient below 8 % in the all plots at Cameroon site. This value is less than half of that for NY and SO (16.0–16.6 %), where the soils are Alfisols. 0–30-cm layer of Alfisols captured rainwater in most of the rainfall events, and soil moisture content in 30–60-cm layer did not frequently increase regardless of the antecedent soil moisture condition as did that of Oxisol at Cameroon site. This contrasting result is probably caused by the larger volume of large-size pores in 0–30-cm layer of Oxisol (0.19–0.34 m^3 m^{-3}) than that of Alfisols (0.07–0.23 m^3 m^{-3}, Table 17.2).

17.1.2.3 Relationship Between Soil Losses and Soil and Rainfall Properties

As a result of statistical analysis on the rainfall events at Tanzanian sites, both sediment concentration and soil loss had significant correlation not with rainfall amount but with average I_{10} within sampling periods at NY (Table 17.4), indicating that soil loss was occurred by storms with high intensity. This also indicated that soil surface at NY had relatively high tolerance against detachment by rainfall and surface runoff because of sealing (Le Bissonnais and Singer 1992), which enhanced runoff generation and consequently increased sediment concentration as the

Table 17.4 Pearson's correlation coefficients between soil losses and environmental parameters at each antecedent soil moisture condition at Tanzanian sites

NY									
	Dry ($n = 4$)			Moist ($n = 4$)			Wet ($n = 4$)		
	Rainfall	I_{10}_max	I_{10}_ave	Rainfall	I_{10}_max	I_{10}_ave	Rainfall	I_{10}_max	I_{10}_ave
Sediment concentration	−0.358	0.531	0.943*	0.878	0.184	−0.123	−0.503	−0.884	−0.425
Soil loss	−0.138	0.733	0.998***	0.591	0.569	0.230	−0.085	0.251	0.462
TA									
	Dry ($n = 6$)			Moist ($n = 7$)					
	Rainfall	I_{10}_max	I_{10}_ave	Rainfall	I_{10}_max	I_{10}_ave			
Sediment concentration	0.926***	0.420	0.619	−0.272	0.448	−0.118			
Soil loss	0.952***	0.255	0.519	0.009	0.451	−0.066			
SO									
	Dry ($n = 14$)								
	Rainfall	I_{10}_max	I_{10}_ave						
Sediment concentration	0.114	0.023	0.065						
Soil loss	0.540*	0.008	−0.192						
MA									
	Dry ($n = 1$) and moist ($n = 10$)								
	Rainfall	I_{10}_max	I_{10}_ave						
Sediment concentration	0.476	0.398	−0.235						
Soil loss	0.594*	0.235	−0.216						

SO consists of only dry condition. MA consists of dry and moist condition, but there was only one dry condition so that statistical analysis was conducted for all together

***Significant correlation at 0.01 level

**Significant correlation at 0.05 level

*Significant correlation at 0.1 level

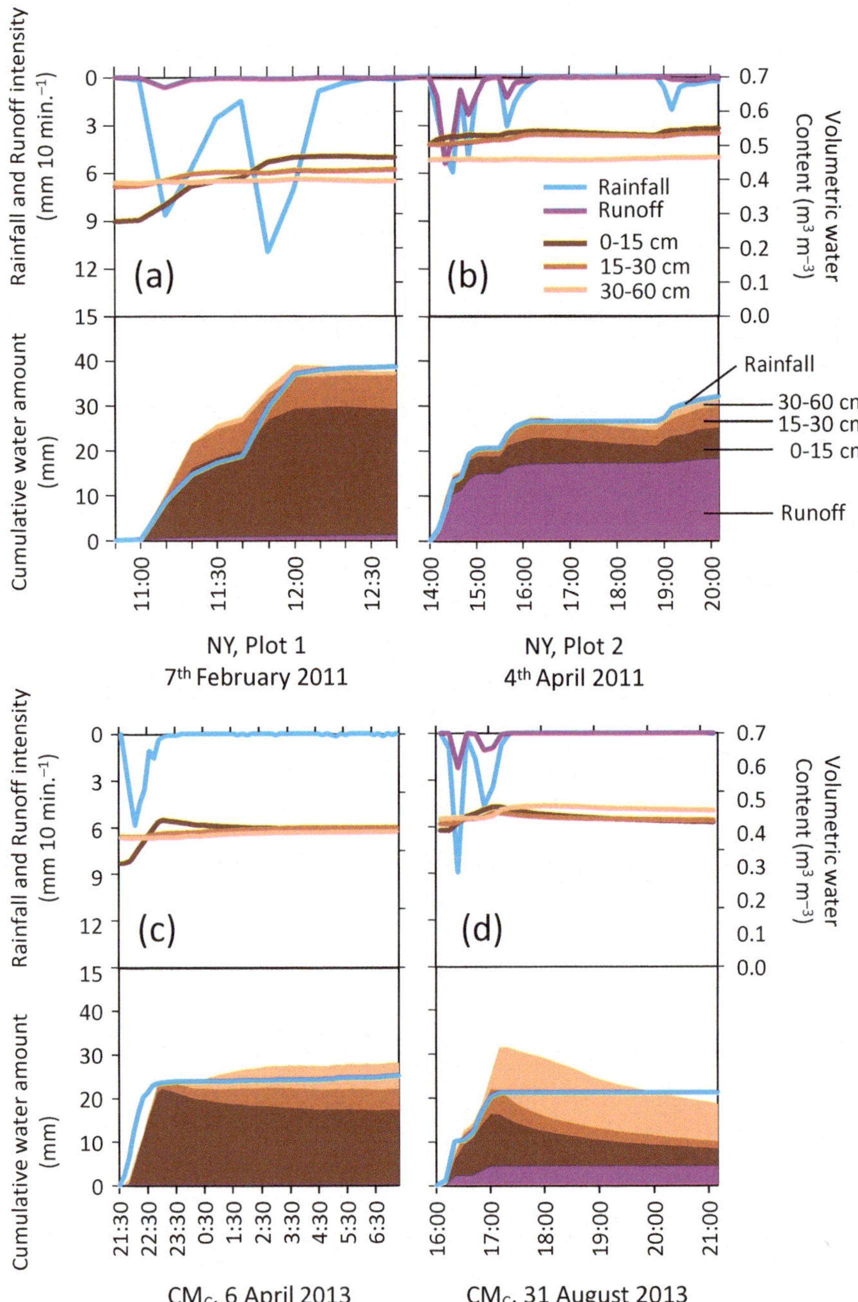

Fig. 17.5 Water budget during rainfall events with different initial surface moisture conditions at NY and CM_C plots. Upper graph shows intensity of rainfall and surface runoff (mm 10 min^{-1}) and volumetric moisture content at each depth. Lower graph shows cumulative water amount derived from rainfall in each proportion. (**a**) Rainfall event when initial surface moisture condition was dry at NY. (**b**) Rainfall event when initial surface moisture condition was wet at NY. (**c**) Rainfall event when initial surface moisture condition was dry at CM_C. (**d**) Rainfall event when initial surface moisture condition was wet at CM_C

erodibility increased. Total sediment concentration was the lowest at SO (17.8 g L^{-1}), indicating the high tolerance probably due to soil crust formation as well as considered with runoff generation at SO. Soil loss at SO and MA was correlated with rainfall amount (Table 17.4), and cumulative soil loss at the both sites mostly increased in the later rainy season in which there was more than twice rainfall amount in the first half (Fig. 17.4). High susceptibility of sandy soil at MA resulted in the highest total sediment concentration regardless of its low I_{10} and runoff ratio (Tables 17.1 and 17.3). At TA, both sediment concentration and soil loss had significant correlation with rainfall amount (Table 17.4). The largest rainfall amount, in addition to the highest I_{10}, through the experimental period at TA (Table 17.1) substantially caused the highest soil loss at all sites (Table 17.3).

Total soil loss during the experimental period at NY and TA was high compared with the foothill sites SO and MA. The former (mountainous) sites had a much higher slope gradient, a higher rainfall amount, and a slightly higher rainfall intensity, generating greater surface runoff water, regardless of runoff ratio (Table 17.1). Despite lower sediment concentrations at NY and TA, runoff amounts were sufficiently large to overcome the soil loss at MA with higher sediment concentration. The soil loss at CM was also larger than that at SO and MA because of high sediment concentration due to high average I_{10} at CM (Table 17.1), although they had similar runoff amounts.

17.1.2.4 Water Erosion Characteristics Under Different Soil and Climate Conditions

NY had the lowest Ks of the surface soil layer (Table 17.2) and that layer was saturated from the middle to the end of the rainy season (Fig. 17.3). The surface layer was gradually saturated with water from 1 February, the beginning of the rainy season. Subsequently, the entire layer from 0- to 60-cm depth was saturated by mid-March (Fig. 17.3), which resulted in the high runoff coefficient despite the relatively low average I_{10} through the experimental period (Table 17.1). In addition, the large rainfall amount at this site substantially caused a large runoff amount (Table 17.3). Although sediment concentration was intermediate, which increased with rainfall intensity of rainfall events (Table 17.4), large amount of runoff resulted in large soil loss (Table 17.3).

At TA site, the high Ks of surface and subsurface layers caused rapid drainage of infiltrated water to deeper layers and caused the smallest runoff coefficient among the Tanzanian sites (Tables 17.2 and 17.3). However, the largest amount of rainfall and the highest average I_{10} through the experimental period generated the greatest runoff amount and soil loss of the all sites (Table 17.3).

Although SO had high K_s, its runoff ratio was the largest of the all sites, and VWC at 30–60 cm depth increased little over the experimental period. Moreover, total soil loss was the least because sediment concentration was the lowest of the all sites (Table 17.3). This unique water erosion characteristic is probably a result of soil crust formed during the rainy season. Soil loss was correlated with rainfall amount (Table 17.4), and the most of soil loss occurred during large rainfall events in the later rainy season. Thus, the importance of soil preservation was highlighted in this period.

The MA site had small rainfall and I_{10} similar to that at SO site. VWC at MA was generally low at all depths because of the lower water holding capacity of sandy soil, and it decreased immediately after rainfall events. The high Ks of surface layer in addition to low rainfall amount resulted in the small runoff amount and low runoff coefficient. However, sediment concentration was the highest of the four sites in Tanzania, where sheet erosion was dominated, and the soil loss had weak correlation with rainfall amount (Table 17.4), indicating that the water erosion risk was potentially high over the rainy season.

The Cameroon site (CM plot) had the highest Ks of surface layer and the largest volume of large-size pores (Table 17.2), resulting in very low runoff coefficient less than half of that in SO and NY. Splash erosion was the dominant erosion process at CM owing to its high I_{10} and low runoff coefficient, resulting in the highest sediment concentration.

Taking into account of each water erosion characteristic, effective land management can be suggested for conservation of soil fertility for each site. At SO, where runoff coefficient was the highest in all plots, it is important to avoid the crust formation with surface mulching to improve the infiltration capacity. At MA, where the sandy soil has the high erosion susceptibility, it is necessary to suppress soil loss with spatial managements, e.g. contour farming (Quinton and Catt 2004), in addition to the surface protection with mulch. At the sites with high rainfall amount, i.e. NY, TA and CM the soil surface should be protected by the mulching against raindrop to protect the soil structure and to suppress splash erosion. Surface mulching would also subsequently preserve the infiltration rate and decrease the runoff coefficient. Therefore, surface mulching is considered to be the most appropriate practice against water erosion loss of soils through the all sites.

17.1.2.5 Effect of Surface Mulching of Runoff Generation and Soil Loss on a Cassava Cropland in Cameroon

The effect of surface mulching on surface runoff and soil loss in a cassava cropland was evaluated in Cameroon site. CM_C plot had lower soil moisture content compared to CM plot likely because of the effect of transpiration of cassava plants, while total runoff coefficient was not different between CM and CM_C plots. On the other hand, soil moisture content was higher in CM_{C+M} plot than in CM_C plot. This is because surface mulching suppressed evaporation from soil surface and remained soil moisture content of surface layer high compared to that in the plot without mulching, as reported by many authors (e.g., Mulumba and Lal 2008; Jordàn et al. 2010).

Total soil loss through the experimental period in CM_{C+M} plot was less than half of that in CM and CM_C plots (Table 17.3). Surface mulching substantially contributed to the protection of soil surface against raindrop with high intensity in this region and suppressed soil loss caused by splash erosion. The amount of runoff water, however, was not affected by the presence of surface mulching in this study. Therefore, it is indicated that the direct impact of rainfall did not decrease the infiltration rate of the surface layer by the formation of crust or clogging in the plots without mulching, CM, and CM_C. This represents the high resistance of the surface soil structure with high drainage capacity of Oxisols (Soil Survey Staff 2014). Considering the high erosion risk of wet soils and in the plots without surface mulching, soils should be kept with some vegetation to avoid constantly high moisture regime and should be covered by mulching during cropping season or weed during the fallow period.

17.1.3   Summary and Conclusion

The results in the current study represented that water erosion characteristic was substantially varied in places in Tanzania and Cameroon, and both soil and rainfall properties in each site were the controlling factors for surface runoff generation and soil losses. High rainfall amount in two mountainous sites, NY and TA, characterized their high surface runoff and soil loss. High stability of soil aggregates in CM resulted in low runoff coefficient. The volume of large-size pores was an important factor influencing the infiltration rate at all sites. Although sandy soils in MA had high infiltration rate and low runoff coefficient, their high susceptibility to transport by surface runoff increased its sediment concentration. Based on the observation of surface runoff and soil loss in five sites in this study, surface mulching is considered to be a widely applicable management to suppress runoff generation and soil loss in tropics, where rainfall intensity is generally high. Surface mulching protects soil surface from the direct impact of raindrops, increases the stability of the aggregates of surface soils and suppresses the crust formation, and remains the infiltration rate. Consequently, splash erosion by raindrops can be reduced and runoff generation is suppressed.

References

Buendia C, Batalla RJ, Sabater S, Palau A, Marcé R (2015) Runoff trends driven by climate and afforestation in a Pyrenean basin. Land Degrad Dev 27(3):823–838. doi:10.1002/ldr.2384

Burgess ND, Butynski TM, Cordeiro NJ, Doggart NH, Fjeldsa J, Howell KM, Kilahama FB, Loader SP, Lovett JC, Mbilinyi B, Menegon M, Moyer DC, Nashanda E, Perkin A, Rovero F, Stanley WT, Stuart SN (2007) The biological importance of the Eastern Arc Mountains of Tanzania and Kenya. Biol Conserv 134:209–231. doi:10.1016/j.biocon.2006.08.015

Cerdà A (1998) Effect of climate on surface flow along a climatological gradient in Israel: a field rainfall simulation approach. J Arid Environ 38:145–159. doi:10.1006/jare.1997.0342

Cerdà A (2002) The effect of season and parent material on water erosion on highly eroded soils in eastern Spain. J Arid Environ 52:319–337. doi:10.1006/jare.2002.1009

Dane JH, Hopmans JW (2002) Pressure plate extractor. In: Dane JH, Topp GC (eds) Methods of soil analysis part 4 physical methods. American Society of Agronomy, Inc., Madison, pp 688–690

Funakawa S, Minami T, Hayashi Y, Naruebal S, Noichana C, Panitkasate T, Katawatin R, Kosaki T, Nawata E (2007) Process of runoff generation at different cultivated sloping sites in north and northeast Thailand. Jpn J Trop Agric 51:12–21

Gabarrón-Galeote MA, Martinez-Murillo JF, Quesada MA, Ruiz-Sinoga JD (2013) Seasonal changes in the soil hydrological and erosive response depending on aspect, vegetation type and soil water repellency in different Mediterranean microenvironments. Solid Earth 4:497–509. doi:10.5194/se-4-497-2013

Gholami L, Sadeghi SH, Homaee M (2012) Straw mulching effect on splash erosion, runoff, and sediment yield from eroded plots. Soil Sci Soc Am J 77:268–278. doi:10.2136/sssaj2012.0271

Greene RSB, Hairsine PB (2004) Elementary processes of soil-water interaction and thresholds in soil surface dynamics: a review. Earth Surf Process Landf 29:1077–1091. doi:10.1002/esp.1103

Grossman RB, Reinsch TG (2002) Bulk density and linear extensibility. In: Dane JH and Topp GC (eds) Method of soil analysis part 4. American Society of Agronomy, Inc., Madison, pp 201–228

Jordán A, Zavala LM, Gil J (2010) Effects of mulching on soil physical properties and runoff under semi-arid conditions in southern Spain. Catena 81:77–85. doi:10.1016/j.catena.2010.01.007

Kinnell PIA (2005) Raindrop-impact-induced erosion processes and prediction: a review. Hydrol Process 19:2815–2844. doi:10.1002/hyp.5788

Klute A (1965) Laboratory measurement of hydraulic conductivity of saturated soil. In: Black CA, Evans DD, White JL, Ensminger LE, and Clark FE (eds) Method of soil analysis part 1. American Society of Agronomy, Inc., Madison, pp 210–221

Lado M, Ben-Hur M (2004) Soil mineralogy effects on seal formation, runoff and soil loss. Appl Clay Sci 24:209–224. doi:10.1016/j.clay.2003.03.002

Lal R (1995) Erosion-crop productivity relationships for soils of Africa. Soil Sci Soc Am J 59:661–667

Lal R (2001) Soil degradation by erosion. Land Degrad Dev 12:519–539. doi:10.1002/ldr.472

Le Bissonnais Y, Singer MJ (1992) Crusting, runoff, and erosion response to soil water content and successive rainfalls. Soil Sci Soc Am J 56:1898–1903

Le Bissonnais Y, Renaux B, Delouche H (1995) Interactions between soil properties and moisture content in crust formation, runoff and interrill erosion from tilled loess soils. Catena 25:33–46. doi:10.1016/0341-8162(94)00040-L

Moore TR (1979) Rainfall erosivity in East Africa. Geogr Ann 61:147–156

Morin J, Benyamini Y (1977) Rainfall infiltration into bare soils. Water Resour Res 13:813–817

Mulumba LN, Lal R (2008) Mulching effects on selected soil physical properties. Soil Tillage Res 98:106–111. doi:10.1016/j.still.2007.10.011

Quinton JN, Catt JA (2004) The effects of minimal tillage and contour cultivation on surface runoff, soil loss and crop yield in the long-term Woburn Erosion Reference Experiment on sandy soil at Woburn, England. Soil Use Manag 20:343–349. doi:10.1079/SUM2004267

Reichert JM, Norton LD, Favaretto N, Huang C, Blume E (2009) Settling velocity, aggregate stability, and interrill erodibility of soils varying in clay mineralogy. Soil Sci Soc Am 73:1369–1377. doi:10.2136/sssaj2007.0067

Salako FK (2006) Rainfall temporal variability and erosivity in subhumid and humid zones of southern Nigeria. Land Degrad Dev 17:541–555. doi:10.1002/ldr.738

Smith GD, Coughlan KJ, Yule DF, Laryea KB, Srivastava KL, Thomas NP, Cogle AL (1992) Soil management options to reduce runoff and erosion on a hardsetting Alfisol in the semi-arid tropics. Soil Tillage Res 25:195–215. doi:10.1016/0167-1987(92)90111-N

Soil Survey Staff (2014) Keys to soil taxonomy, 12th edn. U.S. Department of Agriculture and Natural Resources Conservation Service, Washington

Sutherland RA, Wan Y, Ziegler AD, Lee C-T, El-Swaify SA (1996) Splash and wash dynamics: an experimental investigation using an Oxisol. Geoderma 69:85–103. doi:10.1016/0016-7061(95)00053-4

Tesfaye A, Negatu W, Brouwer R, Van Der Zaag P (2014) Understanding soil conservation decision of farmers in the Gedeb Watershed, Ethiopia. Land Degrad Dev 25:71–79. doi:10.1002/ldr.2187

Wainwright J, Parsons AJ (2002) The effect of temporal variations in rainfall on scale dependency in runoff coefficients. Water Resour Res 38:1271

World Bank (2008) World development report 2008. World Bank, Washington, DC

Zhao G, Mu X, Wen Z, Wang F, Gao P (2013) Soil erosion, conservation, and eco-environment changes in the Loess Plateau of China. Land Degrad Dev 24:499–510. doi:10.1002/ldr.2246

Ziadat FM, Taimeh AY (2013) Effect of rainfall intensity, slope, land use and antecedent soil moisture on soil erosion in an arid environment. Land Degrad Dev 24:582–590. doi:10.1002/ldr.2239

Chapter 18
Utilization of Soil Microbes as a Temporal Nutrient Pool to Synchronize Nutrient Supply and Uptake: A Trial in the Dry Tropical Croplands of Tanzania

Soh Sugihara and Method Kilasara

Abstract In sub-Saharan Africa, soil nitrogen is the most limiting factor for crop production, with considerable nitrogen (N) being lost through leaching. To improve the crop productivity in this region, it is necessary to improve the synchronization of soil N supply and plant N uptake by reducing N loss. In this chapter, based on a field cultivation experiment in the dry tropical croplands of Tanzania, I showed and explained the effectiveness of the utilization of soil microbes as a temporal N pool to synchronize the soil N supply and plant N uptake. Firstly, I revealed that the mismatch of N supply and uptake in this region was due to nitrate leaching during early crop growth, while I also found a clear contribution of soil microbes as an N source for late-season plant growth. Secondly, I found a clear effect of land management (i.e., plant residue application) on soil microbial dynamics, with early season plant residue applications clearly increasing, and maintaining (more than 1 month), the microbial biomass N (MBN). Finally, by applying plant residue 2 weeks before seeding, I assessed the effectiveness of a temporal fluctuation of MBN by analyzing the soil–plant N dynamics and crop productivity. I found that soil microbes contributed to the synchronization of soil N supply and plant N uptake and to the improvement of crop yields.

Keywords Leaching • Nitrogen synchronization • Plant residue • Application • Soil microbial dynamics • Ultisols

S. Sugihara (✉)
Tokyo University of Agriculture and Technology, Saiwai-cho 3-5-8, 183-8509 Fuchu, Tokyo, Japan
e-mail: sohs@cc.tuat.ac.jp

M. Kilasara
Faculty of Agriculture, Sokoine University of Agriculture, Morogoro, Tanzania

© Springer Japan KK 2017
S. Funakawa (ed.), *Soils, Ecosystem Processes, and Agricultural Development*,
DOI 10.1007/978-4-431-56484-3_18

357

18.1 Synchronization of Soil N Supply with Plant N Uptake in Dry Tropical Croplands

In sub-Saharan Africa (dry tropical climate), low soil fertility, and the inability of small farms to afford fertilizers, results in soil nitrogen (N) being the most limiting factor for crop production (Photo 18.1; Bationo and Buerkert 2001; Palm et al. 2001). Therefore, in order to improve crop productivity in this region, it is necessary to improve the synchronization of soil N supply and plant N uptake by reducing N loss (Nyamangara et al. 2003; Adeboye et al. 2006; Gentile et al. 2009). In dry tropical croplands, net N mineralization generally occurs early in the rainy season, during early crop growth, because organic matter (OM) with a low C/N (carbon to nitrogen) ratio accumulates during the dry season (Kushwaha et al. 2000; Singh et al. 2007a). However, during early crop growth, heavy rainfall often leaches this mineralized N, resulting in depletion of available soil N during later crop growth when the need for N uptake is greatest (Fig. 18.1; Hartemink et al. 2000; Chikowo et al. 2004; Shahandeh et al. 2004). Hence, efficient use of the soil N supply in this agroecosystem requires conservation of available soil N at the surface layer during early crop growth and increased soil N supply during later crop growth.

18.2 Importance of Soil Microbes as a Temporary Nutrient Pool in Dry Tropical Croplands

In terrestrial ecosystems, soil microbes act as an important nitrogen (N) pool in soil–plant N dynamics (Singh et al. 1989; Bardgett et al. 2007; Wardle et al. 2004), and recent studies have emphasized their importance for crop N uptake in the dry tropical croplands developed on nutrient-poor soil (Ghoshal and Singh 1995; Cookson et al. 2006; Singh et al. 2007b). To improve the efficiency of N utilization in these nutrient-poor agroecosystems, it is necessary to elucidate the time course for N partitioning between the soil N pool (microbial biomass nitrogen (MBN) and inorganic N in the soil) and plant N uptake during the crop growth period and evaluate the influence of land management on the soil–plant N dynamics (Kushwaha et al. 2000; Herai et al. 2006).

Many studies evaluating the seasonal microbial dynamics in dry tropical ecosystems have reported that MBN is depleted during the rainy season, whereas it remains high during the dry season (Singh et al. 1989; Wardle 1992, 1998). This is because drastically increased soil moisture during the early rainy season depletes soil microbes, due to microbial cell lysis (Fierer et al. 2003; Tripathi and Singh 2007) and enhanced grazing by soil macrofauna (Michelsen et al. 2004). During the rainy season, when plant activity is high, the decrease in MBN leads to nutrient

Photo 18.1 Maize cropland in Tanzania

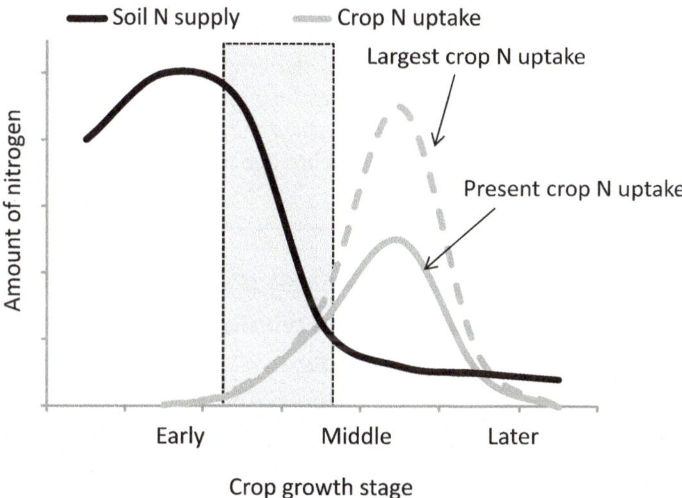

Fig. 18.1 The time course image of soil N supply and crop N uptake in dry tropical croplands. The *shaded area* is the critical N leaching period, and the *dotted gray line* is the potential N uptake pattern of a crop in this region. This figure was generated using the results of Sugihara et al. (2010b). In this study, soil N supply is defined as sum of mineralized N and inorganic N of soil

release from soil microbes, which become an N source for plant growth. On the other hand, the decrease in MBN may result from strong competition for N among plants and soil microbes during the crop growth period (Srivastava and Lal 1994; Kushwaha et al. 2000; Singh et al. 2007b). During a 2-year study in India, Singh et al. (2007b) evaluated temporal variations in MBN by measuring crop root biomass. For rice and barley crops, they observed a negative relationship between MBN and crop root biomass and suggested that strong competition occurred between soil microbes and crops for available nutrients during the seedling and grain-forming stages. However, most studies have focused only on microbial biomass dynamics, and few studies have ever reported the time course of MBN for the plant N uptake pattern in dry tropical croplands. Therefore, the relationship between MBN variation and plant N uptake pattern, as well as the N contribution of soil microbes to plant growth, remains unclear.

Furthermore, land management (e.g., organic matter application, fertilizer application, soil tillage) affects soil microbial dynamics, resulting in a change in N use efficiency (Spedding et al. 2004; Joergensen and Emmerling 2006). Many studies have reported that organic matter application increases MBN in tropical croplands, suggesting that the increased MBN contributes to plant growth by promoting their nutrient source and sink function (Powlson et al. 2001; Singh et al. 2007a, b). In order to synchronize the soil N supply and plant N uptake during the crop growth period, by utilizing the soil microbes as an N pool, it is necessary to understand the effect of land management on the relationship between MBN variation and plant N uptake pattern in the agroecosystem.

In this subsection, I reveal results from a field cultivation experiment in the dry tropical croplands of Tanzania. Firstly, I show a real-life example of mismatch of soil N supply and plant N uptake during the crop growth period in relation to microbial role. Secondly, I discuss the controlling factors of soil microbial biomass in relation to land management. Lastly, I show the applicability of soil microbial biomass as a temporal nutrient pool to synchronize the soil nutrient supply with the plant nutrient uptake.

18.3 Soil–Soil Microbe–Plant N Dynamics in the Dry Tropics: Critical N Leaching

To evaluate the soil–plant N dynamics, in relation to microbial N supply for plant growth in a dry tropical cropland, I conducted a maize cultivation experiment in Tanzania using different land management treatments: no input (control treatment), plant residue application (P), fertilizer application (F), plant residue and fertilizer application (PF), and noncultivated (B). During the 104-days experiment, I periodically evaluated the microbial biomass nitrogen and carbon, plant nitrogen uptake, microbial respiration in situ, inorganic nitrogen in the soil, and environmental conditions.

During the experimental period, total rainfall was approximately 600 mm and approximately 85 % of the total rainfall occurred before 60 days after planting (DAP). As a result, before 60 DAP, soil moisture content was continuously high,

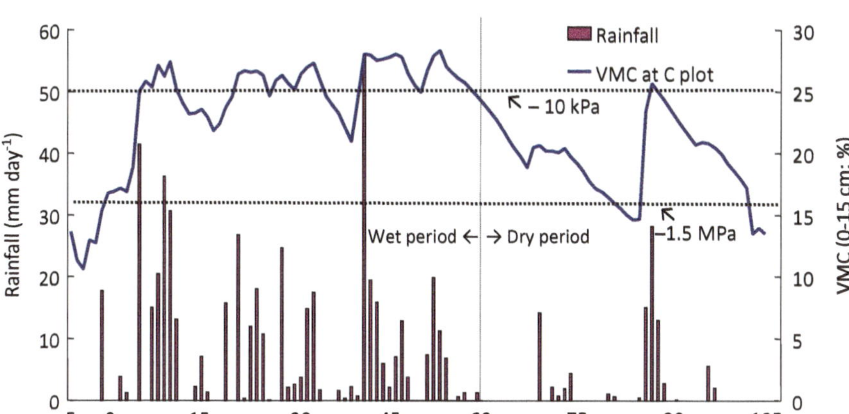

Fig. 18.2 Fluctuations in volumetric moisture content (VMC) in the control plot as a function of daily rainfall. The horizontal *dotted lines* indicate the value of −10 kPa (*above*) and −1.5 MPa (*below*)

around −10 kPa, though it decreased steadily after 60 DAP, and reached −1.5 MPa at 85 DAP (Fig. 18.2). Based on this rainfall and moisture data, the experimental period could be divided into the wet period (0–60 DAP) and the relatively dry period (61–104 DAP).

Soil microbial respiration was high during the wet period (Fig. 18.3). Soil organic matter (SOM) likely underwent substantial decomposition during the wet period, when heavy rainfall maintained high soil moisture content. Shahandeh et al. (2004) showed that rapid N mineralization occurred in the early rainy season in Mali because easily decomposable SOM, which has a low C/N ratio, accumulates during the dry season. Singh et al. (2007a) also observed a high rate of N mineralization during the rainy season in India. In the current study, the cumulative soil respiration, which was estimated by the relationship between the CO_2 efflux rate and soil moisture content (based on Funakawa et al. 2006), was 1.42 Mg C ha^{-1} in the C plot, and more than 70 % of the cumulative respiration occurred during the wet period. Thus, a substantial amount of N was presumably mineralized during the wet period. After 43 DAP, a significant amount of inorganic nitrogen was lost from all of the treatment plots and remained low thereafter (approximately 20–35 kg N ha^{-1} at 0–15 cm of soil depth) (Table 18.1). Considering the substantial N mineralization during the wet period, and the heavy rainfall at 41–43 DAP (91 mm in total), the depletion of NO_3^- at 43 DAP indicated severe N leaching beyond 15 cm of soil depth in all plots. This critical N leaching must have caused the mismatch of soil N supply and plant N uptake (Fig. 18.1).

During the dry period (60–104 days; grain-forming stage), I found a significant effect of plant N uptake on soil microbial dynamics when I compared variations in MBN and the MB C/N ratio between the cultivated plots (control, P, F, and PF

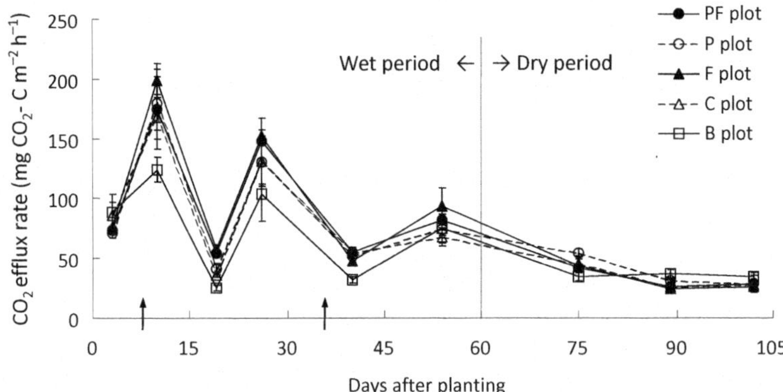

Fig. 18.3 Fluctuations in the CO_2 efflux rate in all plots. The two upward-pointing arrows indicate urea application at 7 and 35 DAP. Bars indicate the standard error. *B plot*, no cultivation plot and control; *F plot*, fertilization plot; *P plot*, plant residue plot; and *PF plot*, plant residue and fertilizer plot (Refer to Sugihara et al. (2010b) for a more detailed explanation of the treatments)

plots) and the noncultivated plot (B plot) (Table 18.1). After 61 DAP, plants continued to absorb substantial amounts of N until 76 DAP, in the control and F plots, and until 92 DAP in the P and PF plots (Fig. 18.4). During this stage (61–92 DAP), I observed a distinct decrease of MBN (from 63–71 to 18–33 kg N ha^{-1}) in all cultivated plots, except the B plot; this MBN decrease was roughly equivalent to the plant N uptake in each plot. Many studies have also reported MBN depletion during the crop growth period and suggested that plants and soil microbes compete for N (Kushwaha et al. 2000; Friedel et al. 2001; Singh et al. 2007b). In India, Singh et al. (2007b) showed a negative correlation between crop (rice and barley) root biomass and MBN for their control, fertilizer, and organic matter application plots. Crop root biomass increased from the seedling stage, to reach a maximum at the grain-forming stage, and declined sharply thereafter, whereas MBN decreased from the seedling stage to the grain-forming stage and increased again at the harvest stage. In my study, MBN continuously decreased until plant N uptake stopped (76 DAP in the control and F plots and 92 DAP in the P and PF plots), and MBN increased again thereafter. This re-increase of MBN at 104 DAP was possibly caused by maize root deposition at the harvest stage (Singh et al. 2007b). In addition, I observed a significant increase in the MB C/N ratio at 76 and 92 DAP in all cultivated plots (excluding the B plot) and layers. An increase in the MB C/N ratio (>10) indicates that N availability is greatly limited in soils (Joergensen and Emmerling 2006). At 104 DAP, when plant N uptake had ceased, the MB C/N ratio in all the cultivated plots and layers decreased to a value similar to that in the B plot, whereas the ratio remained steady in the B plot throughout the dry period (Table 18.1). Similar fluctuations in the MB C/N ratio during crop growth were also observed at a soil depth of 0–27 cm (Germany; Friedel et al. 2001) and also at a soil depth of 0–20 cm (Denmark; Petersen et al. 2003). Due to the drying soil and

Table 18.1 MBN, MB C/N ratio, NH_4^+, and NO_3^- at each treatment plot in each soil layer during the crop growth period in a tropical cropland in Tanzania (Sugihara et al. 2010b)

	Surface soil (0–5 cm)					Subsurface soil (5–15 cm)				
DAP	C plot	P plot	F plot	PF plot	Bare plot	C plot	P plot	F plot	PF plot	Bare plot
MBN (kg N ha^{-1})										
0	18.6 b	17.7 ab	18.6 bc	18.0 ab	18.6 b	26.4 bc	31.4 ab	26.4 bc	32.2 ab	26.4 ab
14	20.6 b	12.8 ab	20.0 ab	26.8 a	21.5 a	32.2 bc	29.7 ab	40.0 ab	38.0 ab	24.7 ab
43	18.3 b	15.7 ab	14.6 bc	16.2 b	15.5 b	38.9 ab	34.6 ab	30.7 bc	32.8 ab	30.3 ab
61	28.7 a	21.4 a	26.1 a	23.0 ab	21.8 a	38.4 abAB	42.0 aAB	39.7 abAB	48.6 aA	25.6 abB
76	13.5 b	14.8 ab	14.7 bc	14.6 b	18.7 b	18.9 c	18.1 b	14.9 c	18.8 b	14.9 b
92	14.4 bA	8.5 bAB	8.2 cAB	6.1 cA	12.6 bAB	20.7 bcAB	13.2 bAB	28.7 bcA	12.5 bB	29.7 abA
104	19.4 b	17.9 ab	19.4 ab	16.8 ab	18.3 ab	49.4 a	46.0 a	48.9 a	51.4 a	35.0 a
MB C:N ratio										
0	8.2 ab	8.0 abc	10.8 ab	10.2 b	10.8	10.2 ab	8.0 bc	11.1 ab	9.0 ab	10.4
14	10.2 a	10.5 ab	10.0 abc	7.7 b	9.6	10.1 ab	10.3 ab	7.2 ab	8.7 ab	10.0
43	7.4 ab	6.0 bc	9.1 abc	8.1 b	7.2	5.0 b	5.8 bc	7.0 ab	6.2 b	5.6
61	5.6 b	5.6 c	6.2 c	5.9 b	6.8	6.2 b	4.9 c	5.8 b	6.3 b	8.2
76	10.4 a	9.3 abc	9.9 abc	8.6 b	8.1	10.7 abAB	8.7 bcAB	11.5 aAB	13.9 abA	7.4 B
92	8.9 abB	12.6 aAB	12.9 aAB	18.1 aA	8.5 B	13.0 aAB	14.2 aAB	7.4 abAB	15.7 aA	7.0 B
104	8.1 ab	7.1 bc	8.6 bc	9.1 b	8.6	5.6 b	6.1 bc	6.2 ab	6.8 b	6.9

(continued)

Table 18.1 (continued)

DAP	Surface soil (0–5 cm)					Subsurface soil (5–15 cm)				
	C plot	P plot	F plot	PF plot	Bare plot	C plot	P plot	F plot	PF plot	Bare plot
NH_4^+ (kg N ha^{-1})										
0	0.6 cB	2.3 bA	1.3 cAB	1.1 cAB	1.6 bAB	1.6 b	8.9 a	2.7 b	2.3 c	3.2
14	1.2 bcB	1.8 bB	37.5 aA	43.9 aA	2.2 bB	4.2 b	6.7 ab	5.0 b	3.6 bc	3.9
43	6.6 a	7.5 a	15.0 b	17.8 b	5.2 a	7.2 a	6.0 ab	9.4 a	8.0 a	6.4
61	2.2 bcA	2.0 bAB	1.5 cAB	2.4 cA	1.1 bB	3.6 b	5.2 ab	3.3 b	3.0 bc	4.2
76	3.0 b	2.2 b	1.6 c	1.8 c	2.1 b	4.3 b	5.0 ab	5.0 b	3.8 bc	5.0
92	2.5 bc	3.1 b	2.8 c	3.0 c	2.6 b	4.3 b	4.1 b	5.4 b	4.0 bc	4.5
104	2.6 bc	2.6 b	2.0 c	2.0 c	2.3 b	3.6 b	3.5 b	3.8 b	4.6 b	4.0
NO_3^- (kg N ha^{-1})										
0	21.7 aAB	27.4 aAB	21.3 bB	35.0 aA	19.2 aB	43.5 aAB	54.7 aAB	42.6 aB	70.0 aA	38.4 aB
14	16.0 abB	13.6 bB	34.8 aA	45.9 aA	19.4 aAB	42.6 aAB	38.9 bB	39.6 abAB	61.1 aA	44.1 aAB
43	1.3 cAB	0.4 eB	0.7 dAB	2.8 cA	0.5 cAB	1.4 c	8.9 c	15.1 c	2.3 c	10.0 b
61	12.6 b	10.6 bc	16.1 bc	14.9 b	10.9 b	21.7 b	16.9 c	22.3 abc	28.1 b	19.8 b
76	10.0 bcAB	5.2 dB	13.6 bcA	11.5 bcA	8.5 bAB	15.7 b	13.0 c	21.4 abc	20.4 b	1.2 b
92	12.0 bA	9.7 cAB	11.9 bcdA	12.7 bcA	8.4 bB	15.9 b	12.6 c	17.8 bc	17.8 b	15.1 b
104	4.8 cAB	3.6 deB	7.0 cdA	6.4 bcA	4.9 bcAB	10.1 bcAB	8.3 cB	10.7 cAB	12.7 bcA	9.5 bAB

Values followed horizontally by a different uppercase letter (A and B) indicate that means are significantly different ($p < 0.05$) between treatment plots (C, P, F, PF, and bare plots) within sampling time; whereas different lowercase letters (a, b, c, d, and e) vertically indicate that means are significantly different ($p < 0.05$) among sampling time within a plot in each soil layer. All values were calculated for area base, by multiplying the soil depth and bulk density (1.21 g cm^{-3})

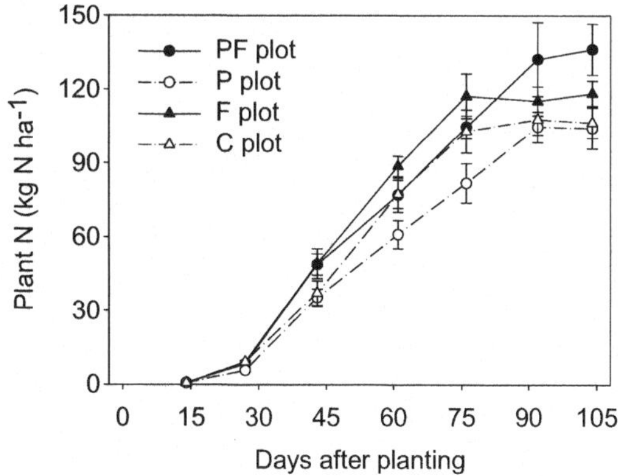

Fig. 18.4 Aboveground plant N uptake in all planted plots. Bars indicate the standard error. Control plot; *F plot*, fertilization plot; *P plot*, plant residue plot; and *PF plot*, plant residue and fertilizer plot (Refer to Sugihara et al. (2010b) for a more detailed explanation of the treatments)

depletion of easily decomposable SOM, the microbial decomposition could only supply a small amount of mineralized N during the dry period. These results indicate that strong competition for nitrogen occurred between soil microbes and plants during this dry period; hence, nitrogen uptake by plants prevented microbial growth (causing high MB C/N ratios). Thus, I concluded that in dry tropical cropland, soil microbes contribute to plant growth by serving as a nitrogen source during the grain-forming stage.

18.4   Controlling Factors of Soil Microbial Dynamics in Dry Tropical Croplands

In this 16-month study in Tanzania, I evaluated seasonal variations in microbial biomass carbon (MBC) and nitrogen (MBN), as well as microbial activity (as qCO_2), with respect to several factors related to soil moisture, land management (i.e., plant residue application, fertilizer application), and cropland type (i.e., 38 % clay soil, 4 % sandy soil). The objectives of this study were to: (1) evaluate the effect of soil moisture on seasonal variations in soil microbial biomass, (2) evaluate the effect of land management on the seasonal variations in soil microbial biomass, and (3) to compare the seasonal variations in soil microbial biomass between the clayey and sandy croplands, with regard to nutrient dynamics.

During the experimental period, there was a clear divide between the rainy and dry seasons (Fig. 18.5). In all treatment plots at both of my test sites, MBC and MBN tended to decrease during the rainy season, whereas they tended to increase, and remain at high levels, during the dry season (Figs. 18.6 and 18.7), although soil moisture did not correlate with MBC or MBN ($P > 0.05$). A couple of explanations have been given for such seasonal variations in MBC and MBN in dry tropical ecosystems: (1) increased soil fauna during the rainy season increases grazing on

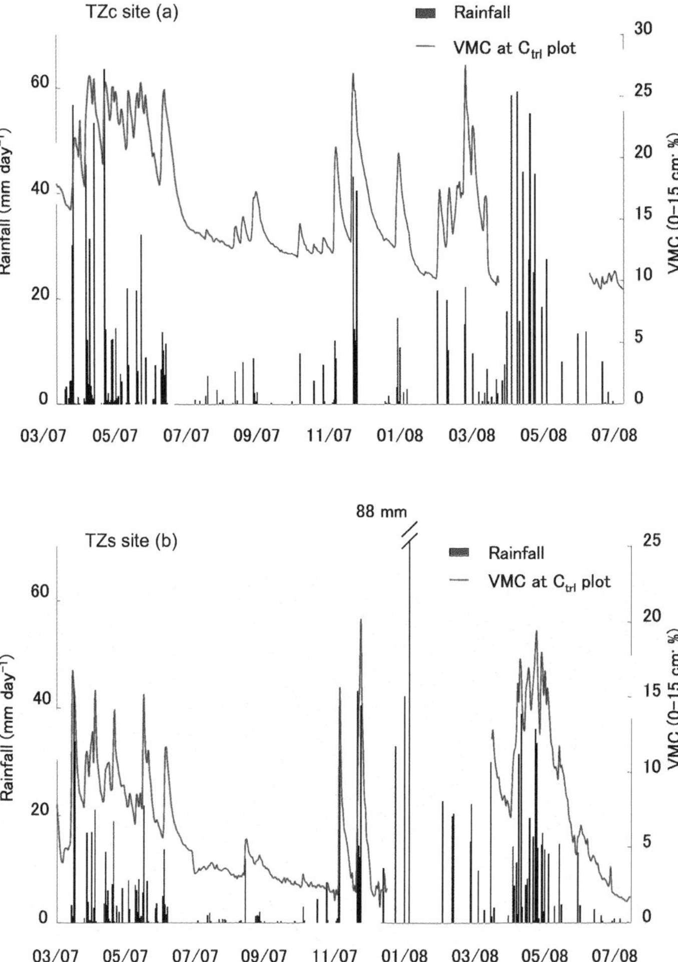

Fig. 18.5 Fluctuation in volumetric moisture content (VMC) at the control plot as a function of daily rainfall at the TZc (**a**) and TZs sites (**b**). VMC data from mid-March to May 2008 at TZc and from December 2007 to February 2008 at TZs could not be obtained because of technical problems with the sensors

soil microbes, thereby reducing their population (Singh et al. 1989; Michelsen et al. 2004), and (2) soil microbes accumulate intracellular solutes during the dry season; as a result, the drastic changes in soil moisture content after the first rainfalls of the rainy season lead to microbe lysis and a dramatic drop in their population (Halverson et al. 2000; Fierer and Schimel 2003). In my study, since MBC and MBN gradually decreased during the rainy season, the increase in soil fauna seems to be a more reasonable explanation for the observed decrease in soil microbes, as I

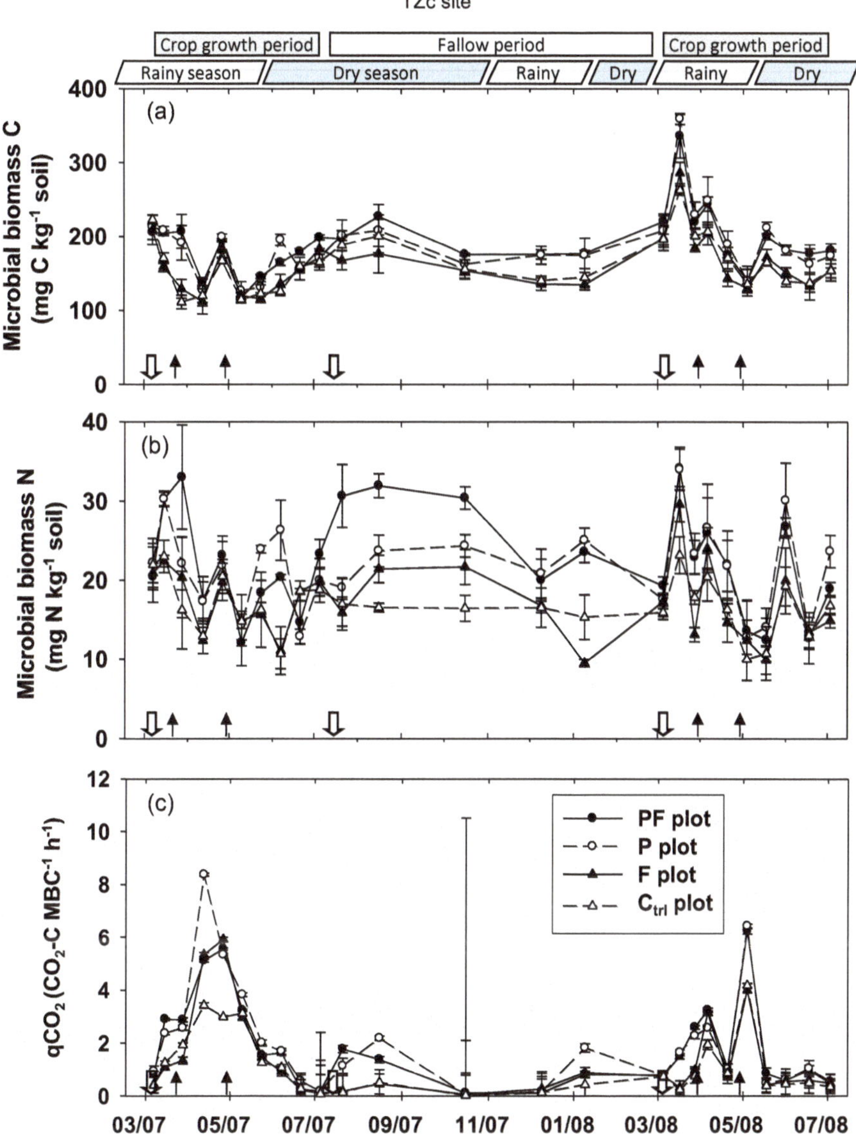

Fig. 18.6 Temporal variations in microbial biomass C (**a**), N (**b**), and in qCO$_2$ (**c**) at the TZc plots. The three *downward-pointing arrows* indicate plant residue application, and the four *upward-pointing arrows* indicate N application. Bars indicate the standard error. Control plot; *F plot*, fertilization plot; *P plot*, plant residue plot; and *PF plot*, plant residue and fertilizer plot (Refer to Sugihara et al. (2010b) for a more detailed explanation of the treatments)

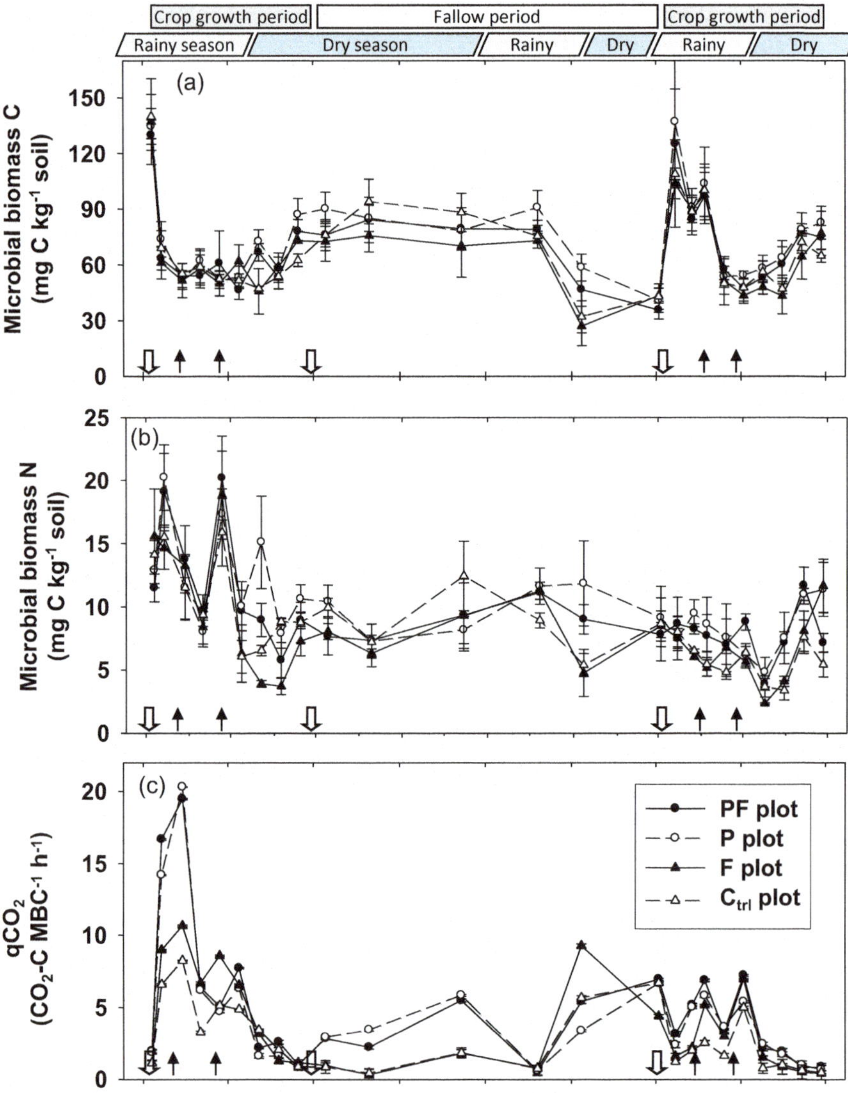

Fig. 18.7 Temporal variations in microbial biomass C (**a**), N (**b**), and in $q\mathrm{CO_2}$ (**c**) at the TZs plots. The three *downward-pointing arrows* indicate plant residue application, and the four *upward-pointing arrows* indicate N application. Bars indicate the standard error. Control plot; *F plot*, fertilization plot; *P plot*, plant residue plot; *PF plot*, plant residue and fertilizer plot (Refer to Sugihara et al. (2010b) for a more detailed explanation of the treatments)

Table 18.2 Summary of effects of land management and seasonal variation on soil microbial factors, such as MBC, MBN, and $q\mathrm{CO}_2$, according to repeated-measure analysis of variance (Sugihara et al. 2010a)

Source	MBC		MBN		$q\mathrm{CO}_2$	
	TZc	TZs	TZc	TZs	TZc	TZs
Plant residue application (P)	NS[a]	NS	0.05	0.02	0.002	0.01
Sampling date (season)	<0.0001	<0.0001	<0.0001	<0.0001	<0.0001	<0.0001
Season*P	NS	NS	NS	0.02	<0.0001	<0.0001
Fertilizer application (F)	NS	NS	NS	NS	0.04	NS
Season	<0.0001	<0.0001	<0.0001	<0.0001	<0.0001	<0.0001
Season*F	NS	NS	NS	NS	<0.0001	<0.0001
PF	0.02	NS	0.03	0.05	<0.0001	0.03
Season	<0.0001	<0.0001	<0.0001	<0.0001	<0.0001	<0.0001
Season*PF	NS	NS	<0.0001	NS	<0.0001	<0.0001

[a]NS means not significant

would have likely seen a dramatic depletion of soil microbes if the trigger was a change in soil moisture (Fierer and Schimel 2003). With regard to the crop growth stage, MBN in both control plots remained low during the crop maturation stage (i.e., the middle to late crop growth period) and the early dry season. As revealed by Singh et al. (2007b) and my previous study (Sugihara et al. 2010a), plant N uptake during the maturation stage caused severe N competition between crop and soil microbes, resulting in low MBN during this period. These results suggest that soil microbes act as a nutrient pool for crop growth, both in clayey and sandy croplands in Tanzania.

In all treatment plots at both sites, the $q\mathrm{CO}_2$ correlated with soil moisture; hence, soil microbes act as decomposers mainly during the rainy season. Compared to the effect of sampling date (i.e., seasonal effect), plant residue and fertilizer applications had little effect on the variations of MBC, MBN, and $q\mathrm{CO}_2$ (Table 18.2). In Canada, Spedding et al. (2004) also found that seasonal variation had a larger impact on soil microbial dynamics than land management (i.e., tillage and residue application). However, at both sites in my study, I observed a clear effect of plant residue application on seasonal variations in MBN (Table 18.2) and also found an increase in MBC and MBN after the addition of plant residue early in the rainy season (beginning of March in 2007 and 2008). Furthermore, at both sites, the increased MBN remained relatively high in the P and PF plots compared with in the control and F plots. These results indicate that seasonal variations in MBC and MBN can be controlled, at least short term, by plant residue application early in the rainy season. During the early rainy season in India, Singh et al. (2007b) also reported a rapid increase in MBC and MBN within a month of the residue application. This suggests that plant residue application not only helps counter the N loss caused by leaching but also synchronizes plant N uptake and N release from soil microbes by utilizing these microbes as an ephemeral nutrient pool during the early crop growth period.

For the clayey (TZc) and sandy (TZs) croplands, there were similar fluctuations in the seasonal variations in MBC, MBN, and qCO_2, but the coefficients of variations of MBC and MBN were higher at TZs than at TZc. This suggests that seasonal variations in MBC and MBN were greater at TZs than at TZc. Soil microbes in clayey soils are protected from macrofauna (i.e., predators) by the clay (Juma 1993) and from dry stress by soil aggregates (Van Gestel et al. 1991); hence, soil microbial biomass is relatively steady in clayey soil compared to that in sandy soil. Similarly, qCO_2 was substantially higher and fluctuated to a greater degree at TZs than at TZc. Many laboratory studies have shown that qCO_2 is higher in sandy soil than in clayey soil (Hassink 1994; Thomsen et al. 2003), and my field results are consistent with those results. Large variations in both MBC and MBN, and a high qCO_2, imply rapid turnover of soil microbes (McGill et al. 1986; Joergensen and Emmerling 2006), resulting in an increased supply of nutrient from the soil microbes. Using the method outlined by McGill et al. (1986) and Friedel et al. (2001), I calculated the turnover rate of soil microbes in the TZc and TZs control plots and found it to be 1.37/year at TZc and 4.90/year at TZs. This means that in the TZc and the TZs control plots, soil microbes should annually supply 41.4 and 91.5 kg N ha^{-1}, respectively. Sakamoto and Hodono (2000) also showed that the turnover rate of soil microbes in sandy soil (4.38 per year; 7 % clay) was more than twice the turnover rate in clayey soil (1.69 per year; 39 % clay). Overall, the nutrient dynamics in tropical agroecosystems appeared to be more substantially affected by soil microbes at TZs than at TZc.

18.5 Utilization of Soil Microbial Biomass to Improve the N Use Efficiency in Dry Tropical Croplands

As aforementioned, in dry tropical agroecosystems, it is important to synchronize plant N uptake with N release from soil microbes during the rainy season in order to minimize N loss from the soil (Singh et al. 2007a). In particular, N leaching during the early part of the rainy season (i.e., seeding and tillering stages of crop growth) leads to critical N loss, because plants are small at this stage and only utilize a small amount of N (Shahandeh et al. 2004). Therefore, in order to synchronize soil N supply with plant N uptake, it is important that available soil N be conserved during early crop growth.

In order to conserve the available soil N until later in the season, when substantial plant N uptake occurs, I assessed the potential role of soil microbes as a temporal N sink–source. During two consecutive years, throughout the ~120-days crop growth period, I evaluated the effect of land management (i.e., no input, plant

Photo 18.2 Experimental field described in this chapter. The *left* site is the plant residue application plot and the right site is the control plot. Two weeks before seeding, plant residue was chopped and plowed into the soil at a depth of 0–15 cm

residue application, plant residue and fertilizer application, fertilizer application, noncultivated) on the relationship between soil N pool (MBN and inorganic N) and plant N uptake. In this study, the application of plant residue was considered the carbon (C) substrate, which should stimulate soil microbial immobilization of potentially leachable N during early crop growth. To prevent severe N competition between soil microbes and crops due to excessive C input (Vigil and Kissel 1991; Hadas et al. 2004), plant residue (2.5 Mg C ha^{-1}; 35 kg N ha^{-1}), which was equivalent to biomass from a mature maize crop (without cobs) and also equivalent to ~50 % of decomposed C during the crop growth period, was applied before seeding (March) (Photo 18.2; Sugihara et al. 2010a, b). The plant residue was applied after the first heavy rainfall event in March, when Tanzanian farmers generally begin to cultivate. To avoid severe N depletion after the excessive C input, seeding was conducted at least 2 weeks after plant residue application and in concert with a rainfall event.

In both years, the plant residue application before planting immediately increased MBN in the P plot to 13.1–19.5 kg N ha^{-1} higher than in the control and F plots, resulting in a larger or similar soil N pool in the P plot during early crop growth compared to in the control and F plots, respectively (Fig. 18.8). This was because the C/N ratio of the applied plant residue was high (~70), and soil microbes immobilized inorganic N that would have potentially leached. According to my estimation, based on the relationship between soil moisture and CO$_2$ efflux rate

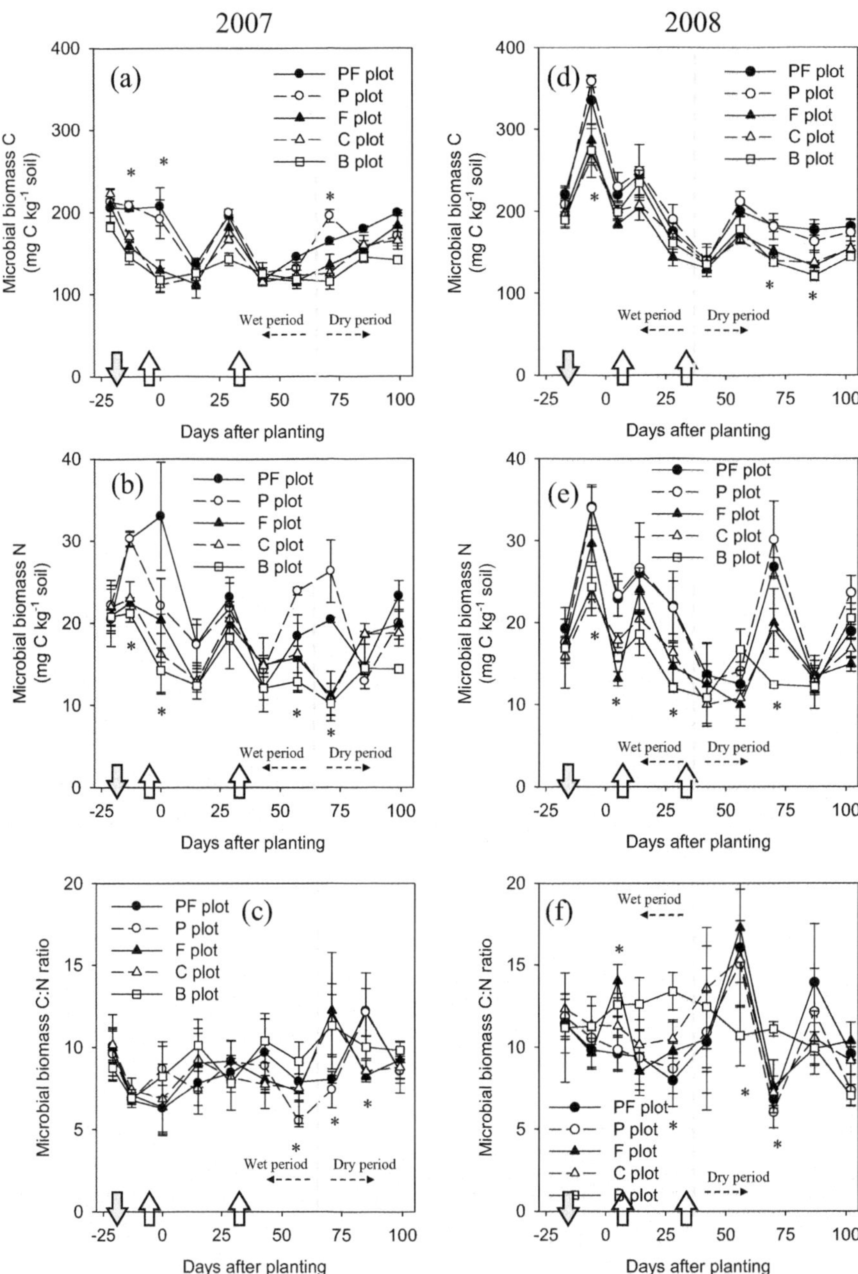

Fig. 18.8 Temporal variations in microbial biomass C (**a, d**), N (**b, e**), and C/N ratio (**c, f**) in 2007 (**a, b, c**) and 2008 (**d, e, f**). The *downward-pointing arrows* indicate plant residue application, and the *upward-pointing arrows* indicate N application. Bars indicate the standard error. (*) indicates a significant difference (one-way ANOVA: $P \leq 0.05$) between treatment plots for each sampling time. Control plot; *F plot*, fertilization plot; *P plot*, plant residue plot; and *PF plot*, plant residue and fertilizer plot (Refer to Sugihara et al. (2012) for a more detailed explanation of the treatments)

(Funakawa et al. 2006), in 2007 and 2008, decomposed C from 20–40 DAP in the P and PF plots was 750–920 kg C ha^{-1} higher than in the control and F plots. This indicates that the C/N ratio of the applied plant residues was gradually reduced during this period, and therefore the net N mineralization of the residue may not have occurred before 40 DAP in the P and PF plots (Vigil and Kissel 1991), when the risk of critical N leaching was high. Thus, plant residue application before planting enhanced the role of soil microbes as a temporal N sink by immobilizing the potentially leachable N, leading to conservation of available soil N for later crop growth.

During the later crop growth (i.e., dry period) in 2007 and 2008, the soil microbes (MBN and MB C/N ratio) and pattern of plant N uptake were closely related, as was previously observed (Sugihara et al. 2010b) (Figs. 18.8, 18.9, and 18.10). The MB C/N ratio also increased significantly, and MBN tended to decrease in all cultivated plots, which was possibly due to increased crop competitiveness for N, driven by an increased root system enhanced assimilation of the soil N pool (Bottner et al. 1999; Mayer et al. 2003). Many other studies have also found an increased MB C/N ratio and decreased MBN, indicating the clear contribution of soil microbial N to plant N uptake during the grain-forming stage in dry tropical croplands (Ghoshal and Singh 1995; Kushwaha et al. 2000; Singh et al. 2007b). Moreover, in the current study at the grain-forming stage, the aforementioned microbial factors did not fluctuate in the B plot as they had fluctuated in the cultivated plots, and this indicates that substantial plant N uptake must underlie the increase in MB C/N ratio and/or the decrease in MBN. Hence, during the later crop growth in all the cultivated plots, soil microbes would have contributed to crop growth as an N source.

In both years, the plant residue treatment clearly affected the relationship between soil microbes (MBN and MB C/N ratio) and plant N uptake during later crop growth because (1) the period of intense N competition (increased MB C/N ratio) was delayed in the P and PF plots (Fig. 18.8c, f) and (2) plant N uptake continued for a longer period in the P and PF plots than in the control plot (Figs. 18.9 and 18.10). The effects of plant residue application likely differed between 2007 and 2008 because of the different rainfall patterns between these years. In 2007, there were few heavy rainfall events, although the overall rainfall level was greater than average. Therefore, N leaching was not excessive, and plant N uptake and yield in 2007 were greater than average. In 2007, the soil N pool in the P plot was 20.0–41.4 kg N ha^{-1} higher than in the control plot and similar in the F plot throughout the experimental period (Fig. 18.9). However, the increased MBN appeared to temporarily suppress plant N uptake (43–57 DAP), because plant N uptake in the P plot (37.0 kg N ha^{-1}) was substantially lower than in the control and F plots (50.0 and 65.3 kg N ha^{-1}, respectively). As previously noted, added C substrate and rainfall (78 mm) in the P plot may have accelerated microbial turnover so that soil microbes could outcompete the crop for N (Bottner et al. 1999; Mayer et al. 2003). As soil dried after 71 DAP, MBN in the P plot decreased

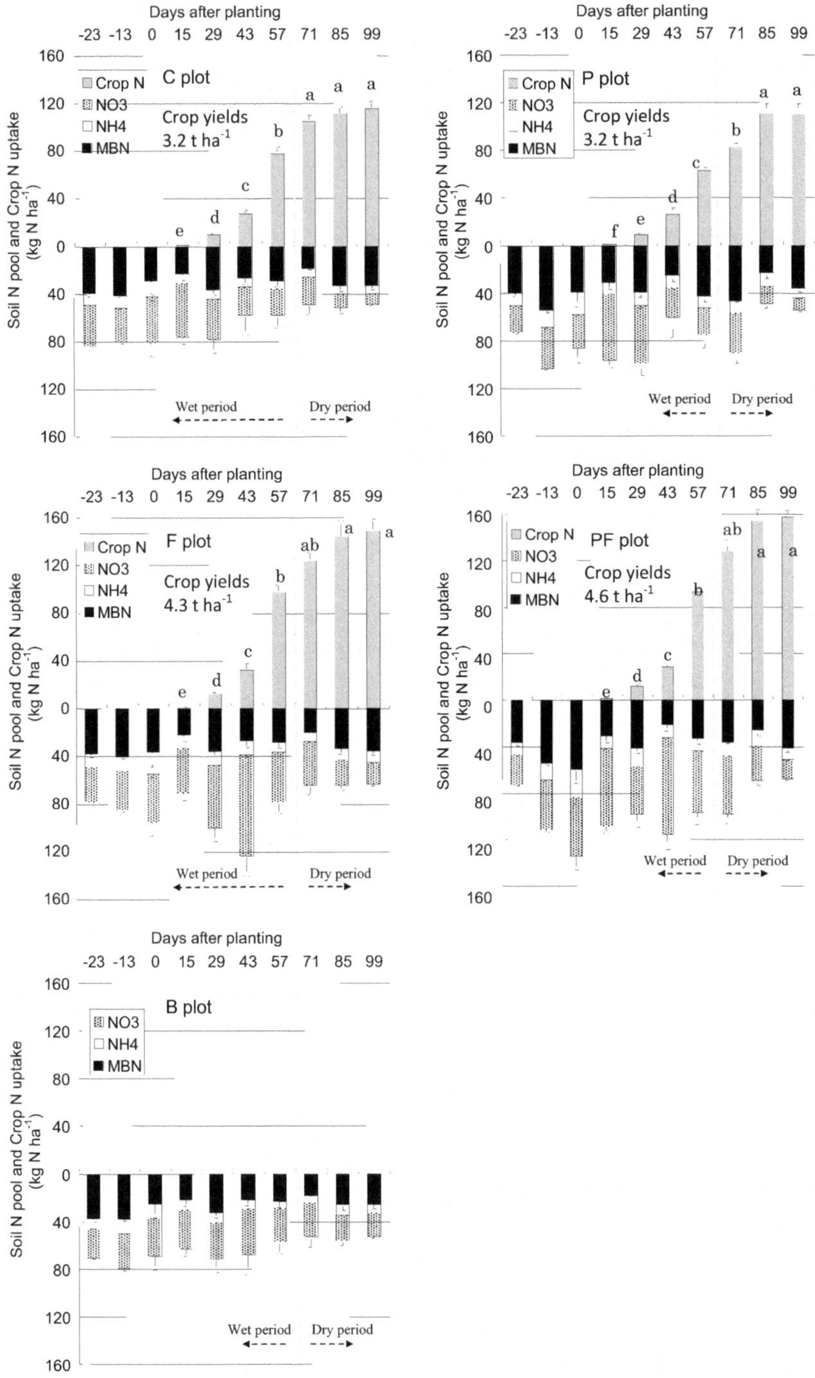

Fig. 18.9 Time course of aboveground crop N uptake and belowground soil N pool (MBN and inorganic N (NH_4^+ and NO_3^-)) in each plot during the experimental period in 2007. Bars indicate the standard error. Different letters (a, b, c, d, and e) indicate that the means of crop N are

with increased MB C/N ratio (Fig. 18.8b,c), and total plant N uptake in the P plot was similar to that in the control plot at harvest. This suggests that the increased MBN finally contributed to crop growth by acting as an N source because of the soil drying or crop maturation (Joergensen and Emmerling 2006; Singh et al. 2007b).

In contrast, in 2008, the rainfall was concentrated at the early crop growth stage, resulting in critical N leaching in all plots (Fig. 18.10). In this relatively normal year, the increased MBN contributed to maintaining the larger soil N pool in the P plot compared with in the control plot during early crop growth (10.5–21.2 kg N ha^{-1}) and later crop growth (16.3–29.9 kg N ha^{-1}). Although the delayed N competition and prolonged plant N uptake period were similar to 2007 (Fig. 18.8f), increased MBN in the P plot did not seem to limit plant N uptake as it had in 2007. Excessive N leaching depressed the soil N pool (especially inorganic N) in the control plot at 42 DAP (43.2 kg N ha^{-1}), resulting in earlier N competition. By contrast, increased MBN in the P plot conserved the larger soil N pool at 42 DAP (60.4 kg N ha^{-1}), thus providing considerable N for crop growth and resulting in the observed delay in soil N deficiency at 56 DAP, which was the same as the chemical fertilized plots (F and PF plots) (Figs. 18.8f and 18.10). Finally, total plant N uptake in the P plot was 19.1 kg N ha^{-1} higher than in the control plot and similar in the F plot. Based on these results, soil microbes certainly supplied the potentially leachable N to plant N uptake during later crop growth, although the contribution of mineralized N from applied plant residue should also be considered. However, from 56–87 DAP in 2008, the decomposed C in the P plot was estimated at 240.1 kg C ha^{-1}, and therefore the amount of net N mineralization would have been 21.2 kg N ha^{-1} (soil C/N $= 11.3$), which was less than plant N uptake in this period (31.9 kg N ha^{-1}). Thus, soil microbes immobilizing the potentially leachable N would have acted as an N source during later crop growth in the P plot, contributing to the synchronization of the soil N supply and plant N uptake (Sugihara et al. 2012).

These results showed that, in order to reduce N leaching and increase the soil N supply during later crop growth, it would be very useful to utilize soil microbes as a temporal N source–sink. In future studies, it would be necessary to evaluate the effect of quality and quantity of applied organic matter, such as high or low C/N ratio (Sakala et al. 2000; Moura et al. 2010), as well as application method (e.g., plowing into the soil or no-till) (Zeleke et al. 2004; Dahiya et al. 2007).

Fig. 18.9 (continued) significantly different ($P \leq 0.05$) among the sampling times within a plot (one-way ANOVA, Tukey's Kramer test). All values were calculated by multiplying the soil depth (0–15 cm) by the bulk density in each plot (1.19–1.27 g cm^{-3}). Control plot; *F plot*, fertilization plot; *P plot*, plant residue plot; and *PF plot*, plant residue and fertilizer plot (Refer to Sugihara et al. (2012) for a more detailed explanation of the treatments)

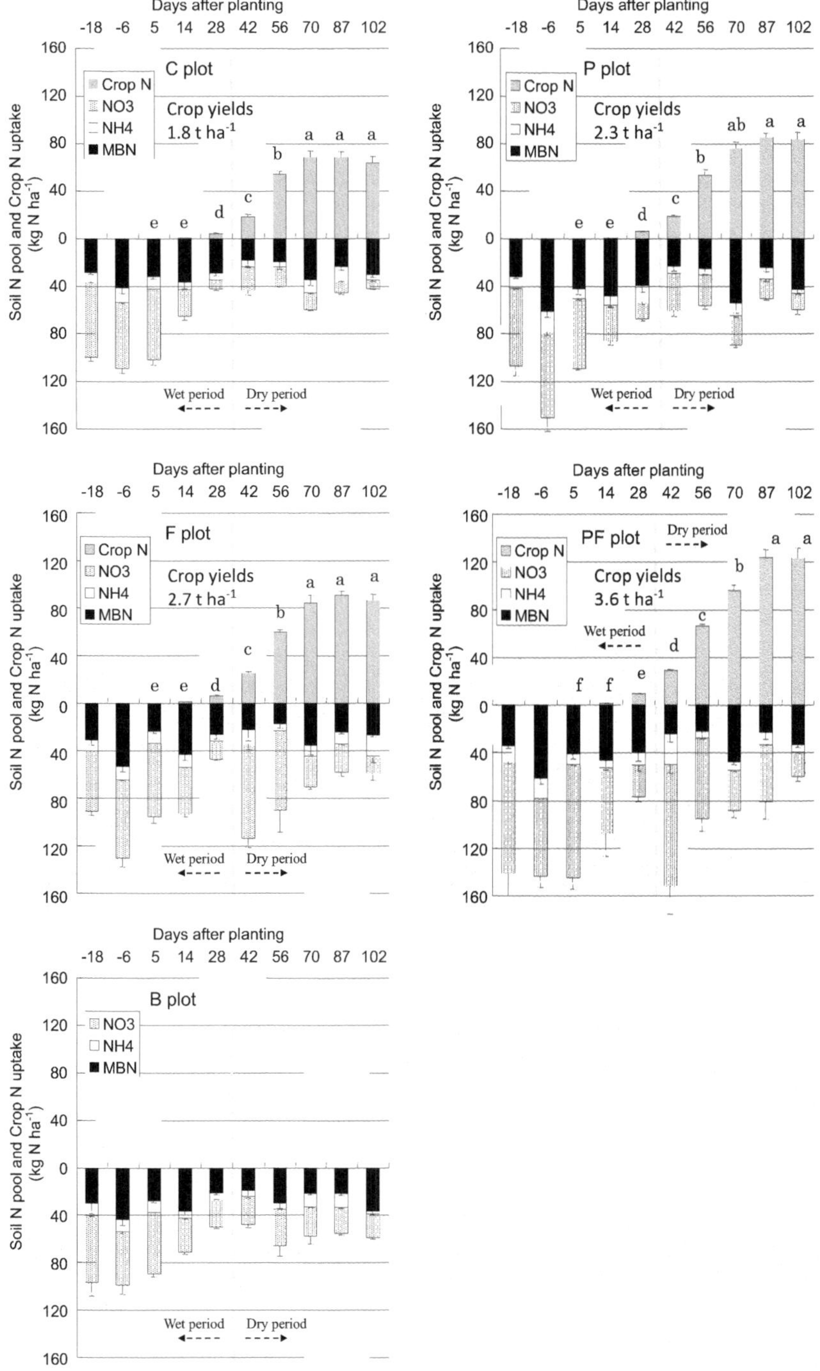

Fig. 18.10 Time course of aboveground crop N uptake and belowground soil N pool (MBN and inorganic N (NH_4^+ and NO_3^-)) in each plot during the experimental period in 2008. Bars indicate the standard error. Different letters (a, b, c, d, and e) indicate that the means of crop N are significantly different ($P \leq 0.05$) among the sampling times within a plot (one-way ANOVA, Tukey's Kramer test). All values were calculated by multiplying the soil depth (0–15 cm) by the bulk density in each plot (1.19–1.27 g cm^{-3}). Control plot; *F plot*, fertilization plot; *P plot*, plant residue plot; and *PF plot*, plant residue and fertilizer plot (Refer to Sugihara et al. (2012) for a more detailed explanation of the treatments)

References

Adeboye MKA, Iwuafor ENO, Agbenin JO (2006) The effect of crop rotation and nitrogen fertilization on soil chemical and microbial properties in a Guinea Savanna Alfisol of Nigeria. Plant Soil 281:97–107

Bardgett RD, van der Wal R, Jonsdottir IS, Quirk H, Dutton S (2007) Temporal variability in plant and soil nitrogen pools in a high-Arctic ecosystem. Soil Biol Biochem 39:2129–2137

Bationo A, Buerkert A (2001) Soil organic carbon management for sustainable land use in Sudano-Sahelian West Africa. Nutr Cycl Agroecosyst 61:131–142

Bottner P, Pansu M, Sallih Z (1999) Modelling the effect of active roots on soil organic matter turnover. Plant Soil 216:15–25

Chikowo R, Mapfumo P, Nyamugafata P, Giller KE (2004) Mineral N dynamics, leaching and nitrous oxide losses under maize following two-year improved fallows on a sandy loam soil in Zimbabwe. Plant Soil 295:315–330

Cookson WR, Marschner P, Clark IM, Milton N, Smirk MN, Murphy DV, Osman W, Stockdale EA, Hisrch PR (2006) The influence of season, agricultural management, and soil properties on gross nitrogen transformations and bacterial community structure. Aust J Soil Res 44:453–465

Dahiya R, Ingwersen J, Streck T (2007) The effect of mulching and tillage on the water and temperature regimes of a loess soil: experimental findings and modeling. Soil Tillage Res 96:52–63

Fierer N, Schimel JP (2003) A proposed mechanism for the pulse in carbon dioxide production commonly observed following the rapid rewetting of a dry soil. Soil Sci Soc Am J 67:798–805

Fierer N, Schimel JP, Holden PA (2003) Influence of drying-rewetting frequency on soil bacterial community structure. Microb Ecol 45:63–71

Friedel JK, Gabel D, Stahr K (2001) Nitrogen pools and turnover in arable soils under different durations of organic farming: II: source-and-sink function of the soil microbial biomass or competition with growing plants? J Plant Nutr Soil Sci 164:421–429

Funakawa S, Yanai J, Hayashi Y, Hayashi T, Watanabe T, Noichana C, Panitkasate T, Katawatin R, Kosaki T, Nawata E (2006) Soil organic matter dynamics in a sloped sandy cropland of Northeast Thailand with special reference to the spatial distribution of soil properties. Jpn J Trop Agric 50:199–207

Gentile R, Vanlauwe B, van Kessel C, Six J (2009) Managing N availability and losses by combining fertilizer-N with different quality residues in Kenya. Agric Ecosyst Environ 131:308–314

Ghoshal N, Singh KP (1995) Effect of farmyard manure and inorganic fertilizer on the dynamics of soil microbial biomass in a tropical dryland agroecosystem. Biol Fertil Soils 19:231–238

Hadas A, Doane TA, Kramer AW, van Kessel C, Horwath WR (2004) Modelling the turnover of 15 N-labelled fertilizer and cover crop in soil and its recovery by maize. Eur J Soil Sci 53:541–552

Halverson LJ, Jones TM, Firestone MK (2000) Release of intracellular solutes by four soil bacteria exposed to dilution stress. Soil Sci Soc Am J 64:1630–1637

Hartemink AE, Buresh RJ, van Bodegom PM, Braun AR, Jama B, Janssen BH (2000) Inorganic nitrogen dynamics in fallows and maize on an Oxisol and Alfisol in the highlands of Kenya. Geoderma 98:11–33

Hassink J (1994) Effect of soil texture on the size of the microbial biomass and on the amount of C and N mineralized per unit of microbial biomass in Dutch grassland soils. Soil Biol Biochem 26:1573–1581

Herai Y, Kouno K, Hashimoto M, Nagaoka T (2006) Relationships between microbial biomass nitrogen, nitrate leaching and nitrogen uptake by corn in a compost and chemical fertilizer-amended regosol. Soil Sci Plant Nutr 52:186–194

Joergensen RG, Emmerling C (2006) Methods for evaluating human impact on soil microorganisms based on their activity, biomass, and diversity in agricultural soils. J Plant Nutr Soil Sci 169:295–309

Juma NG (1993) Interrelationships between soil structure/texture, soil biota/soil organic matter and crop production. Geoderma 57:3–30

Kushwaha CP, Tripathi SK, Singh KP (2000) Variations in soil microbial biomass and N availability due to residue and tillage management in a dryland rice agroecosystem. Soil Tillage Res 56:153–166

Mayer J, Buegger F, Jensen ES, Schloter M, Heb J (2003) Residual nitrogen contribution from grain legumes to succeeding wheat and rape and related microbial process. Plant Soil 255:541–554

McGill W, Cannon WB, Robertson KB, Clark FD (1986) Dynamics of soil microbial biomass and water soluble organic C in Breton L after 50 years of cropping to two rotations. Can J Soil Sci 66:1–19

Michelsen A, Anderson M, Jensen M, Kjoller A, Gashew M (2004) Carbon stocks, soil respiration and microbial biomass in fine-prone tropical grassland, woodland and forest ecosystems. Soil Biol Biochem 36:1707–1717

Moura EG, Serpa SS, Santos JGD, Sobrinho JRSC, Aguiar ACF (2010) Nutrient use efficiency in alley cropping systems in the Amazonian periphery. Plant Soil 335:363–371

Nyamangara J, Bergstrom LF, Piha MI, Giller KE (2003) Fertilizer use efficiency and nitrate leaching in a tropical sandy soil. J Environ Qual 32:599–606

Palm CA, Giller KE, Mafongoya PL, Swift MJ (2001) Management of organic matter in the tropics: translating theory into practice. Nutr Cycl Agroecosyst 61:63–75

Petersen SO, Petersen J, Rubæk GH (2003) Dynamics and plant uptake of nitrogen and phosphorus in soil amended with sewage sludge. Appl Soil Ecol 24:187–195

Powlson DS, Hirsch PR, Brookes PC (2001) The role of soil microorganisms in soil organic matter conservation in the tropics. Nutr Cycl Agroecosyst 61:41–51

Sakala WD, Cadisch G, Giller KE (2000) Interactions between residues of maize and pigeonpea and mineral N fertilizers during decomposition and N mineralization. Soil Biol Biochem 32:679–688

Sakamoto K, Hodono N (2000) Turnover rate of microbial biomass carbon in Japanese upland soils with different textures. Soil Sci Plant Nutr 46:483–490

Shahandeh H, Blanton-Knewtson SJ, Doumbia M, Hons FM, Hossner LR (2004) Nitrogen dynamics in tropical soils of Mali, West Africa. Biol Fertil Soils 39:258–268

Singh S, Ghoshal N, Singh KP (2007a) Synchronizing nitrogen availability through application of organic inputs of varying resource quality in a tropical dryland agroecosystem. Appl Soil Ecol 36:164–175

Singh S, Ghoshal N, Singh KP (2007b) Variation in soil microbial biomass and crop roots due to differing resource quality inputs in a tropical dryland agroecosystem. Soil Biol Biochem 39:76–86

Singh JS, Raghubanshi AS, Singh RS, Srivastava SC (1989) Microbial biomass act as a source of plant nutrients in dry tropical forest and savanna. Nature 338:499–500

Spedding TA, Hamel C, Mehuys GR, Madramootoo CA (2004) Soil microbial dynamics in maize-growing soil under different tillage and residue management systems. Soil Biol Biochem 36:499–512

Srivastava SC, Lal JP (1994) Effect of crop growth and soil treatments on microbial C, N, and P in dry tropical arable land. Biol Fertil Soils 17:108–114

Sugihara S, Funakawa S, Kilasara M, Kosaki T (2010a) Effect of land management and soil texture on seasonal variations in soil microbial biomass in dry tropical agroecosystems in Tanzania. Appl Soil Ecol 44:80–88

Sugihara S, Funakawa S, Kilasara M, Kosaki T (2010b) Dynamics of microbial biomass nitrogen in relation to plant nitrogen uptake during the crop growth period in a dry tropical cropland in Tanzania. Soil Sci Plant Nutr 56:105–114

Sugihara S, Funakawa S, Kilasara M, Kosaki T (2012) Effect of land management on soil microbial N supply to crop N uptake in a dry tropical cropland in Tanzania. Agric Ecosyst Environ 146:209–219

Thomsen IK, Schjonning P, Olesen JE, Christensen BT (2003) C and N turnover in structurally intact soils of different texture. Soil Biol Biochem 35:765–774

Tripathi N, Singh RS (2007) Cultivation impacts nitrogen transformation in Indian forest ecosystems. Nutr Cycl Agroecosyst 77:233–243

Van Gestel M, Ladd JN, Amato M (1991) Carbon and nitrogen mineralization from two soils of contrasting texture and microaggregate stability: influence of sequential fumigation, drying and storage. Soil Biol Biochem 23:313–322

Vigil MF, Kissel DE (1991) Equations for estimating the amount of nitrogen mineralized from crop residues. Soil Sci Soc Am J 55:757–761

Wardle DA (1992) A comparative assessment of factors which influence microbial biomass carbon and nitrogen levels in soils. Biol Rev 67:321–358

Wardle DA (1998) Controls of temporal variability of the soil microbial biomass: a global-scale synthesis. Soil Biol Biochem 30:1627–1637

Wardle DA, Bardgett RD, Klironomos JN, Setala H, van der Putten W, Wall DH (2004) Ecological linkages between aboveground and belowground biota. Science 304:1629–1633

Zeleke TB, Grevers MCJ, Si BC, Mermut AR, Beyene S (2004) Effect of residue incorporation on physical properties of the surface soil in the South Central Rift Valley of Ethiopia. Soil Tillage Res 77:35–46

Chapter 19
Conclusion

Shinya Funakawa

Abstract This chapter provides an overall conclusion and summary of the volume with brief summaries of each chapter. Part I covered soil-forming processes on different continents, focusing on humid regions of Asia and equatorial Africa. Part II presented a comparative analysis of pedogenic acidification processes under different geologic and climatic conditions in humid Asia, where Ultisols predominate, and then provided a comparative analysis of Oxisols/Ferralsols in Cameroon. The analysis in this section suggested that differences in basic soil properties such as clay mineral composition was often accompanied by, or correlated with, different ecosystem processes under natural vegetation, which could be regarded as an "ecosystem strategy for resource (nutrients) acquisition." In Part III, we analyzed the ecological adaptation of traditional agricultural practices to environmental conditions such as climate and soils, to develop a scientific basis for low-input agriculture. These practices, which have proved to be sustainable for some period at least, should be utilized for developing agricultural systems that depend less heavily on external resource inputs. In Part IV, several trials for establishing "minimum-loss" management practices in the tropics were described. In summary, the authors would like to again emphasize that the development of agricultural systems appropriate for use in the tropics needs to be based on the scientific understanding of applied ecological processes, as was outlined in Parts III and IV, in which the loss of soil resources can be minimized by controlling fluxes of carbon, nitrogen, and minerals. As was stated in the introduction, these efforts are justified both from an economic perspective related to the poverty of many of the farmers in tropical countries and the vulnerability of agricultural productivity in tropical countries to changing conditions such as climate change, soil degradation, and worldwide depletion of water and nutrient resources. Further studies on local ecosystem processes and agricultural practices should be encouraged from these perspectives.

Keywords Clay mineralogy • Ecosystem strategy for resource (nutrients) acquisition • Low-input agriculture • Minimum-loss agriculture • Tropics

S. Funakawa (✉)
Graduate School of Global Environmental Studies, Kyoto University, Yoshida-Honmachi, Sakyo-ku, Kyoto 606-8501, Japan
e-mail: funakawa@kais.kyoto.u.ac.jp

© Springer Japan KK 2017

381

S. Funakawa (ed.), *Soils, Ecosystem Processes, and Agricultural Development*, DOI 10.1007/978-4-431-56484-3_19

19.1 Distribution of Clay Minerals in Tropical Asia and Africa, with Special Reference to Parent Materials (Geology) and Climatic Conditions

In Part I of this volume, the soil-forming processes on different continents, humid Asia and equatorial Africa, were discussed in detail, with special reference to our recent findings on clay mineralogy and soil fertility in these regions.

In Chap. 2, secondary mineral distributions in soils from Kalimantan, Indonesia, were investigated to examine the effects of the parent materials and climate at different elevations on the distributions, using B horizon soils from 60 sites. The soil samples were classified from their total elemental compositions using cluster analysis. Secondary minerals were measured by X-ray diffraction and selective extractions. The samples were divided into ferric (high Fe contents), K&Mg (high K, Mg, and Si), and silicic (high Si) groups. The ferric soils were derived from mafic parent materials, whereas the others were derived from felsic or sedimentary parent materials. The K&Mg soils had higher total base contents (suggesting primary minerals) and were less weathered than the silicic soils. Secondary minerals in the ferric soils were characterized by high contents of Fe oxides and gibbsite. The K&Mg and silicic soils had similar secondary mineral (kaolinite and vermiculite) contents, but more mica was found in the former. Only the silicic group soils had secondary mineral contents that changed as the elevation changed (the kaolinite content increased and the vermiculite and poorly crystalline Al and Fe contents decreased as the elevation decreased). Higher temperatures at lower elevations may cause minerals to be altered more. Secondary mineral distributions were primarily controlled by the parent material (mafic or felsic/sedimentary), and secondarily by the climate, which varied with elevation.

In Chap. 3, the impact of moisture regime on clay mineralogy was further discussed using the soils from Indonesia (udic), Thailand (ustic), and Japan (udic). The importance of vermiculitization of illitic minerals from sedimentary rocks and/or felsic plutonic rocks was a key issue. In these soils, mica and kaolin minerals dominated, with lesser amounts of the 1.4-nm minerals in northern Thailand, and significant amounts of 1.4-nm minerals formed in Indonesia and Japan. Based on these findings and a thermodynamic analysis using soil water extracts, the mineral weathering sequences from felsic and sedimentary rocks were suggested for each of the regions. In Thailand, under higher pH conditions associated with the ustic moisture regime, mica is relatively stable, while other primary minerals, such as feldspars, are unstable and dissolve to form kaolinite and gibbsite. Under the lower pH conditions of the udic moisture regime in Japan and Indonesia, mica weathers to form 1.4-nm minerals. Consequently, the soil mineralogical properties are thought to affect the chemical properties of soils, such as the cation exchange capacity (CEC)/clay and pH, and the taxonomic classification of the soils. The CEC/clay of the soils derived from sedimentary rocks or felsic parent materials showed a clear regional trend. The CEC/clay was usually higher than 24 $cmol_c\ kg^{-1}$ (corresponding to Alisols if the argic horizon is recognized) under the udic and perudic soil moisture regimes in Indonesia and Japan. It was predominantly lower than 24 $cmol_c\ kg^{-1}$ (corresponding to Acrisols) under the ustic soil moisture regime in Thailand. The World Reference Base for Soil Resources (WRB) classification is generally consistent with the regional trends of the chemical and mineralogical properties of soils and describes successfully the distribution patterns of acid soils in humid Asia using the criterion of CEC/clay $= 24\ cmol_c\ kg^{-1}$.

In Chap. 4, soil-forming conditions in Tanzania, which is located near the Great Rift Valley, were analyzed with special reference to climatic and geological factors. In total, 95 surface soil samples were collected from croplands, forests, and savannas in different regions across Tanzania, and the physicochemical and mineralogical properties of the samples were analyzed. Soil physicochemical properties varied widely, reflecting the wide variation of climatic conditions and parent materials of soils. Clay mineral composition was essentially similar to the soils developed under the ustic moisture regime of Southeast Asia, in which mica and kaolin minerals usually dominate. Based on a principal component analysis of the collected soil samples, five factors were calculated. From the scores of these five factors, the clay mineralogical composition, and the relationship between geological conditions (or parent materials) and the annual precipitation, the following observations were made: (1) The soil organic matter (SOM) and amorphous compounds were the highest at the volcanic center of the southern mountain ranges, from the east of Mbeya to Lake Malawi. (2) Available phosphorus and potassium were high in the volcanic regions around Mt. Kilimanjaro and in the southern volcanic mountain ranges, presumably due to intensive agricultural practices and fertilizer application. (3) The 1.4-nm minerals were probably formed under conditions of high sodicity and were often observed in the soils near Lake Victoria. (4) The volcanic regions and Great Rift Valley region of Tanzania, where soil is generally more fertile than in other regions, are conducive to modernized agriculture. The semiarid regions of Tanzania suffer from water shortage, while the relatively humid areas have less fertile soil, which predominantly contains kaolin minerals. These conditions are not favorable for agricultural production and should be taken into consideration when studying the feasibility of agricultural development in different areas in the future.

In Chap. 5, soil fertility was analyzed in various regions of sub-Saharan Africa, including Tanzania, Rwanda, western Democratic Republic (D.R.) of Congo, Cameroon, Nigeria, Burkina Faso, Ivory Coast, and others, with special reference to geological and climatic conditions. Comparing soils across the central and western regions of equatorial Africa, the properties of soils in each region can be summarized as follows: soils in Tanzania are affected by the Great Rift Valley movement, including the associated volcanic activity, and were found to be relatively fertile. The soils are generally not intensely acidic, and soil texture is intermediate. The SOM level is moderately high and is partially affected by volcanic activity and the relatively high elevation. The clay mineral composition also suggests that the soil in this region is somewhat less weathered, and possibly supplies more mineral nutrients, than soils in the other regions. Similar advantageous characteristics were found in the soils of the volcanic regions of the highlands of Rwanda and eastern D.R. Congo. The soils in this region are characterized by high base levels, high CEC values (which are influenced by the presence of both SOM and 2:1 clay minerals), intermediate to clayey soil texture, and relatively high SOM levels. The latter two are affected by parent materials and the cool temperatures. In contrast to the Great Rift Valley regions, a large part of Cameroon is situated on the Cameroonian plateau, which is composed of Precambrian basement rocks under humid climates. The soils in this region are characterized by a strong acidic nature, high levels of exchangeable aluminum, fewer base components, moderately low SOM levels, and clayey soil texture, dominated by inactive kaolin minerals. The soils in the western regions of equatorial Africa, such as the Nigeria/Benin and Burkina Faso/Ivory Coast/Liberia regions, are commonly characterized by the presence of sandy soils. The sand content usually exceeds 70 %, while clay content is less than 20 %. As a result, the base reserve is typically low, and SOM levels are less than $10\text{--}15 \text{ g C kg}^{-1}$ soil.

At the end of Part I, the effects of active Al and Fe (acid ammonium oxalate-extractable Al and Fe: Al_o and Fe_o) on preservation of organic carbon (OC) and sorption/release of phosphate in tropical soils are presented in Chap. 6. For OC preservation, Al_o and Fe_o, Fe oxides, and clay fraction have been assumed to be important components stabilizing the organic matter. The $Al_o + Fe_o$ content explained more than 60 % of variation in OC for the all soil groups classified by the degree of soil weathering (weak, moderate, and strong). The quantitative relationships between OC and $Al_o + Fe_o$ were of similar order for all the soil groups. In contrast, Fe oxides and clay contents are less correlated with OC. Effects of Al_o and Fe_o on phosphate sorption capacity and its extractability after sorption were also examined. Furthermore, the retardation effects of phosphate sorption into micro- and meso-pores of Al_o and Fe_o were assessed. The phosphate sorption capacity was strongly correlated with $Al_o + Fe_o$. The proportion of labile phosphate relative to added phosphate and the rate of phosphate released in sequential extractions were negatively correlated with $Al_o + Fe_o$ and porosity of the soils (0.7–4-nm pore SSA relative to the total SSA). These results showed that Al_o and Fe_o are important components of tropical soils that influence OC preservation and phosphate sorption/release, and carbon dynamics and availability of fertilized phosphate will be affected by the amount of Al_o and Fe_o in the soils.

19.2   Ecosystem Processes in Forest–Soil Systems Under Different Geological, Climatic, and Soil Conditions

As stated in Chap. 1, the interaction between terrestrial ecosystem processes and soil development is an essential one. Biological activity is a driving force of soil development, while soil conditions, largely derived from parent material and climate, can either restrict or enhance biological activities and ecosystem processes. Although many processes involve interactions between soil and ecosystems, the link between biological activity and soil acidification/mineral weathering is considered to be one of the most crucial for ecosystem processes in humid climatic conditions, where precipitation exceeds evapotranspiration.

In Part II of this volume, pedogenetic acidification processes under different geological and climatic conditions in humid Asia, where Ultisols (Acrisols and Alisols in the World Reference Base for Soil Resources [IUSS Working Group WRB 2014]) predominate, have been comparatively analyzed in Chap. 7. To analyze dominant processes of soil acidification and plant strategies for nutrient acquisition in tropical forests of Southeast Asia, proton budgets were quantified for plant–soil systems in Indonesia and Thailand. The consistently high acid load found to be transferred by plants to soil suggests that acidification can be a nutrient acquisition strategy for plants that promote cation mobilization in soil through mineral weathering and cation exchange reaction. Soil solution composition indicates that organic acids are the dominant acidifying anions of forest soils developed on sandy sedimentary rocks. The production of organic acids in the O horizons can be enhanced by lignin-rich and P-poor litters and high activity of fungal enzymes (peroxidases) in highly acidic soils. Conversely, bicarbonate regulates cation fluxes in forest soils developed on clayey sedimentary rocks and ultramafic rocks. The spatiotemporal variation in roots and acids can lead to different pathways of pedogenesis and incipient podsolization (aluminum eluviation/illuviation) in sandstone soils and ferralitization (in situ weathering) in iron-rich soils. The processes of nutrient cycles in tropical forests in Southeast Asia are diverse, depending on the levels of soil acidity.

In Chaps. 8 and 9, ecosystem processes that affect on Oxisols/Ferralsols in equatorial Africa (southeastern Cameroon) were analyzed, with special reference to the different vegetation of rainforest and savanna. In Chap. 8, the influence of savannazation on nutrient dynamics was analyzed using soil chemical and microbiological analyses. Distinct differences were observed between the forest and savanna in the soil nutrient stock. Soil carbon stock was usually similar; however, the soil carbon to nitrogen (N) ratio was substantially lower in forest areas than in the savanna, indicating nitrogen-rich conditions in the forest. Moreover, the potassium stock in forest soils was lower than in savanna soils, indicating that potassium deficiency may be one factor limiting afforestation. Soil microbial analysis showed a significant positive correlation between soil moisture and microbial biomass phosphorous (MBP) in the forest, indicating the importance of organic phosphorus (P) mineralization for MBP. In the savanna, a significant positive correlation was found between soil nitrogen availability and MBP, indicating nitrogen limitation for MBP. These results suggest that forest is an N-rich and P-limited ecosystem, whereas savanna is an N-limited ecosystem.

In Chap. 9, nutrient cycling and soil acidification processes in tropical African forests were discussed. Tropical African forests are dominated by Ferralsols (Oxisols) and leguminous species, while Southeast Asian forests are dominated by Acrisols/Alisols (Ultisols) and Dipterocarpaceae. To provide an overview of the carbon (C) and N dynamics, as well as soil acidification processes on Ferralsols and Acrisols, in tropical African forests, we quantified soil respiration and element fluxes through different flow paths (as precipitation, throughfall, litterfall, litter leachate, and soil solutions) and analyzed proton budgets in two secondary forested sites in Cameroon. Our results demonstrate that at an Acrisol site (MV), N was mostly taken up within the O horizon, which has a dense root mat, while half of the input N leached down to the mineral horizon at a Ferralsol site (AD). Nitrification was the main proton-generating process in the canopy and the O horizon of AD, and it caused a large amount of cation leaching, which resulted in the accumulation of basic cations because of the high proton consumption rates in the A horizon. In contrast, because of the dense root mat at MV, the excess cation uptake by plants in the O horizon made the largest contribution to proton generation, which resulted in intensive acidification of the surface soil. Our results suggest that ecosystem processes differ depending on soil type (i.e., soil acidity). Thus, legumes growing on Ferralsols in tropical African forests have unique plant–soil interactions via active nitrification in the O horizon.

The analysis in this section suggested that the difference in the basic soil properties such as clay mineral composition was often accompanied by, or correlated with, different ecosystem processes under natural vegetation.

19.3 Adaptation of Agricultural Practices in Upland Soils Under Different Bioclimatic Conditions in Tropical Asia and Africa

As covered in Chap. 1, agricultural production removes ecosystem resources such as mineral nutrients from the ecosystem as crop yield. This means that replenishment of nutrients is essential for sustaining agricultural productivity over time. Before the establishment of modern agriculture, addition of nutrients from external sources was limited. Farmers depended on natural processes such as mineral weathering and litter input for replenishing mineral nutrients for the next cropping cycle. In Part III of this volume, we analyzed the "ecological adaptation" of traditional agricultural practices to environmental conditions such as climate and soils to understand the scientific basis behind low-input agriculture. These practices, which have proved to be sustainable for some period at least, could be utilized for developing alternative agriculture that depends less heavily on the external input of resources.

In Chap. 10, ecological changes after reclamation of original vegetation (forest or savanna) on Oxisols/Ferralsols in eastern Cameroon were examined. To understand the effects of original vegetation (forest or savanna) on changes of solute leaching and proton budgets after reclamation on Oxisols in Cameroon, we quantified the soil nutrient fluxes in forest, adjacent savanna, and each adjacent maize cropland and compared with the case in Southeast Asia. In forest plot, excess cation accumulation in wood has contributed to soil acidification in the entire soil profile, while soil acidification rates were much lower in savanna plot because of limited plant uptake. As a result of cultivation, NO_3^- fluxes were substantially increased and nitrification was the main process of soil acidification in both croplands. Reflecting the nutrient flux pattern of original vegetation, protons generated by nitrification in forest cropland plot (9.4–10.1 $kmol_c$ ha^{-1} $year^{-1}$) was significantly higher than that in savanna cropland (3.4–4.5 $kmol_c$ ha^{-1} $year^{-1}$). The rate in savanna cropland was comparable to that in Thailand Ultisol (5.0 $kmol_c$ ha^{-1} $year^{-1}$) with moderate soil pH and higher than that in Indonesian Ultisol (1.5 $kmol_c$ ha^{-1} $year^{-1}$) with low pH. Despite low pH of bulk Oxisols of Cameroon, they would provide favorable habitat for nitrifiers with physically well-structured microaggregates, allowing active nitrification in the plots of Cameroon. High rate of nitrification suggests the risk of nutrient deficiency in cropland is more serious in nutrient-poor Oxisols. The effects of reclamation on soil acidification processes would depend on the original vegetation and also on soil pH and physical structure, which affect the nitrification activity. Since the ratio of K^+ concentrations to sum of Mg^{2+} and Ca^{2+} concentrations increased with decreasing soil solution pH, the lower solution pH, which could stem from cultivation, might promote K leaching from cropland.

In Chap. 11, we returned to Ultisols in northern Thailand, whose soil properties and ecosystem processes were extensively investigated in Chaps. 3 and 7. Annual precipitation is approximately 1200 mm at the studied village. Here we analyzed the shifting cultivation system used by the Karen people to understand the functions, such as nutrient replenishment, of the fallow phase. The fluctuation of fertility-related properties of soils through land use stages was analyzed, and the SOM budget was quantitatively calculated, with special reference to soil microbial activities. The factors that have ensured the long-term sustainability of the shifting cultivation system can be summarized as follows: (1) Some soil properties relating to soil acidity improve when SOM increases in the late stages of the fallow phase. The litter input may be supplying bases that are obtained via tree roots from lower down the soil profile. This increase in SOM and bases in the surface soil through forest-litter deposition in the late stages of the fallow phase increases the availability of nutritional elements to crops. (2) The decline in soil organic C during the cropping phase may be compensated by litter input during the 6–7 years of the fallow phase. The organic matter input from the incorporation of herbaceous biomass into the soil system after the establishment of tree vegetation (approximately in the fourth year) was found to be crucial for maintaining the SOM level in the overall nutrient budget. (3) The succession of the soil microbial community from rapid consumers of resources to stable and slow utilizers, along with the establishment of secondary forest, retards further leaching loss of nitrogen and enhances its accumulation in the forestlike ecosystem. The functions of the fallow phase listed above can be considered essential to the maintenance of this forest–fallow system. Agricultural production can therefore be maintained with a fallow period of around 10 years, which is somewhat shorter than widely believed. Traditional shifting cultivation in the study village was shown to be well adapted to its soil-ecological condition.

In Chap. 12, slash-and-burn agriculture under drier conditions in Zambia was investigated. The results of the field study indicated the following: (1) More than

10-year cropping exhausted SOM and woody biomass, SOM and woody biomass were not restored during short fallow. (2) At spots burned with emergent and bush trees, SOM decreased only by burning and could be restored during short cropping, although N contents of SOM and woody biomass were not restored even during fallow. (3) Without burning, 2-year fallow after 3-year cropping is relatively useful management in terms of the maintenance of particulate organic matter, SOM, and grain yield, although woody biomass decreased compared to 2-year fallow after 1-year cropping. The C and N stock could not be fully recovered, especially burned spots. Thus, under rain-fed agriculture without fertilizer in semiarid region, this type of slash and burn was relatively suitable to maintain grain yield and SOM contents although the yield and the SOM were low level compared to other regions where slash and burn is practiced. In the semiarid region of Zambia, the decrease in soil C and N stock during cropping was small with return of plant residue because long dry season constrains loss of SOM through leaching and decomposition and N_2 fixation by free-living N_2 fixation. Therefore, SOM will not decrease drastically under short cropping and short fallow rotation because recent decrease in emergent trees and increase in bush trees brought small loss of SOM during burning.

Based on the results of field studies in different regions of the tropics in Chaps. 10, 11, and 12, land management strategies that use shifting cultivation in different bioclimatic and soil conditions are comparatively discussed in Chap. 13. Four regions with different soil and climatic conditions are included: eastern Zambia (Alfisols; AP, 860 mm), northern Thailand (Ultisols; AP, 1200 mm), East Kalimantan of Indonesia (Ultisols; AP, 2500 mm), and eastern Cameroon (Oxisols; AP, 1450 mm). An increase in N mineralization and resulting increase in N flux, after the original vegetation was removed for cropping, were the highest for the eastern Cameroon forest, followed by northern Thailand, eastern Cameroon savanna, East Kalimantan of Indonesia, and eastern Zambia. This trend, however, is thought to coincide with mineral nutrient loss due to leaching. The loss is likely to be more detrimental for strongly-weathered soils, such as Oxisols in Cameroon. The function of the fallow phase was most clearly observed in the shifting cultivation practiced in northern Thailand; the soil fertility here was reinstated in the later stage of the young fallow phase, around 8 years. A similar trend was found in the soils of eastern Cameroon, where soil fertility was highest in the young fallow forest. This improvement in soil fertility during the fallow stage was not observed for the plots in East Kalimantan, Indonesia, and eastern Zambia, presumably due to the specific climatic conditions. Generally, farming practices and land uses in shifting cultivation in Southeast Asia were well adapted to the soil and climatic conditions. In contrast, the pattern of land use in eastern Cameroon has been affected by human factors, such as the activities of neighboring agropastoralists, as well as by natural conditions. In eastern Zambia, limited and fluctuating precipitation may be the most important factor in agricultural production. Since primary production and soil organic matter decomposition rate are regulated by this, soil fertility status would not change considerably under the present land use practices. Therefore, the classical interpretation of slash-and-burn agriculture is not applicable to the agricultural practices in the latter two African cases.

All of the examples of shifting cultivation studied in Chaps. 10, 11, 12,, and 13 depend on "on-site re-accumulation of mineral nutrients or temporal redistribution of nutrients" that occurs during the fallow phase. In Chap. 14, another approach for nutrient replenishment was investigated. To analyze the resource management approach that involves the import of resources from outside the farm, i.e., "spatial redistribution of resources," an investigation was conducted in southwestern Niger, Africa. It identified livestock management practices, based on the relationship between farmers and sedentary pastoralists, which promote efficient use of livestock excreta through corralling in the Sahel region. Transhumance was found to be essential for household located areas with a high rate of cropping because herds were removed from cropped fields to a drier region in the rainy season. Sedentary pastoralists generally corral and park herds in harvested fields rented from farmers at night. Another type of corralling was performed on a contract basis in farmers' fields, on the condition that the farmers provide food for the pastoralists staying for corralling. About half of the sedentary pastoralists were not engaged in the practice of contract corralling or contracted livestock, suggesting a non-formalized relationship between farmers and sedentary pastoralists for corralling and livestock grazing. In half of the households with contract corralling, the practice changed in two consecutive seasons. This implies that contract corralling is a flexible practice that compensates for the shortage of millet production and forage resources in some years. Considering the amount of livestock excreta applied and the percentage of the area corralled, two-thirds of the households had the potential to promote contract corralling for further efficient use of livestock excreta.

Thus, before the establishment of modernized agriculture when the input of external nutrient resources was limited, farmers depended on natural processes such as mineral weathering and litter input for replenishing nutrients for the next cropping season. Under some circumstances, strategies such as the transport of resources from outside the farm were available. Where the impact of agricultural activities on the natural soil and ecosystem processes is limited, it appears that severe environmental constraints have driven human adaptation of agricultural practices to the natural environments. Most traditional agriculture methods have been developed under the resource management strategy characterized by "low input/minimum loss."

19.4 Strategies for Controlling Nutrient Dynamics in Future Agricultural Activities in the Tropics

The relationship between soil/ecosystem processes and human activities was subjected to considerable change after the introduction of modern technologies into agricultural practices since the early twentieth century, most importantly after chemical fertilizers became available through the development of the Haber–Bosch process. The information and techniques for resource management and sustainable land use, which were important in traditional, low-input agriculture, are now often not adopted. Instead, modern agriculture, characterized by the heavy use of agricultural chemicals and/or intensive irrigation, is the usual approach for agricultural production.

In Part IV, several trials for establishing "minimum-loss" management practices in the tropics were undertaken. In Chap. 15, controlling wind erosion of soils and organic matter using the "Fallow Band System" was analyzed in the semiarid sandy soils of the Sahel region of West Africa. Wind erosion has a significant impact on soil nutrient dynamics in this area. However, these had not previously been accurately evaluated because of the limitations of conventional sediment samplers. In this study, field-scale nutrient flow caused by wind erosion was measured using the Aeolian Materials Sampler (AMS), which was developed to address the limitations of conventional samplers. It was revealed that (i) the annual loss of soil nitrogen from a cultivated field by wind erosion was two to three times greater than the annual absorption by the staple crop. (ii) However, when herbaceous fallow land (more than 5 m wide in the east–west direction) was retained at the west side of the field, most of the blown soil nutrients were captured by the fallow land. On the basis of this high trapping capacity of the fallow land, the "Fallow Band System" was designed to control nutrient loss and improve crop production. A field experiment showed that (a) a single fallow band with a width of 5 m trapped 74 % of annual incoming soil particles and 58 % of annual incoming coarse organic matter, suggesting that the "Fallow Band System" can control the nutrient flow by wind erosion, (b) owing to the deposition of the trapped soil materials by the fallow band, and soil nutrient levels and hydraulic conductivity of the surface soil were greatly improved in the fallow band, which resulted in improved crop yield compared to the field without a "Fallow Band System."

In Chap. 16, surface runoff generation, which is usually strongly related to the process of soil erosion, was investigated in order to establish land use strategies for minimizing water erosion of soils. Surface runoff was measured in three plots at different cultivated sloping sites in north and northeast Thailand using a runoff gauge connected to a data logger. In most cases, rainfall intensity and the surface soil moisture contributed significantly to surface runoff generation. Rainfall intensity at the plot of sandy soils was higher than at the other two plots of fine-textured soils in the northern region. Surface runoff occasionally occurred throughout the rainy season with no clear seasonal trend at the sandy plot in northeast Thailand, unlike at the other two plots in northern Thailand, where surface runoff occurred more frequently during the latter half of the rainy season due to the higher rainfall intensity and/or capillary saturation of surface soils. In all plots, the proportion of surface runoff generated in relation to rainfall increased with the increase of the slope gradient of the plots. The proportion of soil erosion in relation to surface runoff was, however, largest in the sandy plot with the lowest slope gradient, indicating that sandy soils are more easily eroded than clayey soils, presumably due to the weakly organized structure of the soil aggregates. Therefore, the conditions that enhance the risk of surface runoff and soil erosion were found to vary and should be taken into account for agricultural management in the respective regions.

In Chap. 17, a similar approach was adapted in cropping fields in Tanzania and Cameroon, where more strongly weathered soils than Asian soils were found. We installed runoff plots (width 0.8 m × slope gradient 2.4/2.0 m) at four sites (designated NY, TA, SO, and MA) in Tanzania and one site with three treatments in Cameroon: bare plot (CM), cassava plot (CMC), and cassava with mulch plot (CMC+M). Water budgets for rainfall, surface runoff, and soil moisture were created for each rainfall event over a rainy season and the soil losses were measured. High rainfall amounts in NY and TA caused high surface runoff and soil loss. High stability of soil aggregates in CM resulted in low runoff coefficient. Although sandy soils in MA had a high infiltration rate and produced low runoff coefficient, their high susceptibility to transport by surface runoff increased the runoff sediment concentration. Total soil loss in CMC+M was 49 % lower than soil loss in CM and CMC, despite there not being a large difference in runoff volume between the plots, indicating that mulch suppresses the particle detachment by raindrops. Based on the soil erosion observed at the five sites, surface mulching is considered to be a widely applicable management approach to prevent soil loss from water erosion in the tropics, where rainfall intensity is generally high.

In Chap. 18, the possibility of utilizing soil microbes as a temporal nutrient pool to decrease the leaching loss of nutrients, especially nitrogen, was investigated in croplands in Tanzania. Generally, in sub-Saharan Africa, soil nitrogen is the limiting factor for crop production and is often lost through leaching. To improve crop productivity in this region, it is necessary to improve the synchronization of soil nitrogen supply and plant nitrogen uptake. In this chapter, the authors showed and explained the effectiveness of utilization of soil microbes as a temporal nitrogen pool to synchronize the supply of soil nitrogen and plant nitrogen uptake, on the basis of the field cultivation experiment in the dry tropical croplands of Tanzania. The authors first showed that the mismatch of nitrogen supply and uptake in this region was due to nitrate leaching during the early crop growth period. The authors also found a clear contribution of soil microbes as a nitrogen source for plant growth at the later crop growth period. Secondly, the authors found a clear effect of land management, in particular, plant residue application, on the soil microbial dynamics. Plant residue application just before the rainy season can increase the microbial biomass nitrogen (MBN) and maintain it for over a month. Finally, the authors assessed the effectiveness of plant residue application 2 weeks before seeding, by analyzing the soil–plant nitrogen dynamics and crop productivity. We found that soil microbes contribute to the synchronization of soil nitrogen supply and plant nitrogen uptake and to the improvement of crop yields.

As was stated in Chap. 1, I consider that the main direction of development of tropical agriculture in future should focus on:

1. Decreasing the loss of resources, rather than increasing input
2. Controlling the fluxes of carbon, nitrogen, and mineral resources, rather than accumulating them in soils, since the proportion of flux relative to the pool size is usually much larger in tropical ecosystems than in temperate ecosystems

Several trials presented in Part IV were carried out in this line.

At the last of this volume, I would like to emphasize again that the main intention of all these analyses, introduced in Parts III and IV, is derived from our primary concern that we should maintain the knowledge of low-input agriculture with scientific basis in order to develop alternative tropical agriculture, in which the loss of resources should be minimized by controlling the fluxes of carbon, nitrogen, and mineral resources. As was stated in the introduction part, these directions would be justified both from the economic perspective related to the poverty of many farmers in the tropics and instability in the production circumstances specific to tropical countries at the moment or even in the near future, such as climate change, soil degradation, and worldwide depletion of water and nutrient resources. Further studies on local ecosystem processes and agricultural practices should be encouraged from these perspectives.

Printed by Printforce, the Netherlands